T0250618

Lecture Notes in Computer Science 1128

Edited by G. Goos, J. Hartmanis and J. van Leeuwen

Springer
Berlin
Heidelberg
New York
Barcelona
Budapest
Hong Kong
London
Milan
Paris
Santa Clara
Singapore
Tokyo

Jacques Calmet Carla Limongelli (Eds.)

Design and Implementation of Symbolic Computation Systems

International Symposium, DISCO '96
Karlsruhe, Germany, September 18-20, 1996
Proceedings

Series Editors

Gerhard Goos, Karlsruhe University, Germany

Juris Hartmanis, Cornell University, NY, USA

Jan van Leeuwen, Utrecht University, The Netherlands

Volume Editors

Jacques Calmet
University of Karlsruhe, Institute for Algorithms and Cognitive Systems
Am Fasanengarten 5, D-76128 Karlsruhe, Germany
E-mail: calmet@ira.uka.de

Carla Limongelli
University of Rome III, Department of Computer Science
Via della Vasca Navale 84, I-00146 Rome, Italy

Cataloging-in-Publication data applied for

Die Deutsche Bibliothek - CIP-Einheitsaufnahme

Design and implementation of symbolic computation systems : international symposium ; proceedings / DISCO '96, Karlsruhe, Germany, September 18 - 20, 1996. Jacques Calmet ; Carla Limongelli (ed.). - Berlin ; Heidelberg ; New York ; Barcelona ; Budapest ; Hong Kong ; London ; Milan ; Paris ; Santa Clara ; Singapore ; Tokyo : Springer, 1996
(Lecture notes in computer science ; Vol. 1128)
ISBN 3-540-61697-7
NE: Calmet, Jacques [Hrsg.]; DISCO <4, 1996, Karlsruhe>; GT

CR Subject Classification (1991): I.1, D.1, D.2.1, D.2.10, D.3, I.2.2-3, I.2,5

ISSN 0302-9743
ISBN 3-540-61697-7 Springer-Verlag Berlin Heidelberg New York

Typesetting: Camera-ready by author
SPIN 10513495 06/3142 - 5 4 3 2 1 0 Printed on acid-free paper

Foreword

This volume contains the proceedings of the Fourth International Symposium on Design and Implementation of Symbolic Computation Systems, DISCO '96, held in Karlsruhe, Germany, September 18–20, 1996.

The DISCO series of conferences was initiated by A. Miola in 1990 to focus mainly on innovative methodological and technological aspects of hardware and software design and implementation for symbolic and algebraic computation, automated reasoning, geometric modeling and computation, and automatic programming. As illustrated by this volume, the scope of the conference has remained almost unchanged.

However, the successive program committees have modelled what innovative implementation stands for. This trend mirrors the evolution of computer science in the recent years: the paradigm of designing and managing complex system is assuming a more relevant role with respect to the paradigm of algorithms. As a consequence the facets of the conference dealing with hardware and algorithm implementation have faded away.

A feature of DISCO is to demonstrate that it is possible to set the design of symbolic systems on sound theoretical principles. Another theme that spans the proceedings is the trend towards interoperability of software systems. This is rather satisfactory since one of the original goals of this conference was to gather several research communities that used to follow disconnected paths. The contributed papers cover the domains listed as suitable topics for the conference. The invited lectures either present the state of the art of a research domain or point to some new research directions.

As for the previous conferences, the rate of acceptance of submitted papers has been kept around 40%. An innovation of DISCO '96 has been to accept some contributions as short papers. Each submitted paper has been reviewed under the care of at least two program committee members.

The program committee and particularly its chair, L. Carlucci-Aiello, deserve special thanks for efficient and timely processing of the submitted papers. We are very grateful to A. Miola for his help and dedication in setting up this conference. The local organization would not have been possible without the hard work of K. Homann.

All conferences try to attract as many sponsors as possible. It was a very pleasant surprise to find that obtaining sponsorships and/or funding for this conference was much easier than expected. This proves that the conference is now well established and well respected. The sponsors are listed on the following page. However, this list may not be complete at the time this foreword is written. To all of them we extend our gratitude and thanks.

Jacques Calmet – Carla Limongelli

July 1996

Karlsruhe – Roma

Conference Chair

Jacques Calmet

Program Chair

Luigia Carlucci-Aiello

Scientific Committee

L. Carlucci-Aiello (I - Chair)
R. Caferra (F)
J. Calmet (D)
J.A. Campbell (UK)
C.M. Hoffmann (USA)
C. Kirchner (F)
A. Miola (I)
J. Pfalzgraf (A)
F. Pfenning (USA)
A. Salwicki (F/P)

Local Chair

Karsten Homann

In Cooperation with acm SIGSAM SIGART Compulog Net III ECCAI IFIP (TC 12) 

Institute of Algorithms and Cognitive Systems

Contents

Problem-Oriented Applications of Automated Theorem Proving

W. Bibel, D. Korn, C. Kreitz, and S. Schmitt

Fachgebiet Intellektik, Fachbereich Informatik
Technische Hochschule Darmstadt
Alexanderstr. 10, 64283 Darmstadt, Germany
{bibel,korn,kreitz,steph}@intellektik.informatik.th-darmstadt.de

Abstract. This paper provides an overall view of an approach to developing a coherent ATP-system which can deal with a variety of logics and with different applications in a tailored way. The paper also summarizes research results achieved in the course of this development.

1 Introduction

Logical reasoning is an inherent feature of human problem solving. Mathematical reasoning is a particularly precise form of human problem solving which is used especially in scientific applications. Because of its precision it lends itself to automation more easily than reasoning in general. The field of Automated Theorem Proving (ATP) is developing deductive systems which simulate mathematical reasoning. In consequence these systems are of relevance for any application which requires mathematical reasoning.

To mention a few of these applications, finding proofs for mathematical theorems as done in mathematics is of course one of them. A slightly less obvious one is the use of ATP to automatically synthesize computer programs from a given specification. Following the "proofs-as-programs" paradigm [BC85] this task amounts to finding a constructive proof for the existence of a function which maps input elements to output elements of the specified program. A related but more general task is the automated control of the behavior of intelligent agents within a given environment. This application involves the need to automatically generate plans for their actions in that environment. Such plans are a series of actions leading from the initial situation to the desired goal situation. Again this can be modelled as a proof-finding task if these situations as well as the possible actions are modelled in a resource-sensitive logic.

In general, ATP becomes useful whenever the problem to be solved can be formalized in some logical language for which a proof calculus and an efficient implementation is available. In a number of cases, however, we face the problem that such implementations are not available. This is the case in particular if the logic is different from classical first order logic (FOL). Many applications, however, lend themselves to formalizations in logics other than FOL. For instance, trying to formalize the above-mentioned concept of constructive proof within the language of FOL results in a very unnatural formalism which by no means suits to the environments of program development used in concrete systems.

A second problem arises for the use of ATP in applications like those mentioned above. The actual proofs generated by ATP-systems tend to have a very technical look. Before they can be understood by experts of the envisaged application they need to be transformed into a more readable form.

The work done at the Intellectics department of the Technical University in Darmstadt during the last few years has tackled these two problems in the development of ATP-systems. On the one hand we have developed systems, or rather a combined system, which deals with problem formalizations in classical as well as in various non-classical logics. On the other hand we have enhanced their interface towards presenting the generated proofs in a way tailored to the particular application. They are *application-oriented* in the sense that they deal with the logic most convenient for the particular application and also provide an application-specific proof interface. The applications envisaged so far in our approach are the three mentioned above.

Our approach is illustrated in Figure 1. It shows at the top and bottom levels the three areas of application, viz. mathematics, programming, and planning. For each area we mention a particular well-known system of possible use in the respective area (Mathematica, NuPRL, and ISABELLE). The vertical arrows at the top level represent the extraction of proof tasks by the user to be solved by the ATP-system. As far as systems such as Mathematica are concerned one may additionally interpret these arrows as applications of these systems within the proof task. The vertical arrows at the bottom level represent the use of generated proofs within each of these applications. For each of the three areas a particular logic is selected (FOL, intuitionistic logic, modal logics). Their applicability may be characterized more generally as follows.

- **classical logic**: covers all applications which dispense in their proofs with epistemological concepts (e.g. proving mathematical theorems). The availability of a normal form (NF) for classical formulae provides a greater flexibility of problem-specific ATP approaches in terms of the possibility to either work directly on the non-normal form input or to transform it first into normal form. Postprocessing of proofs mainly aims at the readability of the proof steps.
- **intuitionistic logic**: covers applications which require constructive proofs (e.g. program-synthesis). No normal form exists which limits the straightforward application of classical ATP-systems. However, as described later in the paper, the non-classical features in the logical formulae may be regarded as semantic information in addition to the formulae's logical contents in the classical sense. The latter can then again be treated similarly as the classical case. Various approaches to deal with this semantical meta-knowledge have been made as discussed in the paper. Postprocessing of proofs needs also to account for the re-integration of the reasoning about this meta-knowledge into the classical line of reasoning.
- **modal logics**: cover applications which require epistemological concepts (e.g. time, belief). Again, no normal form exists but semantical meta-knowledge can be handled in various ways. Postprocessing of proofs again has

to re-integrate the reasoning about this meta-knowledge into the resulting proof.

This paper provides an overall view of our approach to developing a coherent ATP-system which can deal with each of these logics and with the three areas of applications in a way suitable for each of them. Along with this view the paper summarizes research results achieved in the attempt of mechanizing (classical and) non-classical logics.

In the remaining paper we first review the basic structure of the proof process of ATP-systems in order to be able to describe the need for and the location and details of the extensions which enable the handling of different logics and of application-oriented proof presentations. In the main Section 3 we describe two different ways of dealing with semantical meta-knowledge originating from non-classical logics. For each of these two ways we discuss a number of technical alternatives. We conclude with an outlook to future work.

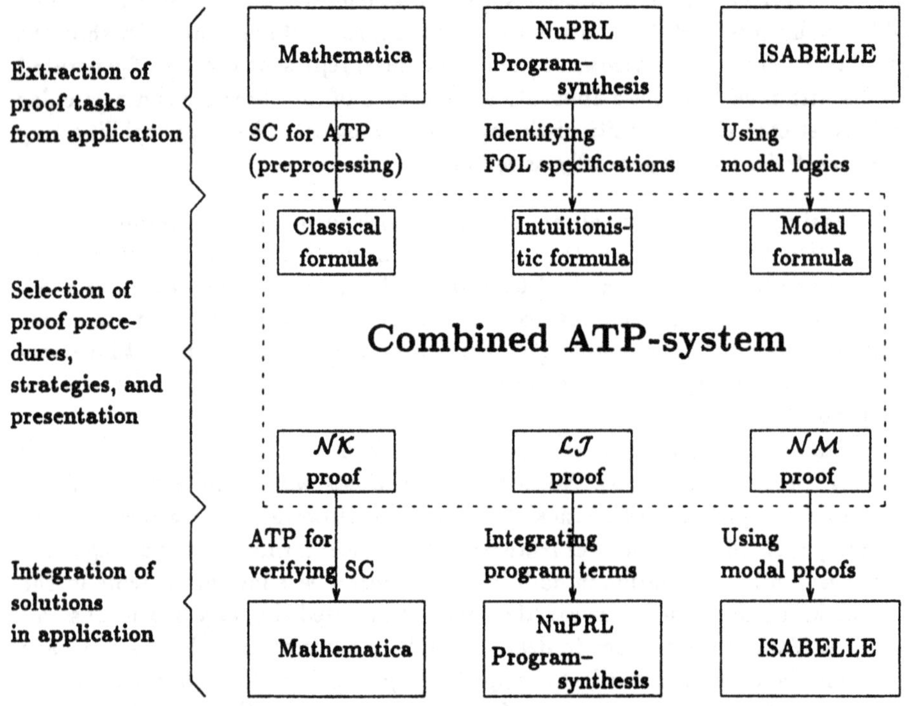

Fig. 1. An application-oriented ATP-system

2 Structuring the Process of Theorem Proving

The core of each ATP-system is the *inference machine* which amounts to sort of a "microprocessor" for theorem proving [Ohl91]. A formula to be proved usually is preprocessed by some input layer in order to transform it into the "machine language" of the inference machine. At the output end the generated proof is possibly postprocessed in order to suit the original input language and to make the result readable for a potential user. Hence, we face the problem of providing mechanisms which deal with three different tasks:

1. **input transformations**: this is the interface between the application-oriented and the inference machine oriented languages. A popular example is the normal form transformation (NFT) which serves to translate arbitrary first order formulae into clause-form which is the preferred language of most inference machines. Another important field of translation is formed by the so-called "logic morphisms" [Ohl91]. They basically aim at encoding semantical meta-knowledge for non-classical logics within the language of classical logic.
2. **inference machines**: they do the actual exploration of the search space which is spanned by the given formula w.r.t. the underlying logic. Usually they consist of a small set of inference rules which are applied to the given formula according to a certain strategy. Popular examples are the various resolution-based strategies [Rob65] as well as the connection-based methods like the extension-procedure [Bib87].
3. **proof presentation**: this is the interface between the logical calculus of the inference machine and the one of the given application. The main goal here is to put the machine-oriented result into a readable form so that the generated proof can be read in terms of the application's own language. Special care has to be taken in order to re-incorporate the necessary information which has been processed during possible input transformations. Examples for such techniques can be found in [And80, Wos90, And91, Pfe87] and [Lin89, Lin90, Gen95]. They try to present resolution- or connection-based proofs within natural calculi such as $\mathcal{LK}$ or $\mathcal{NK}$ [Gen35]. Other approaches try to present them even in natural language [DGH$^+$94]. Similar approaches have been developed for non-classical logics as well [SK95, SK96].

Structuring the proof process in this way provides flexibility for the design of problem-specific ATP-systems. Up to a certain extent, one is given the freedom to distribute the proof process between input transformation and the inference machine. Take, for instance, a classical first order formula which contains equality predicates. Then on the one hand we can deal with the theory of equality by a preprocessing step which generates lemmata that follow from the given formula w.r.t. equality theory [Bra75]. Thus, we would do part of the proof work already within the input transformation layer. On the other hand, we may reason about the equalities by equipping our inference machine with an additional mechanism which allows for reasoning about equality. This amounts to leaving all of the proof work to the inference machine [RW69].

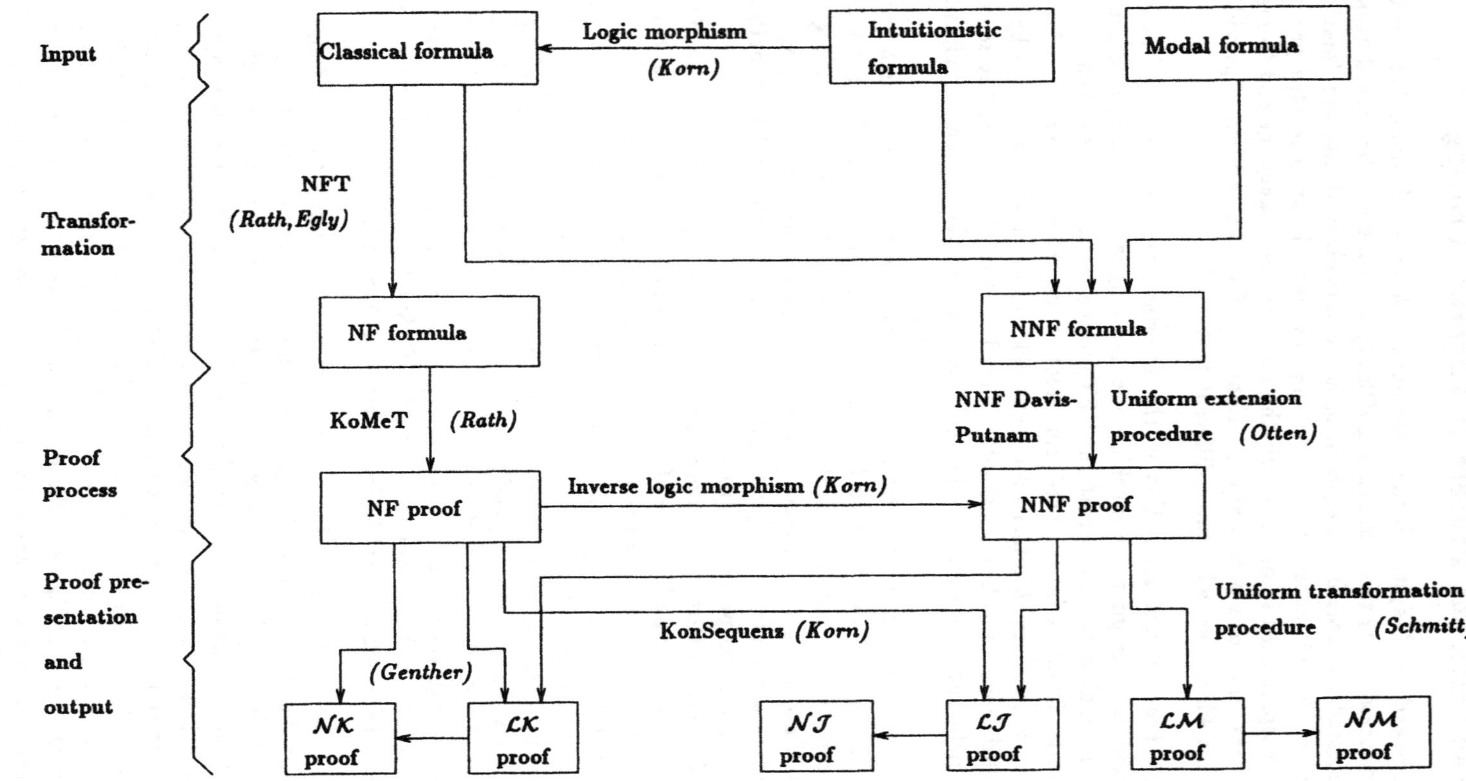

Fig. 2. The concept of combining ATP-systems

Likewise, we might wish to deal with the semantical meta-knowledge for non-classical logics at the input transformations-layer in order to have the transformed formula proved by an inference machine for classical logic. On the other hand we can try to add a logic-specific "co-processor" to the classical inference machine thus obtaining a non-classical inference machine which will directly prove the non-classical formula.

We would like to emphasize that it is not yet clear whether it is more efficient to have a rather simple inference machine which is preceded by a sophisticated input transformation or vice versa. Both approaches have their specific advantages and disadvantages and we have been developing them in parallel. As soon as they both have reached a comparable level of development we will be able to attempt a fair practical comparative evaluation.

Figure 2 provides an overview of the basic techniques which have been developed so far in order to cover the variety of possible approaches. The three above-mentioned layers of the proof process are arranged vertically whereas the three different logics are shown in horizontal order. Beginning at the top we find formulae of the three logics as input objects. From there we can apply various mechanisms which generate suitable inputs for the given inference machines. Among these mechanisms we find the NFT-module providing different techniques for normal form transformation. Practical evaluations have shown that the duration of the proof search strongly depends on the appropriate choice of such a NFT-technique [ER96]. Another example for input transformation is the logic morphism for propositional intuitionistic logic [Kor96] which will be described in detail later on. Below this layer we have three inference machines: the KoMeT-system [BBE+94] which implements various strategic variants of the extension procedure, a non-normal form Davis-Putnam prover for classical propositional logic as well as an inference machine which implements the non-normal form version of the extension procedure thereby allowing for the use of various co-processors for first order non-classical logics [OK96b]. The output layer consists of various techniques which present the generated proofs within natural and logistic calculi for the respective input logic. Obviously their application depends on the way the input-formula was processed. If, for instance, the formula was transformed by a logic morphism then the encoding of the semantical meta-knowledge has to be re-compiled in order to obtain a suitable input for the uniform proof construction mechanism [SK96] shown at the right hand side of the bottom layer. Another approach, the KonSequenz-system [Kor93], was designed to present a classical connection proof for the given formula within an intuitionistic sequent calculus *whenever this is possible.*

In the following section we shall present an example for both of the above-mentioned ways to treat meta-knowledge: a logic morphism mapping propositional intuitionistic logic to propositional classical logic will demonstrate how to handle the meta-knowledge at transformation time. In contrast to that a generic inference machine for dealing with non-normal form formulae will be used as a uniform framework in order to add co-processors for various non-classical logics.

3 Combining Classical and Non-Classical Theorem Proving

ATP in non-classical logics is usually seen as a combination between classical reasoning and reasoning about the semantical meta-knowledge for the given logic. The semantical properties for the logics we have dealt with can be formalized in terms of Kripke-Semantics [Kri63]. The basic idea behind these semantics is the notion of *possible worlds*. Roughly spoken, they represent different interpretations of the predicate symbols and terms within the logical language. Between these worlds an *accessibility relation* **R** is defined in order to describe how the possible worlds interact. The modal operators $\Box$ and $\Diamond$ reflect quantifications over possible worlds which are accessible from a given one via **R**. Thus, given a possible world w and a first order formula F then F is said to be *forced* at w (denoted $w \Vdash F$) if F it is classically satisfied by the (classical) interpretation associated to w. A modal formula $\Box F$ is forced at w if $v \Vdash F$ for every v with $w\mathbf{R}v$. A modal formula $\Diamond F$ is forced at w if $v \Vdash F$ for at least one v with $w\mathbf{R}v$. Various modal logics are determined by combining different properties of **R** (e.g. reflexivity, symmetry, transitivity, etc.).

Proof-theoretically, this world-dependent behavior is reflected in the reduction rules for modal operators within a sequent calculus for the given logic. Applying these rules may cause the deletion of certain formulae within the current sequent. This behavior results in the non-permutability of the respective reduction rules.

From this perspective, reasoning in non-classical logics amounts to classical reasoning about the first order fragment of the non-classical language within single possible worlds combined with reasoning about the modal operators w.r.t. the whole set of possible worlds and the accessibility relation between them. While the former part is usually processed within the inference machine, the latter is not necessarily so. The following two sections present two different approaches for the treatment of this additional reasoning. The first one takes place at preprocessing time. It basically formalizes the Kripke-semantical notions within the language of classical logic. The second one extends the classical inference machine by a co-processor which does the additional reasoning. It basically checks for each classical inference-step whether it is admissible w.r.t. the given modal context.

3.1 Using Classical Theorem Proving for Constructive Logics

In contrast to explicit modal logics, intuitionistic logic does not involve special operators to denote epistemological notions. Rather, the usual logical connectives are interpreted in an epistemological way. For instance, an implicative formula of the form $A \Rightarrow B$ is read as "*we possess a construction which gives us a construction for B, provided we possess a construction for A*" [Hey71]. This notion of already "possessing" a construction can be viewed as being independent of any additional information that might be available somewhere in the future. In this sense $A \Rightarrow B$ is intuitionistically valid w.r.t. some current knowledge, if B is contained within any possible extension of the current knowledge (including no extension at all) which contains A. Likewise, a negated formula $\neg A$ (intuitionistically read as "*there cannot be any construction for A*") is intuitionistically valid w.r.t. some current knowledge, if A is not contained within any possible extension of the current knowledge (again including no extension at all).

Now if we replace "*to possess a construction A*" (i.e. "*A is part of our current knowledge*") with "*A is forced at our current possible world*" then we end up with a modal concept. Possible knowledge extensions become possible worlds and the fact that a set of knowledge entities is a possible extension of another set is denoted by the accessibility relation. The concept of knowledge-extension is ensured by the so-called *heredity condition.* It demands that any atomic formula A which is forced in the current world w is so in any possible world v accessible from w:

$$w \Vdash A \text{ and } w\mathbf{R}v \quad \Rightarrow \quad v \Vdash A$$

Obviously, the accessibility relation must be reflexive and transitive because every possible set of knowledge entities is a (trivial) extension of itself and any extension of an extension of the current set in particular is an extension for the current set as well.

With these insights we are able to define a propositional intuitionistic interpretation I_j as a triple $\langle \mathbf{W}, \mathbf{R}, \iota_j \rangle$ where $\mathbf{W}$ is a (non-empty) set of worlds, $\mathbf{R}$ is a transitive and reflexive relation on $\mathbf{W}$ and ι_j is a function from $\mathbf{W}$ to subsets of the set of atomic propositional formulae such that $w\mathbf{R}v \Rightarrow \iota_j(w) \subseteq \iota_j(v)$ (i.e. the heredity condition holds) for any two $w, v \in \mathbf{R}$. We then call a formula F to be forced at some $w \in \mathbf{W}$ (denoted $w \Vdash F$) iff,

1. F is atomic and $F \in \iota_j(w)$
2. $F = F_1 \vee F_2$ and $w \Vdash F_1$ or $w \Vdash F_2$
3. $F = F_1 \wedge F_2$ and $w \Vdash F_1$ and $w \Vdash F_2$
4. $F = F_1 \Rightarrow F_2$ and $v \Vdash F_2$ provided $v \Vdash F_1$ for any v with $w\mathbf{R}v$
5. $F = \neg F_1$ and $v \nVdash F_1$ for any v with $w\mathbf{R}v$

A formula F is satisfied by an intuitionistic interpretation $I_j = \langle \mathbf{W}, \mathbf{R}, \iota_j \rangle$ (denoted $I_j \models_j F$) iff it is forced at every $w \in \mathbf{W}$. A formula F is intuitionistically valid (denoted $\models_j F$) iff it is satisfied by any intuitionistic interpretation.

A very natural approach to deal with intuitionistic validity within classical logic would be to just formalize the above definition within the language of classical first order logic. An atomic formula A forced at some world w, for instance, would then simply become $\forall v.(wRv \Rightarrow A(v))$, where $A(v)$ encodes $v \Vdash A$ and R encodes the accessibility relation $\mathbf{R}$. Likewise, $w \Vdash A \Rightarrow B$ would become $\forall v.(wRv \Rightarrow A(v) \Rightarrow B(v))$. This technique is known as the *relational translation* [Moo77, Ohl91, vBe84]. Of course, to obtain completeness one has to add some axiom-formulae which encode the properties of $\mathbf{R}$ like, for instance, reflexivity ($\forall w.wRw$). The conjunction of these axiom-formulae then must classically imply the above-mentioned translation of the input formula.

A major disadvantage of this approach lies in the potential undecidability of the resulting formula since it is formulated in full first order logic with non-monadic predicate symbols (R) involving arbitrarily nested combinations of universal and existential quantifications. If we could, however, predict a finite set of accessible worlds $R_F(w)$ for a given formula F and a possible world w, then we could use finite conjunctions over the elements of these sets instead of universally quantifying over them. $w \Vdash A \Rightarrow B$, for example, would become $\bigwedge_{v \in R_F(w)} (A(v) \Rightarrow B(v))$. So the main question to be solved is whether it is always possible to construct a finite countermodel (i.e. a countermodel with a finite set of possible worlds) for every propositional intuitionistic non-theorem. In principle this must be the case, since otherwise propositional intuitionistic logic would not be decidable. Our concern, however, was to find a sufficiently effective construction mechanism for such finite countermodels.

To this end we have been carefully studying the main mechanisms which may yield infinite countermodels: imagine, we would like to find a countermodel for the following intuitionistic non-theorem: $((D \Rightarrow A) \Rightarrow B) \wedge ((C \Rightarrow B) \Rightarrow A) \Rightarrow A$. Then this countermodel must contain a world w_0 where the above formula is not forced. By the definition of forcing for implicative formulae this amounts to the existence of another world w_1 accessible from w_0 with $w_1 \Vdash ((D \Rightarrow A) \Rightarrow B) \wedge ((C \Rightarrow B) \Rightarrow A)$, i.e. $w_1 \Vdash (D \Rightarrow A) \Rightarrow B$ as well as $w_1 \Vdash (C \Rightarrow B) \Rightarrow A$, but $w_1 \not\Vdash A$. Now, if $w_1 \Vdash (C \Rightarrow B) \Rightarrow A$ then any v accessible from there forces A provided it forces $C \Rightarrow B$. In particular this is the case for w_1 itself by the reflexivity of $\mathbf{R}$. On the other hand we know that A is not forced at w_1 so neither can be $C \Rightarrow B$. Therefore there has to be another world w_2 accessible from w_1 where C is forced but B is not. We call w_2 a world which *refutes* the implicative subformula $C \Rightarrow B$. Recall, however, that $(D \Rightarrow A) \Rightarrow B$ was forced at w_1 as well. Hence, any v accessible from w_1 forces B if it forces $D \Rightarrow A$. But then in particular this holds for w_2, so $D \Rightarrow A$ cannot be forced at w_2 since B is not forced there. This, in turn yields the existence of another world w_3 accessible from w_2 which refutes $D \Rightarrow A$, i.e. $w_3 \Vdash D$ but $w_3 \not\Vdash A$. The following picture summarizes this situation:

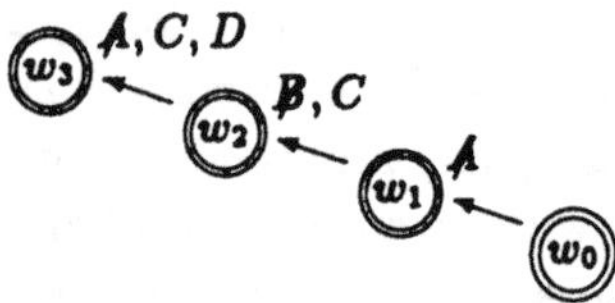

Note that the above-mentioned heredity condition homomorphically extends to arbitrary formulae, i.e. any formula forced at some world remains so for all worlds accessible from there. Hence, $w_3 \Vdash (C \Rightarrow B) \Rightarrow A$ as well, since this was the case for w_1. Now, we face about the same situation for w_3 which we had for w_1: A is not forced at w_3, thus there must be a w_4 accessible from w_3 forcing C but not B, hence neither $D \Rightarrow A$, again yielding another w_5 that forces D but not A, and so on:

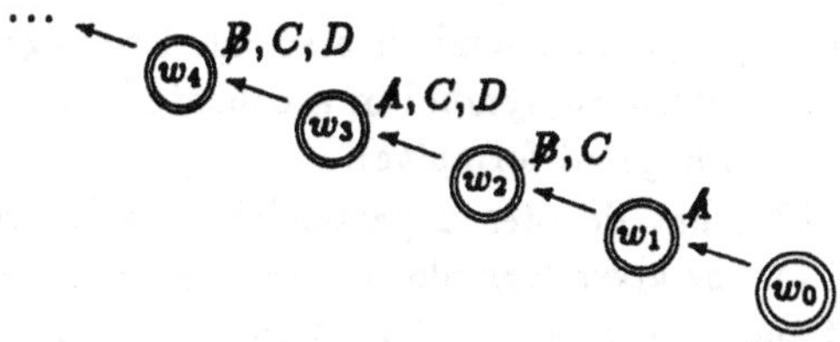

Although this model is potentially infinite, we can replace it by the finite one in the previous figure. The reason for this again stems from the heredity condition. At w_2 we find C forced for the first time. By the heredity condition this remains so for any world accessible from there. But then $C \Rightarrow B$ is forced if and only if B is. Thus, in any world accessible from w_2 $C \Rightarrow B$ becomes logically equivalent to B. Hence, in the above example we would not have to consider w_4 anymore, since knowing $w_3 \not\Vdash A$ we only need to infer $w_3 \not\Vdash B$ instead of $w_3 \not\Vdash C \Rightarrow B$. From this example we may conclude that we do not need to add any worlds refuting a certain implication as soon as our current world is accessible from one which already refutes this implication.

Besides implicative formulae also for negated formulae possible worlds have to be added to a potential countermodel. Imagine that during the attempt to construct a countermodel for a propositional formula F we need to show the existence of a possible world w_0 which does not force a specific negated subformula of F, say $\neg A$. According to the definition of intuitionistic forcing for negated formulae this amounts to the existence of at least one possible world w_1 accessible from w_0 such that $w_1 \Vdash A$. But by the heredity condition A will remain forced at every world accessible from w_1. So let us advance through an arbitrary chain of possible worlds following w_1 via the accessibility relation. Again by the heredity condition we know that every formula which becomes forced along our way will remain so from there on. But then sooner or later we must have reached a world $\max_F(w_1)$ where no other world accessible from there forces more subformulae of F than $max_F(w_1)$ itself (recall that the set of subformulae of F is finite). We shall call such a world $\max_F(w_1)$ *F-maximal*. Now the crucial feature of $\max_F(w_1)$ is that for any subformula F' of F we have $\max_F(w_1) \Vdash F'$ if and only if F' is classically satisfied by the set $\iota_j(\max_F(w_1))$. Hence, when we wish to argue that a negated subformula $\neg A$ of our input-formula F is not forced at some world w_0 we simply check if there is an F-maximal world $\max_F(w_1)$ forcing A, i.e. if A is classically satisfied by the model $\iota_j(\max_F(w_1))$. Thus, it becomes unnecessary to consider any further possible world accessible from $\max_F(w_1)$.

In order to obtain a suitable encoding of a potential intuitionistic countermodel for a given input formula F we first of all associate a unique function symbol $\mathtt{w}_i$ to each implicative or negated subformula of F which has polarity 0 (i.e. is a *positive part* in the terminology of [Sch77]). Then, beginning with a "root" possible world $\mathtt{w}_0$, we construct a set W of terms that are obtained by applying a proper choice of concatenations of the $\mathtt{w}_i$ to $\mathtt{w}_0$. Roughly spoken, the order of the $\mathtt{w}_i$ within these concatenations is determined by the tree ordering between the associated subformulae. No function symbol will occur more than once within a single such concatenation (reflecting the above-mentioned insights on implicative subformulae). Function symbols associated to negated subformulae will only occur as the outermost symbol of a concatenation (reflecting the above-mentioned insights on negated formulae).

Given a term $t \in W$ (that denotes a particular possible world within a potential countermodel for the given formula) the set $R_F(t)$ of accessible worlds will then be the set of those terms $t' \in W$ that contain t as a subterm. The translation procedure itself will come in the form $\Psi(t, F')$ meaning to encode that the subformula F' is forced at the possible world denoted by the term t. It basically reflects the definition of intuitionistic forcing as mentioned above. If, for instance, $F' = F_1' \Rightarrow F_2'$ and F' has polarity 1 (i.e. F' is possibly forced at t) then $\Psi(t, F') = \bigwedge_{t' \in R_F(t)} \Psi(t', F_1') \Rightarrow \Psi(t', F_2')$, meaning that if F' is forced at t then F_2' is so at any t' accessible from t where F_1' is forced. Conversely if F' is of polarity 0 (i.e. F' is possibly not forced at t) then $\Psi(t, F') = \Psi(\mathtt{w}_i(t), F_1') \Rightarrow \Psi(\mathtt{w}_i(t), F_2')$ where $\mathtt{w}_i$ is the unique function symbol associated to F'. This reflects that if F' is not forced at t then there is a possible world $\mathtt{w}_i(t)$ accessible from t where F_1' is forced but F_2' is not. Special care has to be taken if $\mathtt{w}_i$ already occurs somewhere in t. According to our above expositions we then know that F_1' is already forced at t. Thus the translation will evaluate to $\Psi(t, F_2')$ only in this case, indicating that F_2' is not forced at t. Hence t is a possible world accessible from itself by the reflexivity of $\mathbf{R}$ where F_1' is forced but F_2' is not. Likewise, we do not have to extend t via $\mathtt{w}_i$ if the outermost function symbol in t is associated to a negated subformula (hence denoting a F-maximal world).

To sum up we have achieved a morphism from intuitionistic to classical logic that maps propositional input formulae to propositional output formulae (note that no quantifiers or uninstantiated terms occur in the output formula). The translation result can be directly used as an input for any classical theorem prover. No theory reasoning is required at proof time anymore — neither implicitly (through an encoding within the object-language) nor explicitly (through a modification of the inference machine).

In practice, this translation technique has turned out to be quite successful. Not only it preserves decidability, hence allowing the use of a Davis-Putnam-style prover. Compared to other techniques it reduces the search space considerably and does so the more negated subformulae occur within the given input formula. Moreover, we are quite confident to be able to lift the underlying ideas to full first order intuitionistic logic. Once we have achieved this goal we then would have to provide a technique that reconstructs the meta-knowledge reasoning from a given classical proof of a transformed intuitionistic formula in order to eventually complete our translation approach. This technique, shown in figure 2 as "Inverse logic morphism", is currently under development.

3.2 A Uniform Framework for Non-Normal Form Theorem Proving

In the previous section we have presented the first approach for dealing with additional reasoning which is caused by non-classical semantics of the logical connectives. Its basic idea was to encode these semantics into a classical first order formula at preprocessing time. In the second approach which will be described below this additional reasoning is achieved by an extension of the classical inference machine, i.e. by using a specialized co-processor.

The difference to classical normal form theorem proving essentially consists of proving theorems within a non-normal form environment. On the one hand the input formula can be used in a direct way without any translations and normal form transformations. On the other hand when considering non-classical logics the semantical information (depending on the selected logic) may be processed during the proof procedure itself which makes a single inference step more complicated. Finally this additional information is contained in the resulting proof and makes it easier to construct a readable output, i.e. to present the proof within sequent or natural deduction calculi.

For this purpose we have developed a uniform framework for non-normal form theorem proving. The resulting method basically consists of a two step algorithm, i.e. a uniform procedure finding the proofs [OK96b] and a uniform transformation procedure converting these proofs into sequent-style systems [SK96]. During the development of the algorithm we have put particular emphasis on the uniformity of our environment. First of all we have to unify all logics under consideration into one system of matrix- and sequent-style calculi, i.e. fixing the *invariant* and *variant* parts of these logics when searching for matrix proofs and transforming into sequent proofs. From this we obtain the basic structure of the corresponding procedures using the invariant parts as uniform steps. The variant parts are designed in a modular way to define the special properties depending on the selected logics. This leads to generalized invariant proof and transformation procedures where the variant parts are included as a kind of *drawer–technique* for specializing to the single logics. Furthermore this approach realizes a great degree of flexibility and can be extended to other logics in an easy way.

Non-normal form Automated Theorem Proving

The theoretical foundation for our proof procedure is a matrix characterization of logical validity developed in [Wal90] where a given formula F is logically $\mathcal{L}$–valid iff each *path* through F has at least one *complementary connection* according to the selected logic $\mathcal{L}$. This characterization can be seen as a generalization of Bibel's connection method [Bib87] to capture also non-classical logics which do not have a normal form. This is achieved by introducing the syntactical concept of *prefixes* to encode the Kripke semantics of the logics explicitly into the proof search. As a result we obtain an extension of the notion of *complementarity* depending on the logic under consideration. Whereas in the classical case two atomic formulae are complementary if they have different polarity but can be unified by some substitution on their sub-terms, for non-classical logics also the prefixes of the two atomic formulae (i.e. descriptions of their modal context in the formula tree) have to be unifiable. Thus in addition to the conventional unification algorithms our procedure requires a special *string-unification* procedure [OK96a] for making two prefixes identical. Moreover, the absence of normal

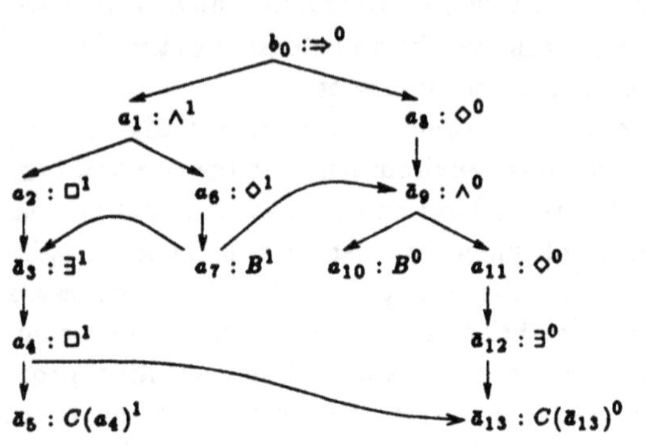

x	$pre(x)$
b_0, a_1, a_2, a_6, a_8	b_0
$\bar{a}_3, a_4$	$b_0\bar{a}_3$
$\bar{a}_5$	$b_0\bar{a}_3\bar{a}_5$
a_7	b_0a_7
$\bar{a}_9, a_{10}, a_{11}$	$b_0\bar{a}_9$
$\bar{a}_{12}, \bar{a}_{13}$	$b_0\bar{a}_9\bar{a}_{12}$

$\sigma_L(\bar{a}_3) = \sigma_L(\bar{a}_9) = a_7$
$\sigma_L(\bar{a}_{12}) = \bar{a}_5$
$\sqsubset_L = \{(a_7, \bar{a}_3), (a_7, \bar{a}_9)\}$
$\sigma_Q(\bar{a}_{13}) = a_4$
$\sqsubset_Q = \{(a_4, \bar{a}_{13})\}$
$\sigma_L^{\#}(pre(a_4)) = b_0a_7$
$\sigma_L^{\#}(pre(\bar{a}_{13})) = b_0a_7\bar{a}_5$

Fig. 3. Formula tree, prefix table, and substitutions of the formula from Example 1

forms requires a proof procedure which operates on the formula tree instead of a clause set. The notions of *current clauses* and *active paths* which are used to guide the development of a classical matrix proof in normal form have to be extended as well. The following example illustrates the appropriately extended characterization of logical validity based on non-classical connections.

Example 1. (Non-classical validity) Consider the modal formula $F \equiv \Box\exists x.\Box C(x) \wedge \Diamond B \Rightarrow \Diamond(B \wedge \Diamond\exists x.C(x))$. Its formula tree (denoted by an ordering $\ll$) is shown in figure 3 where we use *positions* for a unique indexing of all subformulae. Additionally we have associated each position x with the main operator of the corresponding subformula and its polarity. For a uniform treatment of quantifier unification σ_Q and the special string unification σ_L (specifying the logic $\mathcal{L}$) we distinguish variable positions (marked with an overbar) and constants. The table next to the tree shows the prefixes corresponding to the positions of $\ll$. Basically a prefix $pre(x)$ assigned to a position x is a string consisting of the root b_0 and all positions which dominate x in $\ll$ and are successors of positions associated to a modal main operator. For $S5$ the prefixes consist only of the last positions in these strings. Using the well-known tableaux

$$\left[\begin{bmatrix} b_0\bar{a}_3\bar{a}_5 : C(a_4)^1 & b_0a_7 : B^1 \end{bmatrix} \begin{bmatrix} b_0\bar{a}_9 : B^0 \\ b_0\bar{a}_9\bar{a}_{12} : C(\bar{a}_{13})^0 \end{bmatrix}\right]$$

Fig. 4. Matrix representation of the formula from Example 1

classification of subformulae [Wal90, Fit83] we obtain the corresponding *matrix* (two-dimensional representation) shown in figure 4 where each atom is represented next to its prefix. The vertically arranged atoms correspond to a branching at some β–formula in an analytic tableau. The horizontally written atoms are of formula-type α which does not cause a split in a tableau proof. The matrix has two paths (defined on the set of positions associated with the atoms)

$\{\bar{a}_5, a_7, a_{10}\}$ and $\{\bar{a}_5, a_7, \bar{a}_{13}\}$. The corresponding connections are $\{a_7, a_{10}\}$ and $\{\bar{a}_5, \bar{a}_{13}\}$. A *combined substitution* $\sigma = \langle\sigma_Q, \sigma_L\rangle$ can be computed as shown below the prefix table. The induced relations $\sqsubset_Q$ and $\sqsubset_L$ have already been integrated into the tree ordering $\ll$ and result in a relation $\lhd = \ll \cup \sqsubset_Q \cup \sqsubset_L$. The irreflexivity of $\lhd$ is *one* required condition for σ to be $\mathcal{L}$*–admissible*. Furthermore σ_L respects semantical conditions w.r.t. the $\mathcal{L}$–accessibility relation R_0 for the modal logics $D, D4, T, S4, S5$. The interaction between σ_Q and σ_L satisfies the cumulative and constant domain conditions for these logics, too (see [Wal90] for detailed definitions). Summarizing *all* these conditions we obtain σ being $\mathcal{L}$*–admissible* for these cases. It is easy to verify that σ makes the two connections σ–complementary and hence the corresponding paths as well. Thus F is shown to be $\mathcal{L}$–valid.

F can also be proven $\mathcal{L}$–valid in $T, S4$ and $S5$ under varying domains by extending σ_L. For $T, S4$ we must add $\sigma_L(\bar{a}_5) = \emptyset$, and for $S5$ $\sigma_L(\bar{a}_5) = a_7$ satisfying the corresponding condition.

A uniform and compact presentation of Wallen's matrix characterizations for classical, intuitionistic, and various modal logics can be found in [OK96b] or [SK96]. The proof procedure working on these compactly represented matrix characterizations has been developed in [OK96b] involving extended concepts of paths, (open) subgoals, and multiplicities for non-normal form matrices. The basic mechanism for proof search is similar to Bibel's extension procedure [Bib87]. It consists of a short and uniform (invariant) algorithm for connection driven path checking of a formula F. By following connections we avoid the notational redundancies occurring in sequent or tableaux-based proof procedures (driven by the logical connectives) whereas by considering non-normal form we reduce the size of the matrix to be searched through by a connection based proof search. The variant component of the procedure determining the complementarity of the connections strongly depends on the underlying logic $\mathcal{L}$. It has been successfully realized using the string-unification procedure [OK96a, OK96b] on the corresponding prefixes of the connected atoms. The modular design of this unification procedure and its integration into the search algorithm allows us to treat a rich variety of logics in a uniform, efficient, and simple way. We could even extend it to logics not yet considered simply by providing a component for checking complementarity according to a matrix characterization of validity in this logic.

Converting non-normal form matrix proofs into sequent-style systems

Uniform methods for proof search in a non-normal form environment are only the first step in solving problems from various application domains via ATP-methods. In a second step we also have to develop a uniform algorithm which transforms non-normal form matrix proofs into a comprehensible form. This step is necessary because the efficiency of automated proof methods based on matrix characterizations strongly depends on a compact representation of a proof. This makes it very difficult to use an automated generated proof as guideline for solving application problems.

Since the development of automated theorem provers attempts have been made to convert machine proofs into a humanly comprehensible form. Since a classical non-normal form proof explicitly preserves structural information (in contrast to a matrix proof in clause form) a corresponding sequent proof can be extracted in a direct way. For non-classical logics this easy approach becomes rather complicated since semantical information (also contained in the output "proof code") has to be integrated. In [SK96] unified and compact representations of the matrix characterizations [Wal90], *conventional* sequent calculi, and *prefixed* sequent calculi have been presented for all logics under consideration. For the representation of the sequent calculi we use a uniform system of tables to specify the variant parts of the logics. The tables reflect conditions on rule applications depending on different Kripke-semantics of the logics. For this reason these tables can be seen as the counterpart to the string unification procedure, i.e. variant parts in the matrix characterizations. The invariant parts of the sequent systems are given by one basic system of inference rules for the logical connectives.

In [SK96] we have developed a transformation procedure working on this uniform framework of logical calculi. Starting with a matrix proof of a formula F we obtain a combined substitution $\sigma = \langle \sigma_Q, \sigma_L \rangle$, which induces, together with the tree ordering of F, an ordering $\triangleleft$ on the nodes of the formula tree. This ordering efficiently encodes the fact that within a (top-down) sequent or tableaux proof there are certain restrictions on the order of rule applications which are necessary to "reduce" the formula F in order to reach the axioms of the calculus. Thus the basic idea of our algorithm is very simple. Essentially we have to traverse $\triangleleft$, to select appropriate sequent rules according to the subformula represented by each node and its polarity, and to keep track of all the subgoals already solved. No search will be involved in the transformation. From the uniform representation of the sequent calculi we obtain a set of tables to be consulted by the algorithm when determining an appropriate sequent rule corresponding to a given node. Positions of type β cause a sequent proof to split into two independent subproofs and force us to determine which sub-relations of $\triangleleft$ will be relevant in each subproof. Finally, since $\triangleleft$ does not uniquely determine the order of rule applications, we have to identify proof-relevant positions in the formula tree which have priority over others. These may again depend on the underlying logic and force us to insert *wait*-labels to certain nodes in order to keep them from being reduced too early.

Prefixed sequent systems. Because of a strong similarity between the matrix characterizations and Fitting's *prefixed tableaux systems* [Fit69] we have developed our algorithm in a first phase to convert a matrix proof into a *prefixed* sequent proof. From the prefixes computed during proof search via string unification we obtain the prefixes for the sequent proof in a direct way. Hence, the semantical information is preserved for prefix construction in the prefixed sequent proof. The input of the uniform transformation algorithm $TOTAL$ is given by a reduction ordering $\triangleleft$ and a logic $\mathcal{L}$ such that $\triangleleft$ represents a matrix proof in $\mathcal{L}$. $TOTAL$ traverses $\triangleleft$ to compute a *totalization* of the partial ordering

$$
\dfrac{\dfrac{\dfrac{\dfrac{\dfrac{\dfrac{\dfrac{b_0a_7\bar{a}_5 : C(a_4),\ b_0a_7 : B \vdash b_0a_7 : B \qquad \dfrac{\dfrac{b_0a_7\bar{a}_5 : C(a_4),\ b_0a_7 : B \vdash b_0a_7\bar{a}_5 : C(a_4)}{b_0a_7\bar{a}_5 : C(a_4),\ b_0a_7 : B \vdash b_0a_7\bar{a}_5 : \exists x.C(x)}\ \exists r\ (\bar{a}_{12}, \bar{a}_{13})}{b_0a_7\bar{a}_5 : C(a_4),\ b_0a_7 : B \vdash b_0a_7 : \Diamond\exists x.C(x)}\ \Diamond r\ (a_{11})}{b_0a_7\bar{a}_5 : C(a_4),\ b_0a_7 : B \vdash b_0a_7 : B \wedge \Diamond\exists x.C(x)}\ \wedge r\ (\bar{a}_9, a_{10})}{b_0a_7\bar{a}_5 : C(a_4),\ b_0a_7 : B \vdash b_0 : \Diamond(B \wedge \Diamond\exists x.C(x))}\ \Diamond r\ (a_8)}{b_0a_7 : \Box C(a_4),\ b_0a_7 : B \vdash b_0 : \Diamond(B \wedge \Diamond\exists x.C(x))}\ \Box l\ (a_4, \bar{a}_5)}{b_0a_7 : \exists x.\Box C(x),\ b_0a_7 : B \vdash b_0 : \Diamond(B \wedge \Diamond\exists x.C(x))}\ \exists l\ (\bar{a}_3)}{b_0 : \Box\exists x.\Box C(x),\ b_0a_7 : B \vdash b_0 : \Diamond(B \wedge \Diamond\exists x.C(x))}\ \Box l\ (a_2)}{b_0 : \Box\exists x.\Box C(x),\ b_0 : \Diamond B \vdash b_0 : \Diamond(B \wedge \Diamond\exists x.C(x))}\ \Diamond l\ (a_6, a_7)}{\vdash b_0 : \Box\exists x.\Box C(x) \wedge \Diamond B \Rightarrow \Diamond(B \wedge \Diamond\exists x.C(x))}\ \Rightarrow l\ (b_0),\ \wedge l\ (a_1)
$$

Fig. 5. Prefixed sequent proof of the formula from Example 2

$\lhd$. From this we obtain a linear sequence of sequent rules which forms a proof in the prefixed sequent system corresponding to $\mathcal{L}$. The specific sequent rules and their parameters (i.e. prefixes and terms) are determined by the main operator of the currently visited node in $\lhd$ and by the substitutions σ_L and σ_Q. The *correct* application of sequent rules and construction of prefixes will be guaranteed by using our table system according to the selected logic $\mathcal{L}$. For the details we refer again to [SK96].

Example 2. (Conversion into prefixed sequent systems) From Example 1 we take the formula $F \equiv \Box\exists x.\Box C(x) \wedge \Diamond B \Rightarrow \Diamond(B \wedge \Diamond\exists x.C(x))$, the substitutions σ_L, σ_Q, and the reduction ordering $\lhd$ generated from the matrix proof. For *all* positions we obtain the prefixes under σ_L shown in the following table:

x	b_0, a_1, a_2, a_6, a_8	$\bar{a}_3, a_4, a_7, \bar{a}_9, a_{10}, a_{11}$	$\bar{a}_5, \bar{a}_{12}, \bar{a}_{13}$
$\sigma_L^{\#}(pre(x))$	b_0	b_0a_7	$b_0a_7\bar{a}_5$

We start our algorithm *TOTAL* traversing the reduction ordering $\propto^*$ for the logics $D, D4, T, S4$ with cumulative domains. The total ordering of positions encoding the sequent rules is given by

$$[b_0, a_1, a_6, a_7, a_2, \bar{a}_3, a_4, \bar{a}_5, a_8, \bar{a}_9, a_{10}, a_{11}, \bar{a}_{12}, \bar{a}_{13}]$$

The corresponding prefixed sequent proof respecting this ordering looks like shown in figure 5. The goal sequent of the prefixed sequent proof is given by $\vdash b_0 : F$. For explanation of the traversing process we use the picture of $\lhd$ from Example 1. The conversion algorithm starts with positions b_0, a_1 where we construct the the rules $\Rightarrow r : b_0$ and $\wedge l : b_0$ (in the following we write the constructed prefix and quantifier parameter behind the rule name). After skipping a_8 we are blocked at $\bar{a}_9$ because of the reduction ordering $(a_7, \bar{a}_9) \in \sqsubset_L$. We continue the conversion at a_6 which can be seen as a goal-oriented selection in order to remove the effect of blocking at $\bar{a}_9$. We obtain the rule $\Diamond l : b_0a_7$ and the atom $B^1 : b_0a_7$ solving a_7. Now we can delete the blocking arrow pointing to $\bar{a}_9$. Next we skip a_2 and, reaching $\bar{a}_3$, construct the two rules $\Box l : b_0a_7$ and $\exists l : b_0a_7; a_4$ (introducing the eigenvariable a_4 and deleting the arrow blocking $\bar{a}_{13}$). Visiting a_4 and $\bar{a}_5$ the atom $C(a_4)^1 : b_0a_7\bar{a}_5$ can be isolated in the sequent proof. We make up the reduction $\Diamond r : b_0a_7$ at a_8 and afterwards, the application of $\wedge l : b_0a_7$ at β–position $\bar{a}_9$ forces a split of the reduction ordering into two independent suborderings $\lhd_1, \lhd_2$ (detailed definitions given in [SK96]). For $\lhd_1$, we solve a_{10} having the atom $B^0 : b_0a_7$, and hence an axiom rule with $B^1 : b_0a_7$ (a_7 already solved). For reducing $\lhd_2$ we visit $a_{11}, \bar{a}_{12}$ and $\bar{a}_{13}$ obtaining $\Box r$: $b_0a_7\bar{a}_5$ and $\exists r : b_0a_7\bar{a}_5; a_4$, where $\sigma_Q(\bar{a}_{13}) = a_4$ has been used at $\bar{a}_{13}$. Finally, the sequent proof will be finished solving $\bar{a}_{13}$, i.e. applying the axiom rule with $C(a_4)^0 : b_0\bar{a}_7\bar{a}_5$.

The eigenvariable a_4 is associated with the prefix b_0a_7 and the prefix at $\bar{a}_{13}$ is given by $b_0a_7\bar{a}_5$. We use a_4 for the quantifier reduction $\exists r$ with $\sigma_Q(\bar{a}_{13}) = a_4$ satisfying the cumulative domain condition on the semantics of the considered logics, i.e. $b_0a_7 \, R_0 \, b_0a_7\bar{a}_5$ (where R_0 denotes the accessibility relation on prefixes [Fit83]). The sequent proof can be extended to $T, S4$ for varying domains by extending the modal substitution $\sigma_L(\bar{a}_5) = \emptyset$.

Conventional sequent calculi. The transformation algorithm $TOTAL$ developed in the first phase realizes conversion into prefixed sequent systems only. The advantage lies in an easy re-integration of semantical meta-knowledge by using the prefixes from the matrix proof in a direct way. Furthermore we are able to capture modal logics for which there are no cut-free conventional sequent calculi (for example $S5$ and all modal logics under consideration with constant domains). The disadvantage of converting into prefixed sequent systems is given by little practical use of these calculi in real application systems.

The latter fact leads to the second phase of developing a uniform conversion procedure, i.e. to extend the algorithm $TOTAL$ into $TOTAL^*$ which transforms non-normal form matrix proofs into conventional sequent calculi [Gen35, Wal90]. For non-classical logics we have to consider the fact that prefixes cannot be used explicitly. The non-permutabilities of inference rules are now encoded by the structure of the rules themselves which may delete sequent formulae during a reduction. In this case the absence of prefixes makes it necessary to deal with β-splits and to consider the additional priorities of proof-relevant positions while the essential algorithm remains unchanged.

Because of the compact representation of matrix proofs it is not trivial to extract the sequent proofs in a direct way. We have to develop a concept for re-building the notational redundancies which were eliminated in the matrix proof but still have to occur in a sequent proof. The main problem occurs when splitting at β–positions in the transformation process. In this case one has to decide which sub-relations of $\lhd_1, \lhd_2$ are still relevant in the matrix (sub-)proof and which subformulae are necessary for further reductions closing the corresponding branch in the sequent proof.

In our general transformation algorithm a more detailed consideration of β–nodes can have two different effects. When creating prefixed sequent systems (denoting a *lower* transformation level) all semantical informations concerning the specific logic are encoded by the prefixes of the sequent formulae. In this case deleting *irrelevant sub-relations* after a β–split can be viewed as an optimization which eliminates redundancies. In contrast to that a conversion into conventional sequent calculi without explicit prefixes has to take into account that the semantics of a logic, in particular the non-permutabilities of rule applications, is now completely contained in the structure of the inference rules. Thus the reduction ordering $\triangleleft$ has to be extended by additional ordering constraints (called *wait*–labels) which cause deletion of sub-relations after β–splits to be more than an optimization feature. Now these operations will be essential for preserving the completeness of our transformation procedure.

In [Sch95] we have developed a concept of reductions on non-normal form matrix proofs for intuitionistic logic which respects these optimization- and completeness features. In [SK96] we have extended this concept to modal logics, especially the definitions of required *wait*–labels for making the reduction ordering $\triangleleft$ complete. These *wait*–labels depend on the selected logic $\mathcal{L}$ and, hence can be seen as an additional variant part for conversion in conventional sequent calculi. The whole concept of β–splits is one of the theoretical foundations for the uniform transformation algorithm $TOTAL^{\star}$. One basic principle is the concept of proof relevant positions, where a position x is said to be *relevant* if some successor leaf position c is part of a connection $\{c, c_2\}$. However, we were able to show that for some of the logics K, $K4$ and $D4_{co}$ (*co* means in *constant domains*) this principle fails. There are theorems which have *pure* relevant sub-formulae not involved in any connection of the matrix proof [1]. For all other logics under consideration one should use the deletion of reductions after splitting even when creating prefixed sequent proofs.

Both parts, the proof search and the transformation procedure can be integrated into an application system (like a proof development system) by providing appropriate interfaces similar to the ones described in [KOS95]. We hope that this general methodology will make ATP useful in a rich variety of application domains.

4 Future Work

This paper has provided an overall view of the development of an ATP-system which is able to handle three different types of applications in a tailored way. In this last section a short outlook into future work will be given. We restrict this outlook mainly to the application of automated theorem proving techniques for the development of correct programs from formal specifications which is one of the three areas of applications discussed in the paper.

[1] $\Diamond B$ in the K-theorem $\Box A \wedge \Diamond B \Rightarrow \Diamond A$, and $\Diamond\Diamond B$ in the $D4_{co}$–theorem $\forall x.\Box\Box A(x) \wedge \Diamond\Diamond B \Rightarrow \Diamond\forall x.A(x)$.

Currently, we are developing a C-implementation of our connection-based proof procedure for intuitionistic logic [OK95, OK96b] and of our method for converting matrix proofs into intuitionistic sequent proofs [SK95, SK96]. We also implement an interface which integrates the resulting software with the interactive proof/program development system NuPRL [C+86]. This will enable to use an efficient proof procedure interactively [KOS95] as a tactic within the NuPRL system for solving sub-problems from first order logic, the need for which arises during the process of deriving programs with the NuPRL system.

Furthermore we intend to investigate how inductive proof methods can be integrated into program synthesis systems by use of the same technology as just described. Further research focuses on providing a rigorous formalization of the meta-theory of programming in order to raise the level of formal reasoning in formal program synthesis from low level calculi to one comprehensible for programmers and to implement program development strategies on this level [Kre96]. These are considered to be steps on the way towards enabling a user of a program development system to focus on the key ideas in program design while being freed from formal details which in addition are needed to ensure correctness.

It seems very likely that our techniques can be generalized to some subset of linear logic and other calculi in use in artificial intelligence and computer science. In fact we are currently working on a matrix-characterization for validity in linear (and other) logics in order to provide the basis for extending our combined ATP system to a larger variety of application domains.

The last topic of future research mentioned here in the context of the material presented in this paper concerns the enhancement of efficiency of the resulting system. We plan to investigate how the techniques, used in systems like Setheo [LBB92] and KoMeT [BBE+94] and resulting in the high performance of these systems, can be imported into our combined system. Examples of the kind of techniques we have in mind are preprocessing techniques and the use of typing information during unification. As mentioned in Section 3.1 we also aim at using these classical proof systems directly as inference machines via semantics-based translation techniques [Kor96]. We plan to lift the existing approach to full intuitionistic predicate logic. In order to re-integrate the generated proofs into the NuPRL-environment we are currently developing an appropriate postprocessing mechanism.

Experiments with examples from applications will eventually have to demonstrate whether after such provisions non-normal form proof techniques have an advantage over conventional clause-form theorem provers which is an open question of practical relevance up to this day.

References

[And80] P. ANDREWS. *Transforming matings into natural deduction proofs.* In *CADE-5*, LNCS 87, pp. 281–292. Springer, 1980.

[And91] P. ANDREWS. *More on the problem of finding a mapping between clause representation and natural-deduction representation. Journal of Automated Reasoning*, Vol. 7, pp 285–286, 1991.

[BC85] J. L. BATES, R. L. CONSTABLE. *Proofs as programs. ACM Transactions on Programming Languages and Systems*, 7(1):113–136, January 1985.

[Bib87] W. BIBEL. *Automated Theorem Proving.* Vieweg Verlag, [2]1987.

[BBE⁺94] W. BIBEL, S. BRÜNING, U. EGLY, T. RATH. *Komet.* In *CADE-12*, LNAI 814, pp. 783–787. Springer, 1994.

[Bra75] D. BRAND *Proving theorems with the modification method.* In J. Hopcroft e. a., eds. *SIAM Journal of Computing Vol. 4*, pp. 412–430. Philadelphia, 1975.

[C⁺86] R. L. CONSTABLE ET. AL. *Implementing Mathematics with the NuPRL proof development system.* Prentice Hall, 1986.

[DGH⁺94] B. I. DAHN, J. GEHNE, T. HONIGMANN, L. WALTHER, A. WOLF. *Integrating Logical Functions with ILF*, Preprint 94-10, Humboldt University Berlin, 1994.

[ER96] U. EGLY, T. RATH. *On the practical value of different definitional translation to normal form.* To appear in *CADE-13*, Springer, 1996.

[Fit69] M. C. FITTING. *Intuitionistic logic, model theory and forcing.* Studies in logic and the foundations of mathematics. North–Holland, 1969.

[Fit83] M. C. FITTING. *Proof Methods for Modal and Intuitionistic Logic.* D. Reidel, 1983.

[Gen95] K. GENTHER. *Repräsentation von Konnektionsbeweisen in Gentzen-Kalkülen durch Transformation und Strukturierung.* Tech. Rep. AIDA-95-12, TH Darmstadt, 1995.

[Gen35] G. GENTZEN. *Untersuchungen über das logische Schließen. Mathematische Zeitschrift*, 39:176–210, 405–431, 1935.

[Hey71] A. HEYTING. *Intuitionism — An Introduction.* North Holland Publishing Company, Amsterdam, [3]1971.

[Kor93] D. KORN. *KonSequenz — ein Konnektionsmethoden-gesteuertes Sequenzenbeweisverfahren.* Master-thesis, TH Darmstadt, 1993.

[Kor96] D. KORN. *Efficiently Deciding Intuitionistic Propositional Logic via Translation into Classical Logic.* Tech. Rep. AIDA-96-09, TH Darmstadt, 1996.

[Kre96] C. KREITZ. *Formal Mathematics for Verifiably Correct Program Synthesis. Journal of the Interest Group in Pure and Applied Logics (IGPL)*, 4(1):75–94, 1996.

[KOS95] C. KREITZ, J. OTTEN, S. SCHMITT. *Guiding Program Development Systems by a Connection Based Proof Strategy.* In M. Proietti, editor, 5th *International Workshop on Logic Program Synthesis and Transformation*, LNCS 1048, pp. 137–151, Springer Verlag, 1996.

[Kri63] S. A. KRIPKE. *Semantical analysis of modal logic I. Normal modal propositional calculi.* Zeitschrift f. mathematische Logik u. Grundlagen d. Mathematik 9, pp 67–96, 1963.

[LBB92] R. LETZ, J. SCHUMANN, S. BAYERL, W. BIBEL. SETHEO: A high-performance theorem prover. *Journal of Automated Reasoning*, 8:183–212, 1992.

[Lin89] C. LINGENFELDER. *Structuring computer generated proofs*. *IJCAI-89*, 1989.

[Lin90] C. LINGENFELDER. *Transformation and Structuring of Computer Generated Proofs*. PhD thesis, Universität Kaiserslautern, 1990.

[Moo77] R. C. MOORE. *Reasoning about Knowledge and Action* IJCAI-77, pp 223–227, Stanford, California 94305, 1977.

[Ohl91] H. J. OHLBACH. *Semantics-Based Translation Methods for Modal Logics*. *Journal of Logic and Computation*, Vol. 1, no. 6, pp 691–746, 1991.

[OK95] J. OTTEN, C. KREITZ. *A connection based proof method for intuitionistic logic*. *TABLEAUX-95*, LNAI 918, pp. 122–137, Springer, 1995.

[OK96a] J. OTTEN, C. KREITZ. *T-string-unification: unifying prefixes in non-classical proof methods*. *TABLEAUX-96*, LNAI 1071, pp. 244–260, Springer, 1996.

[OK96b] J. OTTEN, C. KREITZ. *A Uniform Proof Procedure for Classical and Non-Classical Logics*. To appear in *KI-96*, Springer, 1996.

[Pfe87] F. PFENNING. *Proof Transformations in Higher-Order Logic*. PhD thesis, Carnegie Mellon University, January 1987.

[Rob65] J. A. ROBINSON., *A machine-oriented logic based on the resolution principle Journal of ACM*, Vol. 12, pp 23–41, 1965.

[RW69] G. ROBINSON AND L. WOS., *Paramodulation and theorem proving in first order theories with equality*. In B. Meltzer and D. Michie, eds., *Machine Intelligence 4*, pp. 135-150, Edinburgh University Press, 1969.

[Sch95] S. SCHMITT. *Ein erweiterter intuitionistischer Sequenzenkalkül und dessen Anwendung im intuitionistischen Konnektionsbeweisen*. Tech. Rep. AIDA-95-01, TH Darmstadt, 1995.

[SK95] S. SCHMITT, C. KREITZ. *On transforming intuitionistic matrix proofs into standard-sequent proofs*. *TABLEAUX–95*, LNAI 918, pp. 106–121, Springer, 1995.

[SK96] S. SCHMITT, C. KREITZ. *Converting Non-Classical Matrix Proofs into Sequent-Style Systems*. To appear in *CADE–13*, Springer, 1996.

[Sch77] K. SCHÜTTE. *Proof theory*. Springer-Verlag, New York, 1977.

[vBe84] J. VAN BENTHEM. *Correspondence Theory*. In: *D. Gabbay and F. Guenther (eds.): Handbook of Philosophical Logic*, Vol. II, pp. 167–247, D. Reidel Publishing Company, Dordrecht, 1984.

[Wal90] L. WALLEN. *Automated deduction in non-classical logics*. MIT Press, 1990.

[Wos90] L. WOS. *The problem of finding a mapping between clause representation and natural-deduction representation*. *Journal of Automated Reasoning*, Vol. 6, pp 211–212, 1990.

Σ^{IT} – A Strongly-Typed Embeddable Computer Algebra Library

Manuel Bronstein

Institute for Scientific Computation
ETH Zentrum IFW
CH-8092 Zürich, Switzerland

Abstract. We describe the new computer algebra library Σ^{IT} and its underlying design. The development of Σ^{IT} is motivated by the need to provide highly efficient implementations of key algorithms for linear ordinary differential and (q)-difference equations to scientific programmers and to computer algebra system users, regardless of the programming language or interactive system they use. As such, Σ^{IT} is not a computer algebra system per se, but a library (or substrate) which is designed to be "plugged" with minimal efforts into different types of client applications.

1 Introduction

Σ^{IT} is a new computer algebra library under development at the ETH Zürich. It is not designed to evolve a general purpose computer algebra system, but to the contrary, it is designed to provide reusable and efficient implementations of highly specialized primitives and algorithms for linear ordinary differential and (q)-difference equations, either to any other computer algebra systems, or to scientific programmers coding in more traditional compiled languages. The need for providing such primitives was recognized by the EC CATHODE working group, whose solution to this problem was to provide primitives with compatible naming schemes and identical semantics written in all of `axiom`, `maple` and `reduce`. This allows users of those primitives to program higher level algorithms, e.g. the Kovacic algorithm for second order linear ordinary differential equations [6], on top of those primitives in a portable way accross those three systems. Our approach is to provide a single, compiled version of those algorithms, and make it both linkable into applications, or callable from any interactive shell. This single-source approach allows for easier maintenance, tuning, and integration of future algorithmic improvements. Thus, our emphasis is not on wide coverage of algebra, but on providing selected state of the art algorithms in a "plug-and-play" fashion to a wide range of different client applications. This is what we mean by the term "embeddable" when it appears in this paper. The closest previous attempt to provide such services to the scientific programming community is the Weyl computer algebra substrate [14], which can be seen as a set of algebraic datatypes that can be embedded into a Lisp system, providing it with computer algebra services. We describe in this paper the features of Σ^{IT} that make it significantly different from other libraries, namely:

- its use of the Aldor[1] programming language [12] and of its OOP features;
- its emphasis on skew-polynomials and the other ground mathematical objects of pseudo-linear algebra [4];
- its smooth merging of both a strongly-typed object-oriented and functional interface for low-level linking, and a typeless string-based interface for high-level process to process communication. This efficient merging is the key to efficient embeddability at all levels without type inference.

2 The use of Aldor

Σ^{IT} is entirely written in the Aldor programming language [12]. Aldor is a new compiled language, which can be described as a second generation successor of the language used to program the axiom library [5]. We refer to [12, 13] for more details about Aldor. Σ^{IT} has been designed from the start to take advantage of the following language features, whose benefits for programming algebraic algorithms are well known:

- Object-oriented programming with templating, subclassing and multiple inheritance. This allows for generic algorithms over mathematical categories.
- Aldor functions produce C-linkable object code. This allows for linking together Aldor modules with C, C++ and Fortran libraries.
- Aldor provides an abstraction level similar to axiom, so there is no need to use two different languages, one for the kernel (typically C/C++ for efficiency) and one for the high level library code (typically interpreted).
- Similarly to Lisp, Aldor provides built-in big integers, transparent dynamic allocation and a garbage collector, freeing the programmer from such tasks as memory management.

We illustrate the high-level of mathematical abstraction provided by Aldor and Σ^{IT} in Section 4, where we describe the use of Σ^{IT} within Aldor programs. A side-benefit of using Aldor, is the use of the asdoc LaTeX2e document class [7], which has been developped specially for the purpose of processing documentation in Aldor source files. Thanks to this class, each Σ^{IT} exported type and function is documented in the source code, producing a reference manual with a unix-like manual page for each exported name.

3 The algebraic facilities

As explained in the introduction, the goal is to provide primitives and algorithms for the manipulations of skew-polynomials [4]. Those objects form a common abstraction of linear ordinary differential and (q)-difference operators, so for example the Weyl algebra $\mathbb{Q}[x, d/dx]$ is a particular skew-polynomial ring. Since those rings are noncommutative, the basic category hierarchy provided by Σ^{IT} includes several noncommutative classes (Fig. 1). It was chosen to provide only categories

[1] Aldor is mostly known under its internal working names $A^{\#}$ and AXIOM-XL

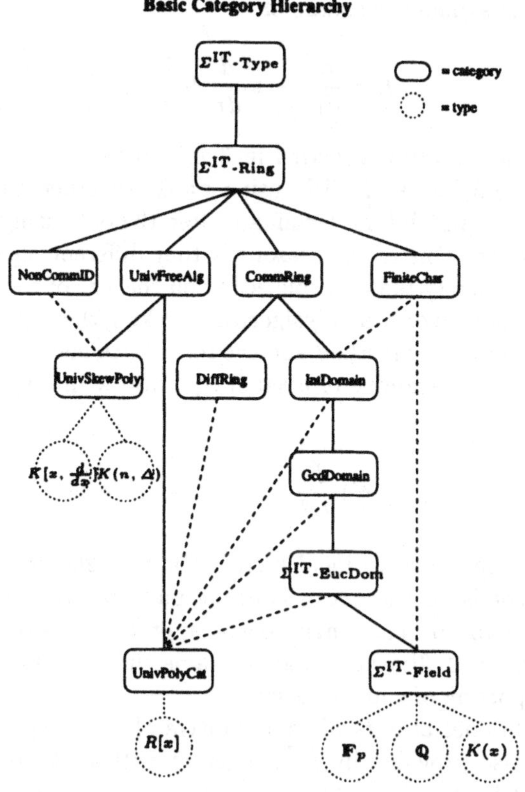

Fig. 1. The Σ^{IT} basic category hierarchy

for which there are existing types and implemented generic algorithms in the library, rather than to attempt as wide a coverage as in the `axiom` category hierarchy [5]. The resulting category structure is a rooted DAG. Σ^{IT} provides implementations of all the CATHODE primitives for arbitrary skew-polynomials, *i.e.*:

- ring arithmetic, plus left and right Euclidean division;
- left and right greatest common divisors and least common multiples;

In addition, adjoints and symmetric powers [9] are provided for differential operators. The only significant higher-level operation provided for a differential operator L is the computation of its rational kernel, i.e. $\ker(L) \cap \mathbb{Q}(x)$. This operation is a key primitive of differential Galois theory [11], and by using polyalgorithms and recent improvements [1], our implementation of rational kernel is significantly faster than the existing alternatives that are distributed with commercial computer algebra. For example, on a single-processor SPARC 10/41, Σ^{IT} was able to compute the symmetric square of

$$L = \frac{d^7}{dx^7} - x\,\frac{d}{dx} + \frac{1}{2} \qquad (1)$$

and its one-dimensional rational kernel in 3 minutes of CPU time. The same computation took 3.5 hours of CPU using `maple` together with its `ratlode` share-library package, and did not terminate after 12 hours using the `ratDsolve` `axiom` library function. One of the reasons is that different algorithms for this problem have a complexity that is cubic either on the size of the operator, or on the size of its solutions, so only a polyalgorithm making the right choice between the undetermined coefficient method and the recurrence method of [1] can be efficient on all inputs. An example of a small equation with large solution is the legendre operator

$$L_n = (1-x^2)\frac{d^2}{dx^2} - 2x\,\frac{d}{dx} + n(n+1)\,,$$

which has a one-dimensional rational kernel. For $n = 200$, Σ^{IT} took 3.8 CPU seconds vs. 6.3 seconds for the `maple ratlode` package, while for $n = 500$ this ratio becomes 18.6 to 40. The single-source approach makes algorithmic improvements such as [1, 2, 3] available at the same time to all computer algebra systems, with no porting of code necessary.

The more traditional objects of commutative algebra, e.g. univariate polynomial rings are also provided in Σ^{IT}, together with a standard set of basic algorithms, e.g. modular gcd's, polynomial remainder sequences, resultants, and factorization in $\mathbb{F}_p[x]$, $\mathbb{Z}[x]$ and $\mathbb{Q}[x]$. Types for matrices and vectors are also included. All the relevant types for polynomials, operators and linear algebra are provided in both dense and sparse versions. As of January 96, the current count of exported operations is over 100 type constructors, exporting over 500 different names (each name, e.g. `+` corresponds to many different functions).

4 Low-level embeddability

By low-level embeddability, we mean the use of Σ^{IT} as a strongly-typed library to be linked in an application program. As for any other compiled language, Σ^{IT} is provided as an archive of object files, together with one include file `sumit.as`, and including it into any `Aldor` program gives access to the full Σ^{IT} type system and its algorithms. Since Σ^{IT} is a compatible extension of the `axllib` `Aldor` standard library, one just needs to replace the `#include "axllib"` line by `#include "sumit"`. The following sample program, which computes the rational kernel of the symmetric square of the operator (1) illustrates the abstraction level provided by `Aldor` and Σ^{IT}. Note the use of calls to `monom` to create the independent variable and derivation, and the required coercion of the rational number 1/2 to a differential operator. Those strong-typing "annoyances" for non-programming users will not appear in the typeless interface presented in the next section.

```
----------------------- Katz' G2 example -------------------------
#include "sumit"
macro {
     Z   == Integer;                            Q  == Quotient Z;
     QX  == DenseUnivariatePolynomial(Q, "x"); Qx == Quotient QX;
     QXD == LinearOrdinaryDifferentialOperator(QX, "D");
}
main():() == {
     import from SingleInteger, Z, Q, QX, Qx, QXD;
     import from Vector Qx, LODORationalSolutions(QX, QXD);
     x:QX   := monom;                   -- create independent variable
     dx:QXD := monom;                   -- create derivation
     L := dx^7 - x*dx + (1/2)@Q::QX::QXD;
     V := rationalKernel symmetricPower(L, 2);
     print<<#V<<"-dimensional space of solutions found:"<<newline;
     for i in 1..#V repeat print << V.i << newline;
}
main();
```

Since `Aldor` is a general purpose programming language, Σ^{IT} can be used as a computer algebra engine inside programs whose main emphasis is not computer algebra. In fact, the only published research use of Σ^{IT} to date is of this form: it was used as a skew-polynomial engine inside a symbolic-numeric implementation of Balser's multisummation algorithm for divergent series [8]. The resulting program is a symbolic-numeric software that links in addition to a C-graphics library for rendering its results. The linked use of Σ^{IT} is of course its most flexible and efficient use.

A direct benefit of the use of the `Aldor` language is that Σ^{IT} can be linked in the above fashion to C/C++ and Fortran programs, through compiled stubs for each function used. Those compiled stubs only need to convert the arguments between C and `Aldor` types, so binary data can still be shared. Here is an example of a stub exporting polynomial factorization over $\mathbb{F}_p$ to C/C++ programs. Note the `export` statement making its arguments available to client programs, as well as the storage of the coefficient prime field into the POLY structure, which is itself exported to client programs.

```
----------------------------- dupfp.as --------------------------
-- C interface to dense Fp[X] factorization
#include "sumit"
macro {
    Z      == SingleInteger;      DUP == DenseUnivariatePolynomial;
    POLY   == Record(F:PrimeFieldCategory, Fx:DUP F);
    FACTOR == Record(factor:POLY, exponent:Z);
    FPOLY  == Record(lc:Z, size:Z, factors:PrimitiveArray FACTOR);
}
export {
    FpXfactor: POLY -> FPOLY;
```

```
    FpXmake:   (Z, PrimitiveArray Z, Z) -> POLY;
    FpXprint:  (POLY, String) -> ();
} to Foreign C;

FpXprint(p:POLY,x:String):() == print0(explode p)(x);
local print0(Fp:PrimeFieldCategory,p:DUP Fp)(x:String):()==print(p,x);

FpXmake(deg:Z, c:PrimitiveArray Z, charac:Z):POLY == {
    macro Fp == SmallPrimeField charac;
    import from Integer, Fp, DUP Fp;
    p:DUP Fp := 0;
    for i in 0..deg repeat p:=add!(p,monomial(c(i+1)::Fp,i::Integer));
    [Fp, p];
}

FpXfactor(p:POLY):FPOLY == factor0 explode p;
local factor0(Fp:PrimeFieldCategory, p:DUP Fp):FPOLY == {
    import from Z, Product DUP Fp, FACTOR, PrimitiveArray FACTOR;
    import from PrimeFieldUnivariateFactorizer(Fp, DUP Fp);
    import from Integer, Record(lcoeff:Fp, factors:Product DUP Fp);
    r := factor p;    v:PrimitiveArray FACTOR := new #(r.factors);
    for term in r.factors for i in 1@Z.. repeat {
        (f, n) := term;   v.i := [[Fp, f], retract n];
    }
    [lift(r.lcoeff), #(r.factors), v];
}
```

To the above stubs corresponds the C header file **sumit_dupfp.h**, which declares the 3 functions that are exported from the stub, but using C types as arguments:

```
/* sumit_dupfp.h: header for sumit interface to Fp[x] factorizer */
typedef void *FPX;

typedef struct single_factor {
    FPX factor;
    unsigned int exponent;
} FACTOR;

typedef struct factored_poly {
    int leadingCoefficient;
    unsigned int numberOfFactors;
    FACTOR **factors;
} FACTORED;

extern FPX  FpXmake(unsigned int deg, int *coeffs, unsigned int charac);
extern void FpXprint(FPX poly, char *varname);
extern FACTORED *FpXfactor(FPX poly);
```

The above header file is the only visible part of the stub for client programs. For example, the following C program computes the factorization of $x^7 + 5x^6 + 7x^5 + 9x^4 + 3x^3 + 4x^2 + x + 6$ in $\mathbb{F}_{11}[x]$:

```
#include <stdio.h>
#include "sumit_dupfp.h"
main() {        /* x^7+5x^6+7x^5+9x^4+3x^3+4x^2+x+6*/
    int i, coeff[8] = {6, 1, 4, 3, 9, 7, 5, 1};
    FPX p = FpXmake(7, coeff, 11);    /* F_11[x] */
    FACTORED *fp = FpXfactor(p);
    printf("The factorization of $");
    FpXprint(p, "x");
    printf("$ in $F_11[x]$ is: $$\n %d ", fp->leadingCoefficient);
    for (i = 0; i < fp->numberOfFactors; i++) {
        printf("(");
        FpXprint(fp->factors[i]->factor, "x");
        printf(")^%d ", fp->factors[i]->exponent);
    }
    printf("\n$$\n");
}
```

Embedding Σ^{IT} into Fortran programs is done in a similar fashion, with stubs and `export` statements. Because the **Aldor** compiler can produce Lisp code, it is also possible to load the entire Σ^{IT} type system into a Lisp system, for example into **axiom**. Although the linked use of Σ^{IT} into other compiled languages is less flexible than into **Aldor** programs, this makes computer algebra algorithms available with minimal effort to general purpose programs. Such computer algebra services are already being provided by the many C-based computer algebra kernels available (e.g. PARI, SACLIB, GB, KANT, SIMATH, GANITH, KAN), but none of them provides algorithms for linear operators, with the exception of KAN [10], which provides however a different algorithm (Gröbner bases).

5 High-level embeddability

The previous section described how Σ^{IT} can be used as an engine by programmers. Since most computer algebra system systems provide a higher level style of programming in their interface language, Σ^{IT} is also designed to be used transparently from almost any interactive shell. The main problem to solve for such interaction is the strong typing of all Σ^{IT} objects. For example, the **axiom** interpreter [5] provides an interface between a mostly typeless user world and a strongly-typed algebra library, but with the heavy cost of type-inferencing. A different design concept was selected for Σ^{IT}: the use of light-weight servers providing typeless access to selected functionalities (in contrast to the **axiom** interpreter, which provides access to all the library). In some sense, servers can be seen as string-based stubs for process to process communication. Since the set of provided services is fixed once a server is compiled, it is crucial that designing new servers or extending existing ones can be done very easily. This is made possible by the object-oriented design of **Aldor**: generic subclassable parsers and interpreters are provided, so new servers with additional or different semantics can be derived easily from existing ones. We describe this process in more details in this section.

5.1 Expression Trees

In order to provide a typeless face to the outside world, an expression tree type is implemented in Σ^{IT}. A significant difference with Lisp S-expressions, or the parse and output trees provided by most systems, is that the operators (internal nodes) in the Σ^{IT} expression trees are full-fledged types (classes) rather than names or placeholders. The common category (abstract class) for all the operators enforces the export of various methods for communication with the outside world (e.g. conversion to TEX, `axiom` or `maple` input, generation of C code). Should a new conversion be desired to a new format or protocol (e.g. OpenMath), it is sufficient to add a corresponding method to each operator, without modifying any of the types in the algebra engine. In several cases, adding a default implementation to the operator category is enough (for example most prefix operators generate TEX input by adding a \ in front of their name). The root category in the algebra hierarchy (Fig. 1) enforces a conversion into expression trees, so every exported type is convertible to a typeless tree, thereby allowing its values to be exported to the outside world. For the reverse direction, Σ^{IT} provides an expression parser that produces expression trees from any input stream. This results in a two-sided design (Fig. 2), with a strongly-typed algebra side and a typeless expression tree side that shields the algebra completely from the outside world. It is possible

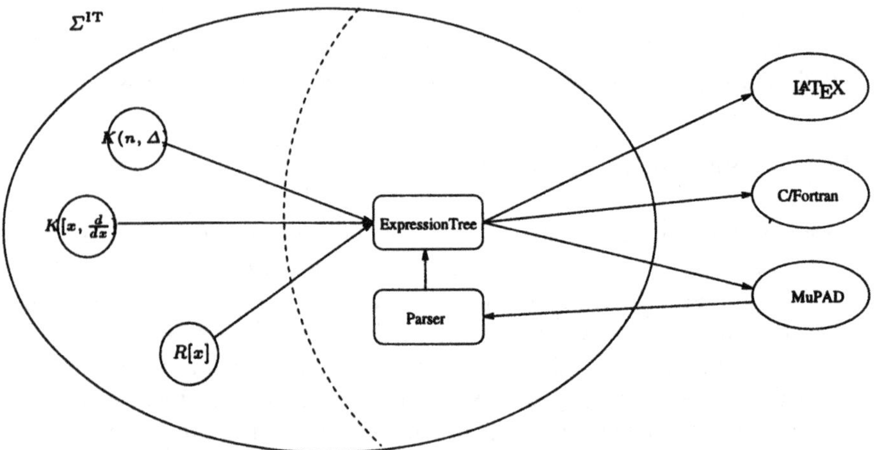

Fig. 2. The strongly-typed/typeless 2-sided design

to program mathematical algorithms operating on the typeless expression trees (e.g. numerical evaluation or pattern matching), and to take advantage of their object-orientedness in the process. In contrast to several systems, Σ^{IT} uses the same tree type for both input and output.

5.2 Partial rings

The only remaining problem in order to complete a server is the interpreting of an expression tree, i.e. converting an expression tree produced by the parser into a strongly-typed object. The semantics of this operation depend of course on the target type, and cannot be carried out without type-inference in the absence of information. In order to avoid the cost of type-inference, a server provides access only to a limited set of types. To support such a set of types, Σ^{IT} provides the category `PartialRing`, which is an abstract class for generic unions of rings. Any partial ring (∂-ring for short) must export at least all the ring and field arithmetic $(0, 1, +, -, *, /, \hat{})$ but all the exported operations are allowed to fail on some combinations of inputs, for example if the operands are not in compatible components of the union. Failure is reported through the use of the `Partial` type constructor of `Aldor`, although raising an exception will also be a possibility when exceptions are supported by `Aldor`. An implemented example is the `bernina` server for linear ordinary differential operators. This server exports to the outside world the following ∂-ring:

$$\mathbb{Z} \cup \mathbb{Q} \cup \mathbb{Q}[x] \cup \mathbb{Q}(x) \cup \mathbb{Q}[x, \frac{d}{dx}] \cup \mathbb{Q}(x, \frac{d}{dx}) \cup \mathbb{Q}(x)^{\mathbb{N}} \tag{2}$$

where $\mathbb{Q}(x)^{\mathbb{N}}$ means vectors of fractions of arbitrary length. Because of the various natural embeddings, the above ∂-ring is implemented as the 2-branch union $\mathbb{Q}(x, d/dx) \cup \mathbb{Q}(x)^{\mathbb{N}}$, but the knowledge of all other components is used when deciding whether an operation (e.g. exponentiation) can be carried out. Note that a ∂-ring exports only signatures involving itself (% in `Aldor`), so the components of its union are completely hidden from the outside. Each exported operation from a ∂-ring is programmed by checking the components of the arguments and dispatching the call to the appropriate Σ^{IT} types.

5.3 Evaluators

Once we have a specific ∂-ring R in mind, the need for type-inferencing is removed, since the interpreter only needs to convert expression trees to R. Since R is expected in addition to export operations corresponding to all the functions exported by the server under consideration, the interpreter only needs to traverse the tree, calling the appropriate function from R at each node. To make this process easy, Σ^{IT} provides a category `Evaluator(R:PartialRing)`, which can be seen as an abstract interpreter. But since any ∂-ring must provide ring and field arithmetic as well as an embedding from $\mathbb{Z}$ into itself, the default generic interpreter is able to process correctly all the integer and symbolic leaves, arithmetic operators, as well as assigment operators. Thus, a server only needs to provide function for handling operators not covered above, usually only the exported prefix functionalities. For example, the following code is the complete `bernina` interpreter, where `MultiLodo` is an implementation of the ∂-ring given by (2). All the other interpreter functionalities are provided by the generic behaviour of `Evaluator(R)`.

```
--------------------- evalodo.as ---------------------------
#include "sumit"
#library pring "multlodo.ao"
import from pring;
macro R == MultiLodo(QX, Qxd);

LODOEvaluator(QX:UnivariatePolynomialCategory Quotient Integer,
 Qxd:LinearOrdinaryDifferentialOperatorCategory Quotient QX):
  Evaluator R == add {
    evalPrefix!(name:String, nargs:SingleInteger,
        args:List R, tab:SymbolTable R):Partial R == {
            nargs = 1 => eval1(name, first args);
            nargs = 2 => eval2(name, first args, first rest args);
            failed;
    }
    eval1(name:String, x:R):Partial R == {
        name = "adjoint" => adjoint x;
        name = "diff" => D(x, 1);
        name = "kernel" => kernel x;
        failed;
    }
    eval2(name:String, x:R, y:R):Partial R == {
        name = "apply" => x y;
        name = "diff" => D(x, y);
        name = "symmetricPower" => symmetricPower(x, y);
        name = "kernel" => kernel(x, y);
        failed;
    }
}
```

5.4 Servers

Together with the built-in parser, a combination of a ∂-ring and an evaluator for it can be turned trivially into a read-eval loop that takes its input from a stream, processes it, and writes the result to another stream in any supported format when requested. Tools for standard read-eval loops (timing, prompting, output format conversions) are provided by the `Shell` type, which is itself parametrized by both a ∂-ring and an evaluator for it, allowing for rapid coding of a top-level server. The end-product is either a library or a stand-alone process that provides services while appearing to the outside like a typeless expression-tree

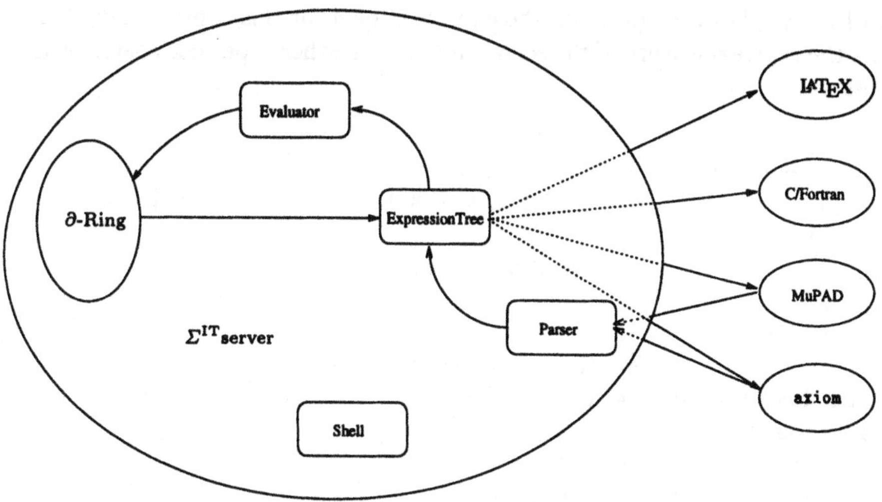

Fig. 3. A typical Σ^{IT} server

based computer algebra server (Fig. 3). Two sample servers are currently implemented, one for linear ordinary differential operators, and one for univariate polynomial factorization. A server can be used in 2 slightly different ways: linked into an application, in which case it provides a single entry-point that parses and execute a command string, or as a stand-alone process communicating via files or sockets. We illustrate the latter use of `bernina` from `maple`: the function below is a `maple` function that computes the kernel of the symmetric square of an operator L given as a polynomial in x and D, by calling a Σ^{IT} service via files:

```
quadraticInvariants := proc(L)
    readlib(write); open(`bernin`);
    writeln(`L := `); writeln(lprint(L)); writeln(`;`);
    writeln(`maple(kernel(symmetricPower(L,2)));`); close();
    open(`bernout`); write(`sumitresult := `); close();
    system(`bernina -q < bernin >> bernout`);
    read bernout; sumitresult; end;
```

Σ^{IT} can then compute the kernel of the symmetric square of the operator (1):

```
> v := quadraticInvariants(D^7 - x*D + 1/2);
> print(v);
                                  [ x ]
```

Note that the above function can be used within further `maple` programs, providing for a short implementation of the completely reducible cases of [11]. Of course, the above function can be easily translated to any interactive shell providing file I/O and linear output. In addition, each server provides a basic terminal interface, and can thus be used interactively directly. We conclude with the above computation performed directly in the `bernina` server. Comparing with the sample Aldor program shown in Section 4 for this computation, highlights the typeless feature of the server interface (neither types nor coercions are necessary):

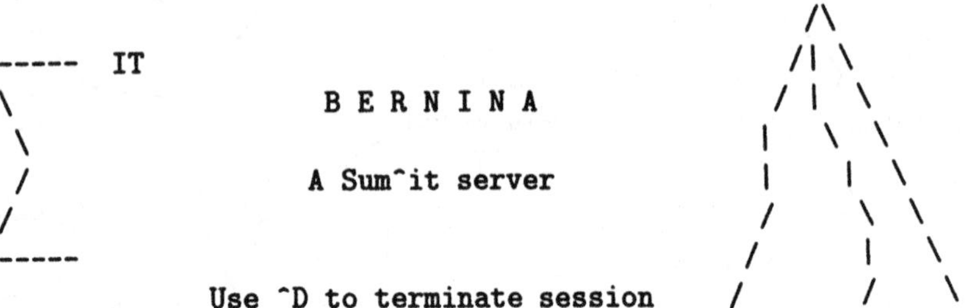


```
Independent variable = x
Derivation = D
1 --> L := D^7 - x*D + 1/2;
2 --> L2 := symmetricPower(L,2);
3 --> V := kernel(L2);
4 --> axiom(V);
vector [x]
```

References

1. S. A. Abramov, M. Bronstein & M. Petkovšek (1995): On polynomial solutions of linear operator equations, *Proceedings of ISSAC'95,* ACM Press, 290–296.
2. S. A. Abramov & K. Yu. Kvashenko (1991): Fast Algorithms to Search for the Rational Solutions of Linear Differential Equations with Polynomial Coefficients, *Proceedings of ISSAC'91,* ACM Press, 267–270.
3. M. Bronstein (1992): On Solutions of Linear Ordinary Differential Equations in their Coefficient Field, *Journal of Symbolic Computation* **13,** 413–439.
4. M. Bronstein & M. Petkovšek (1996): An introduction to pseudo–linear algebra, *Theoretical Computer Science* **157**, 3–33.
5. R. D. Jenks & R. S. Sutor (1992): *Axiom – The Scientific Computation System,*
6. J. J. Kovacic (1986): An Algorithm for Solving Second Order Linear Homogeneous Differential Equations, *Journal of Symbolic Computation* **2** 3–43.
7. N. Mannhart (1995): *asdoc* – Aldor *documentation class for LaTeX2e,* `ftp://ftp.inf.ethz.ch/org/cathode/asharp/asdoc/`
8. F. Naegele (1995): *Autour de quelques équations fonctionnelles analytiques,* Thèse de Mathématiques Appliquées, INPG, Grenoble.
9. M. Singer & F. Ulmer (1993): Galois Groups of Second and Third Order Linear Differential Equations, *Journal of Symbolic Computation* **16,** 9–36.
10. N. Takayama (1992): *Kan,* `ftp://ftp.math.s.kobe-u.ac.jp/pub/kan/kan96.tgz`
11. F. Ulmer & J. A. Weil (1994): *Note on Kovacic's algorithm,* Prépublication 94-13, IRMAR, Rennes.
12. S. M. Watt & al. (1994): A first report on the $A^{\#}$ compiler, *Proceedings of ISSAC'94,* ACM Press, 25–31.
13. S. M. Watt & al. (1994): *The $A^{\#}$ user's guide,* NAG Ltd.
14. R. Zippel (1990): *The Weyl computer algebra substrate,* Technical Report 90-1077, Dept. of Computer Science, Cornell University.

$\text{Disc}_{\text{Atinf}}$: A General Framework for Implementing Calculi and Strategies

Christophe Bourely and Nicolas Peltier

LIFIA-IMAG
46, Avenue Felix Viallet
38000 Grenoble, France
Nicolas.Peltier@imag.fr, Christophe.Bourely@imag.fr

Abstract. The underlying ideas, design principles and capabilities of a system called $\text{DISC}_{\text{ATINF}}$(Defining and Implementing Strategies and Calculi), allowing to implement a large class of computational systems (proof procedures, constraints solving algorithms...) are presented. Systems are described by sets of rules along with a strategy to guide the application of these rules. This formalism has been chosen because it fits well to the standard mathematical practice.
$\text{DISC}_{\text{ATINF}}$ is user-oriented and very flexible. It allows to specify the source objects and the transformation process in a very easy way. Moreover it is very general.
We describe the main features of $\text{DISC}_{\text{ATINF}}$ and gives examples of applications — showing evidence of the usefulness of our system. Finally, we briefly compare our system with similar ones.

Mathematicians, logicians and computer scientists usually spend a lot of time in implementing the computational systems they define. Such systems (typically proof procedures, rewriting systems, and more recently constraints solving procedures...) are very often described by set of *rewriting rules*, along with *a strategy* guiding the application of these rules.

When defining such systems, two different steps must be considered. During the first one, the user defines the system, and proves its fundamental properties (termination, soundness, refutational completeness...). Usually this is unfortunately not enough for most applications due basically to combinatory explosion. Then the user must make it more efficient, for example by defining some *strategies* or *heuristics* in order to improve its practical performances. Some of the rules may also have to be improved or modified. A lot of work is often needed before the defined system is well understood, and before a "final" version of the algorithm can be written. For example, efficient unification algorithms have only been proposed a long time after the definition of the first unification algorithm.

The second step consists in designing an efficient implementation of the system. A careful study of the behavior of the system is often required in order to find data structures and algorithms that are well adapted to the problem at hand.

Nevertheless, during the first step, an implementation of a prototype of the system is often needed to facilitate the theoretical study: for example even when proving the theoretical properties of a system, it is necessary to study its behavior on particular examples. Moreover, a lot of experimentation are usually needed before "good" strategies or heuristics can be defined. All these experiments are very difficult (and in fact impossible) to do by hand, hence the implementation of a prototype of the system is unavoidable. Even using a high level programming language such as LISP, PROLOG, or CAML designing such an implementation can be very time consuming.

The starting point of our work was to design a software to do what we *do not want to do by hand*, i.e. to *simulate the application of the rules* on an object, using a particular strategy. This "system interpreter" has to provide the following facilities:

It must allow to specify the underlying language and the rules in a very natural way. The aim is to be as close as possible to the way mathematicians (logicians, computer scientists,...) write it in a paper in order to minimize the amount of time spend in implementing the particular system. In other words, the system must be able to understand the system as it stands in a paper. It must be general enough to allow to express a very large class of computational systems. It should allow to express as many strategies as possible in order to guide the application of the rules. Again, these strategies should be expressed using a very *natural* and *general* syntax.

In this work, we describe such a system called DISC$_{\text{ATINF}}$ (standing for **D**escribing and **I**mplementing **S**trategies and **C**alculi). This software is a very useful tool that was developed within the ATINF project (see [2, 3] for a precise description of ATINF). We describe precisely the features of the system and give some examples, in order to illustrate its power and generality.

1 What is Disc$_{\text{Atinf}}$ good for ?

Roughly speaking DISC$_{\text{ATINF}}$ is based on the transformation of a given *object* by *rules* restricted by a set of *application conditions* and following a particular *strategy*.

DISC$_{\text{ATINF}}$ needs 3 different kinds of input: the definition of the language, i.e. the abstract and concrete syntax of the objects manipulated, described by using a set of **type constructors**. The **source object** of the language on which the method will be applied. This object is specified using the language previously defined. The specification of the transformation process. It is described by a **set of rules** (using the language defined above) and possibly with the **strategy**. The strategy may consist in a set of application conditions associated to the rules of the system and/or in a set of procedures. They are both specified by using a LISP-style programming language.

The transformation process can be divided into 2 steps: First, select a rule ρ, an object o and a strategy $\mathcal{S}$. This selection can be made automatically (using some pre-defined criteria), or by a user-defined function. Second, transform the object o with the rule ρ. The transformation process is guided by the strategy $\mathcal{S}$ and *possibly* by the user for eliminating the remaining indeterminism.

DISC$_{ATINF}$ also includes useful commands to communicate with the user or another system (for example to display the results of these transformations): a typical working session can be decomposed into a sequence of *selections* and *transformations*, separated by some commands to interact with the user, or with another system.

The following sections give a brief but precise overview of the features of our system.

2 The underlying language

First we must define the syntax of the language. In this aim, we have to define a set of *types*, with the command:

`(define types <type>`$_1$ `<type>`$_2$ `...` `<type>`$_n$`)`

The set of types contains some predefined types such as **string** or **integer** (other types such as **real** will also be implemented in the near future). There also exist some non standard types: **position** (denoting the set of positions in an object), **substitution** (denoting the set of all substitutions), and **set of substitutions**.

Types can also be specified by using some *type constructors* allowing to build sets, lists or products of existing types. Each definition can be completed by a condition restricting the form of the object belonging to this type.

```
<type> = <type_definition> where <condition>
```

It is worth mentioning that the definition of a type not only specifies the *structure* of the type (set, list, tuples of types) but also the concrete syntax of the type, i.e. the way the elements of the type will be written. For example, we can specify that lists of integers must be written on the form: $(x_1, \ldots, x_n)$ or $[x_1, \ldots, x_n]$, etc. In this aim, we simply specify 3 *separators* s_1, s_2, s_3, that must be of type "connective" (see below). s_1 and s_3 show respectively the beginning and the end of the list. s_2 is used to separate two elements x_i, x_{i+1} of the list.

Let us detail some of the possible constructors:

Connective. `<type> = '<string>'`: this type contains exactly one element and can be used for some particular constant or as a separator for other types.

List. `<type> = ListOf <separ_option> <type>`: This type is the set of (possibly empty) sequences of objects of same `<type>`. `<separ_option>` allows to specify the way the object will be written, as pointed out above.

`<separ_option> = [ <connect>`$_1$ `-- <connect>`$_2$ `-- <connect>`$_3$ `]`

The elements of type `<type>` will be written and read on the form:

`<connect>`$_1$ `<objects>`$_1$ `<connect>`$_2$ `...<connect>`$_2$ `<objects>`$_n$ `<connect>`$_3$

Of course, some of the connectives $1, 2, 3$ may be empty. Moreover `<separ_option>` is only optional: if nothing is specify, default setting will be used.

Set. `<type> = SetOf <separ_option> <type>`: a (possibly empty) set of objects of same type. Again `<separ_option>` allows to specify the concrete syntax of the object.

Product. `<type> = ProductOf <separ_options> (<name>`$_1$ `: <type>`$_1$`,` ..., `<name>`$_n$ `: <type>`$_n$ `)`: This type is simply the product of `<type>`$_1$ ...`<type>`$_n$. The operator "." allows to get a particular component of the tuple.

Commutative Product.
`<type> = StructOf <separ_option> (<name>`$_1$`, ...,` `<name>`$_n$ `: <type>)`: this type can be seen as the quotient of the product of the type `<type>`^{n} by the relation of commutativity of each argument.

Union. `<type> = <type>`$_1$ `OR <type>`$_2$ `OR ...OR <type>`$_n$: the resulting type is the union of the types `<type>`$_i$.

The definition of the variables and objects are specified by the following commands:

`(define variable <idf> : <type> )`

`(define objects <idf> : <type> <value> )`

It is also possible to find the term at a position `p` in a given object `T`, by using the constructor ".": `T.p`. `p` is an object of type position.

Example (first order unification):

In order to illustrate the use of DISC$_{\text{ATINF}}$ we will take an very simple example: we will use DISC$_{\text{ATINF}}$ to implement a very simple unification algorithm for first order terms. Let us recall the problem considered.

We consider the set of terms $\tau(\Sigma, \mathcal{X})$ inductively defined on a signature Σ and a set of variables $\mathcal{X}$.

We call *unification problem* a conjunction of equations of the form $t = s$, where t and s are terms. The aim is to define an algorithm allowing to find the *most general unifiers* of a given problem $\mathcal{P}$ that is to say the most general solution of $\mathcal{P}$ (it is well known [9] that this m.g.u. exists).

For defining our unification algorithm we have to define the terms of first order logic. This is done by the following command:

```
(define type

TERM =  ProductOf (F : string, Arguments : ListOf ( SubTerm: TERM))
             OR  X : variable where (X in SetVars)
             OR  A : string.)
```

This command defines the type `TERM` as either a string or the product of a string and a list of objects of type `TERM`. As the reader can see, the type `TERM` is parameterized by the variable `SetVars`. The set `SetVars` will contain the set of variables of the problem (the unknowns)[1]. The name `SetVars` may be undefined at this time (Check will be done latter, when the function called).

[1] We could also have defined the set of function symbols Σ (with their arity) or used some syntactic criterion to distinguish constant from variables.

Then we define the type "Unification Problem". First we define the type "Equation" as a commutative product of two terms, then a unification problem as a set of equations.

```
EQUATION = Eq : StructOf [ -- ' = ' -- ] ( T,S : TERM ).
       OR FALSE

PROBLEM = SetOf [ '( ' -- ' and ' -- ' )' ] ( E : EQUATION ).
```

The expression `[ -- ' = ' -- ]` and `[ '( ' -- ' and ' -- ' )' ]` respectively allow to specify that an equation (t, s) must be written `t = s` and that a set of equation $\{e_1, \ldots, e_n\}$ must be written ($\mathtt{e}_1$ **and** ...**and** $\mathtt{e}_n$)

If we enter the command:

```
(define object SetVars: SetOf( string ) (x, y, z, u, v))
(define object P : PROBLEM
    (f(x,x) = g(y,z) and f(u,v) = f(h(y),g(h(y),z))))
```

P will be parsed as an object of type "PROBLEM" and will be associated to the formula $f(x,x) = g(y,z) \wedge f(u,v) = f(h(y), g(h(y), z))$.

3 Rules and Methods

A *method* is either a sequence of methods or a sequence of rules.

```
(define method <idf> : <type> <list_of_idf>)
(define method <idf> : <type> <list_of_rules>)
```

3.1 Rules

A rule allows to transform an object belonging to a given type into another object, when some conditions are satisfied. The syntax of the rules is the following:

```
<rule>      ::= <idf> : <type><rule_def>
<rule_def>  ::= <left_side> |--> <right_side> .
                if <condition>              % optional
                where <condition>           % optional
```

The left and right side of a rule are *generic objects* also called *meta-objects* that will be either unified with a given object or instantiated by a substitution.

Meta-objects Meta-objects are defined exactly like standard objects (see Section 2) except that we add 3 new options, very useful for defining rules:

If a meta-object (of type T) is an identifier, then it is considered as a *variable* of type T. These variables can be unified with any object.

It is possible to represent a list of n objects (where n is not explicitly specified) by an expression of the form: `<name>i ... <name>j`, where `<name>` is *the common part* of the objects of the set and i, j are *integer variables* denoting the bounds of the set. The expression: `<name>{k}` (where k is an integer) denotes the k-th object of the sequence.

It is also possible to explicitly denote a term `T` containing a term `T'` at a position `p` by: `T [ p : <type> T' ]`. `p` denotes a position in `T` (it can be either a variable or an explicit position $i_1.i_2.\ldots$). Similarly, it is possible to denote the object `T` where the object at position `p` is replaced by `T'` by: `T [ p <- <type> T' ]`, or to denote the object `S` obtained by replacing each occurrence of `T'` in `T` by `T''`: `T { T' <- <type> T'' }`. (this last construction is mainly useful for the right side of the rule).

Semantic The semantic of a rule is the following: the left side of the rule is unified with the given object. This unification is of course performed modulo the particular properties of the type constructors (commutativity, associativity...). The result is a *set of substitutions*. The aim of the condition `if` (pre-condition) is to restrict the admissible substitutions, hence to restrict the possible applications of the rule. This verification may — as a side effect — modify the substitutions considered. The right part of the rules is instantiated by the chosen substitution then the initial object is replaced by the result of the transformation. The condition `where` (post-condition) allows a post-treatment before instantiation, allowing for example to enrich the substitutions (by using external data or functions).

Example Let us define the rules of our unification algorithm. We will use the usual rules of *decomposition*, *occur check*, *replacement* and *clash*... The unification procedure will be defined as a method containing all these rules.

```
(define method Unification: PROBLEM
     Decompose(PROBLEM) :
       f(t_1,...,t_n) = f(s_1,...,s_n) |-->
       t_1 = s_1 and ... and t_n = s_n.

     Clash(PROBLEM) :
       f(t_1,...,t_n) = g(s_1,...,s_n) |--> FALSE.
         if (f != g).

     Replacement(PROBLEM) :
       x = t and P |--> x = t and P [ x <- t ] .
         if (and (x in SetVars) (not (x in t))
                 (or (not (t in SetVars)) (t in P))).

     OccurCheck(PROBLEM) :
       x = t [ p : TERM x ] |--> FALSE.

     Simplify (PROBLEM) :
       FALSE and P |--> FALSE.)
```

4 Strategies

As already evidenced by the above example, the syntax of the functional language used by DISC$_{ATINF}$ is very close to LISP. This kind of syntax has been chosen because it is very simple and widely used in the computer scientists community. Expressions are of the form: **idf**, **integer**, or **(function arg$_1$ ...arg$_n$)** where **arg$_i$** is an expression. These terms are evaluated recursively as usual in the current environment. This language offers all the features of a standard programming language. Classical LISP functions (**if, while, let** ...) as well as all the usual operators (and, or, +, - ...) can of course be used. Functions can be defined, using the command:

define function <idf> (<args>) <func_def>

Moreover, this language also allows to handle the objects and the rules previously defined. New functions are indeed provided in order to have access to the components of a given object or rule, to apply a rule on an object... This feature allows to define a large class of strategies with little effort. Objects of the source language are considered exactly as standard LISP objects, and can be handled in the same way. When defining the application conditions of a rule, one can of course have access to the value of the variables of this rules, that are part of the environment of the condition. They can be handled *exactly* as standard variables. This allows to check whether a given variable satisfies some condition (for instance, in the above example, the expression **(x in SetVars)** allows to check whether the meta-variable **x** of the rule **Replace** belongs to the set of variables).

We list below the main functions of our language, together with examples in order to illustrate their application[2].

(<object>$_1$ == <object>$_2$) check whether two objects are equal.

(apply <method> on <object> with <strategy> using <substitution>)

Apply a **<method>** on an **<object>** with a **<strategy>** and a **<substitution>**.

The strategy can be either defined by the user or selected in a set of *predefined strategies*. It allows to select a method, a rule, a particular application of a rule or the evaluation of a condition.

There exist 5 predefined strategies:

first: apply the first rule find on the first object. **saturate**: apply the rules of the system until the object cannot be reduced any more. **random**: choose the rule and the object randomly. **top**: only try to apply the rules in the position root, and not in subterms. **all**: apply all the possible rules: the result is a set of objects.

For example we can use the command: **(apply Unification on P with saturate)** in order to transform the problem P into a solved one.

(ForAll (<var> in <set>) <expression>)

---

[2] This list is not exhaustive

Evaluate the expression **<expression>** by replacing **<var>** by each element of the set **<set>**.

A *set* can be any object of type **SetOf**, **ListOf**, **StructOf** or a set of integers of the form: `[ v1 .. v2 ]`.

(Select <var> <object>$_1$ in <object>$_2$ with <strategy>)

Return the set of possible substitutions σ, such that σ is an unifier of **<object>**$_1$ and a sub-term of **<object>**$_2$.

The command: **(Select sigma T in T' with top)** return the set of unifiers of the meta-objects **T** and **T'**.

Defining new strategies: the user can define new strategies that will be used as the predefined strategy **saturate**, **first**... in the functions **apply**, **select**...

The syntax is the following: **define strategy <strategy> : <method> <expression>**

This command associates the **<expression>** to the **<method>**.

When the method is called, by the command **apply**, ..., with the strategy **<strategy>**, the expression **<expression>** is evaluated. The keyword **object** denotes the object on which the strategy is applied.

Example (continued) Assume we want to define a new strategy for our unification system: we want to apply the rule *replacement* only if no other rules are applicable.

Then we have just to enter the command:

```
(define strategy new_strategy : Unification
      (eval
            (define method Unification_2 : PROBLEM
                 Decompose Clash OccurCheck)
            (while
              (apply Unification_2 on object with saturate))
              (apply Replace on object with first)))
```

5 Implementing a Constraint Solver

We give an example illustrating the use of our software. We use DISC$_{\text{ATINF}}$for defining a procedure for solving first order constraints in the empty theory.

A Constraint Solving Algorithm

An *equational problem* is a formula of the form: $\exists\overline{x}.\forall\overline{y}.\mathcal{M}$, where $\mathcal{M}$ is a *system*, i.e. a purely equational formula without quantifiers.

A substitution σ is said to be a *solution* of a problem $\mathcal{P}$ iff $\mathcal{P}\sigma$ is valid in the empty theory (i.e. in the Herbrand universe).

Equational problems have a lot of interesting application, for example in inductive theorem proving, specification, model building...[5, 4]. In [5, 6] an algorithm has been proposed to find the solutions of any equational problem. This algorithm is given as a set of rules, allowing to transform any equational problem into a set of problems in "solved form".

This algorithm can be very easely implemented by using the system $\text{DISC}_{\text{ATINF}}$. First we have to define the syntax of the source language.

```
(define type

ATOM =  'TRUE' or
        'FALSE' or
        Eq : StructOf [ -- ' == ' -- ] ( T,S : TERM ) or
        DisEq : StructOf [ -- ' != ' -- ] ( T,S : TERM ) .

SYSTEM =   E : ATOM or
           Conjunction : SetOf [ '( ' -- ' and ' -- ' )' ] ( S : SYSTEM ) or
           Disjunction : SetOf [ '( ' -- ' or ' -- ' )' ] ( S : SYSTEM ) .

SET_VARS = SetOf( x : variable ).
AUXILIARY_UNKNOWN = 'exists' W : SET_VARS .
PARAMETERS = 'forall' Y : SET_VARS .

PROBLEM = S : SYSTEM or
          X : AUXILIARY_UNKNOWN Y : PARAMETERS '.' S : SYSTEM .)
```

Because of lack of space, we do not give here the definition of all the rules of the method. Most of them can be directly translated into the formalism of $\text{DISC}_{\text{ATINF}}$, without changes. We give below one representative rule.

```
( define method EXPLOSION ( CONSTRAINT ) :

%% Explosion

  EXPLOSION ( CONSTRAINT ) :

      exists X. forall Y. P[ pos : ATOM x != u ]
                     |-->
            [ P1 ou ... ou Pn ] .
   if (and
      ( IsVar x )
      ( not ( x in Y ) )
      ( not ( IsVar u ) )
      ( not( IsEmpty ( Inter Y ( vars_term u ) ) ) )
   )
   where ( eval
      ( let n 0 )
      ( ForAll ( F in EnsFoncs )
            ( let m ( F . Arity ) )
            ( let f ( F . Name ) )
            ( AddTo n 1)
            ( ForAll ( i in 1 .. m )
```

```
                ( let w (new_var) )
                ( let w{i} (copy w) )
                ( let w'{i} ( assign TERM w . ) )
                ( PushSet EnsVars w )
            )
            ( let X'' ( assign SET_VARS (w1, ..., wm) .))
            ( let X' ( Union X X'' ) )
            ( let Q{n} ( assign PROBLEM
                exists X'. forall Y. ( P and x == f( w'1, ..., w'm ) ) .
                ))
              ( let P{n} ( assign CONSTRAINT Q{n} . ) ))
              ( ForAll ( A in EnsCstes )
                 ( let a ( assign TERM A . ) )
                 ( AddTo n 1 )
                 ( let Q{n} ( assign PROBLEM
                     exists X. forall Y. ( P and x == a ) .))
                  ( let P{n} ( assign CONSTRAINT Q{n} . ) ))))
```

We have experimented the constraint solving system implemented with DISC$_{\text{ATINF}}$ on some problems proposed by H. Comon. All the problems were solved in a very short time (less than 1 second on a SUN 4 ELC). We give below some examples of such "benchmarks" and the result of the solving procedure.

```
(define object Pb3 : CONSTRAINT
  (exists (). forall (y).
    (s(x) != s(y) or x == y)))

(apply SOLVE on Pb3 with saturate)

-> True

(define object Pb15 : CONSTRAINT
  (exists (). forall (y1,y2,y3,y4). (y2 != f(y1,y1) or y3 != f(y2,y2) or
  y4 != f(y3,y3) or x != f(y1,y1)))).

(apply SOLVE on Pb15 with saturate)

-> exists (v1,v2) forall ().x == f(v1,v2) and v1 != v2)
or exists (v3,v4) forall (). x == g(v3,v4)
or exists (v5,v6,v7).forall ().x == h(v5,v6,v7)
or exists (v8) forall ().x == s(v8) or x == zero
```

6 Comparison with other approaches

A large amount of excellent work has been done on definition of computational systems, specially in what is usually called *logical frameworks*. A logical framework is a kind of implementable meta-logic in which all useful logics can be represented. They are usually based on the use of higher order logics, that allow to use the well known Curry-Howard isomorphism: the *propositions* of the source logic correspond to the *types* of the logical framework, a *proof* of this proposition corresponds to a *term* of the corresponding type.

The systems AUTOMATH [1], ELF [11], COQ [7]...and many others deserve to be mentioned.

Though $\text{DISC}_{\text{ATINF}}$ shares some caracteristics with logical frameworks, clearly it does not belong to this category of tools. Our system does not aim at *universality* but at *ease to use* and *efficiency*. It uses a very weak logic to code the source language: each notion must be expressed by first order terms. On the other hand, powerful tools are made available to make this coding user-friendly and flexible.

Besides, our system focus rather on the definition of the *strategies* used to apply the rules of the system: one of the main aims of our system is to provide a formalism allowing to define a large class of strategies. In contrast to the notion of *tactics* used for example in COQ, those strategies are expressed in a specific (internal) formalism.

Therefore $\text{DISC}_{\text{ATINF}}$ is closer to systems like ELAN [10] or ECOLOG [8] which aim is to specify constraint logic programming languages. It differs from them is several respects: ECOLOG is an environment for constraint logic programming developed in CAML and based on rewriting logic. Some parts of the system — for example the definition of the abstract syntax — cannot be specified in the formalism of ECOLOG and must therefore be developed directly in CAML. In contrast to this, $\text{DISC}_{\text{ATINF}}$ allows to specify all aspects of the system in a very natural formalism.

The system ELAN is very close to $\text{DISC}_{\text{ATINF}}$. The main differences between our system and ELAN are the following: $\text{DISC}_{\text{ATINF}}$ allows more flexibility in the definition of the rules of the system. For example the notion of position and subterm are not present in ELAN, which restricts the class of rules that can be specified. The formalism used by ELAN to specify strategies is weaker and less general than the one used by $\text{DISC}_{\text{ATINF}}$. Roughly speaking, strategies are specified in ELAN by *regular expressions* using the operators **while** (iterate a strategy until irreducible normal forms are reached) **iterate** (iterate a strategy as long possible), and the composition in sequence. The operator **don't known choose** allows to urge the system to explore all the possible branches while **don't care choose** explores only the first branch.

It is not difficult to see that this formalism can be easely simulated by defining functions in the language of $\text{DISC}_{\text{ATINF}}$, using functions like **if**, **while** and **apply**.

In some sense, our formalism operates at a *lower level* than the one of ELAN. Our expressive power is therefore bigger, but on the other hand the formalism of ELAN is simpler and more concise.

7 Conclusion and Future Work

We have described a system called $\text{DISC}_{\text{ATINF}}$ for implementing computational systems described by sets of rules and showed it at work on non-trivial examples.

It has been implemented in C++. The two main characteristics of this system are the following:**User-oriented**: The system is very flexible and allows to specify the system in a very natural and user-friendly way. **Generality**: $\text{DISC}_{\text{ATINF}}$ allows to express easely a very large class of systems. A programming language is provided for the definition of strategies and heuristics.

We hope that our system will make the definition of prototypes of computational systems easier (particularly for non computer scientists) and therefore will provide useful help to the theoretical and practical studies in various fields of research.

Acknowledgements

We would like to thank Ricardo Caferra and Gilles Défourneaux for a careful reading of an earlier version of this paper.

References

1. N. D. BRUIJN. *Essays on Combinatory Logic, Lambda Calculus and Formalism*, chapter A Survey of the project AUTOMATH. Academic Press, 1980.
2. R. CAFERRA and M. HERMENT. $\text{GLEF}_{\text{ATINF}}$: A graphic framework for combining provers and editing proofs. In *Design and Implementation of Symbolic Computation Systems*, pages 229–240. Springer-Verlag, 1993.
3. R. CAFERRA and M. HERMENT. A generic graphic framework for combining inference tools and editing proofs and formulae. *Journal of Symbolic Computation*, 19(2):217–243, 1995.
4. R. CAFERRA and N. ZABEL. A method for simultaneous search for refutations and models by equational constraint solving. *Journal of Symbolic Computation*, 13:613–641, 1992.
5. H. COMON. *Unification et Disunification. Théorie et Applications.* PhD thesis, INPG, Grenoble, 1988.
6. H. COMON and P. LESCANNE. Equational problems and disunification. *Journal of Symbolic Computation*, 7:371–475, 1989.
7. T. COQUAND, C. PAULIN-MOHRING, G. DOWEK, and G. HUET. The calculus of constructions. documentation and user's guide 110. INRIA-ENS, Rocquencourt, 1989., 1989.
8. M. HABERSTRAU. ECOLOG: An Environment for COnstraints LOGics. In E. Domenjoud and C. Kirchner, editors, *CCL'92*, Le Val d'Ajol, 1992.
9. J. JOUANNAUD and C. KIRCHNER. Solving equations in abstracts algebra: a rule based survey of unification. In J.-L. Lassez and G. Plotkin, editors, *Essays in Honor of Alan Robinson*, pages 91–99. The MIT-Press, 1991.
10. C. KIRCHNER, H. KIRCHNER, and M. VITTEK. Designing constraint logic programming using computational systems. In F. Orejas, editor, *2nd CCL Workshop*, September 1993.
11. F. PFENNING. *Logical frameworks*, chapter Logic programming in the LF logical framework. Cambridge University Press, 1991.

Equality Elimination for the Tableau Method

Anatoli Degtyarev* and Andrei Voronkov**

Computing Science Department, Uppsala University
Box 311, S-751 05 Uppsala, Sweden

Abstract. We apply the *equality elimination method* to semantic tableaux with equality. The resulting logical system is a combination of a goal-directed tableau calculus with a basic superposition calculus. Unlike most other known methods of adding equality to semantic tableaux, equality elimination does not use simultaneous rigid E-unification or its modifications. For controlling redundancy, we can use powerful strategies of subsumption and simplification. We also make an extensive comparison with related works in the area.[1]

1 Semantics tableaux and equality

The problem of extending tableaux with equality rules is crucial for enriching the deductive capabilities of the tableau method. This problem is attacked by a growing number of researchers during the last years.

When we studied literature on tableau-based automated reasoning techniques, we found two tendencies in this area. The first tendency is to search for a proof by methods bringing us back to classical papers in proof theory, mainly to Prawitz [36] and Kanger [22] and also to results on normalization of proofs in sequential calculi (Orevkov [34], Maslov, Mints and Orevkov [29]). Since many notions have been (for good or for bad) reinvented in papers on tableau methods, we would like to recall fundamental ideas of earlier papers.

Prawitz and Kanger proposed the method of applications of proof rules ($\forall \rightarrow$) and ($\rightarrow \exists$) (later called γ-rules in Smullyan [39] or Fitting [18])[2]. The idea was to introduce a new kind of variables (called "dummies" in both Prawitz [36] and Kanger [22]), and to delay instantiation of these variables until necessary information for it is obtained. Comparing this approach to an earlier work of Beth [5], Prawitz [36] noted: "the solution proposed here is quite different but well-suited for mechanical use". This method was independently proposed in Russia by N. Shanin and has been characterized as the "metavariable method" in Maslov, Mints and Orevkov [29]. Information for instantiation is provided by

* Supported by a grant from the Swedish Roval Academy of Sciences

** Supported by a TFR grant

[1] Due to the space restrictions, we omit proots in this paper. Proofs may be found in the technical report [10].

[2] There are differences in *representation* of inferences between Kanger [22] and Fitting [18] — Kanger uses sequent proofs. Sequents in Kanger [22] correspond to branches in tableaux Fitting [18].

constructing an uninstantiated proof, and checking from time to time whether one can find values for dummies which make it a valid proof. In Kanger [22] this check is reduced to verifying that the top sequents are "directly demonstrable", i.e. can be obtained from axioms by applications of equality rules. The possibility of applying all equality rules before all other rules has also been demonstrated in Orevkov [34] and used in Maslov [28].

Dummies or metavariables have later been called "free variables" in Fitting [18]. Free variables are substituted for existentially quantified variables in the goal, when γ-rules are applied. Directly demonstrable sequents become closed branches of a tableau, the procedure for a "direct demonstration" becomes the check for the rigid E-unifiability (Gallier, Raatz and Snyder [19]). The simultaneous rigid E-unifiability for closing all branches of a tableau has recently been proved undecidable (Degtyarev and Voronkov [13]). Thus, the problem arises to find methods not based on the simultaneous rigid E-unifiability[3].

The second tendency in tableaux-based theorem proving with equality is that of "globality". The main idea of using metavariables — to provide a smallest instantiation — comes out through the notion of a most general unifier. But another advantage of post-Kanger procedures is that the idea of metavariables is combined with the principle of *locality* (Maslov, Mints and Orevkov [29]). Among these procedures are resolution, hyperresolution, the inverse method and their numerous modifications (see e.g. surveys Maslov, Mints and Orevkov [29], Degtyarev and Voronkov [9, 8]).

The tableau method achieves goal-orientation by sacrificing the principle of locality. A substitution in tableau methods is applied to the whole tableau (see e.g. the "MGU replacement rule" from Fitting [18]), which corresponds to a set of sequents, while in the inverse method (see Voronkov [42], Degtyarev and Voronkov [11]) a substitution is applied to a sequent. In Section 4 we show that the use of the functional reflexivity in Fitting [18] is a consequence of the global substitution application. In order to reduce high non-determinism caused by applications of functional reflexivity and MGU replacement rule, we have to search for methods using local rules.

In this paper we propose a method of handling equality in tableaux which tries to solve the above mentioned problems. Our method is based on extending a tableau prover by a bottom-up equation solver using basic superposition. Solutions to equations are generated by this solver and used to close branches of a tableau. Thus, the method combines (non-local) tableau proof search with the (local) equation solving. Only completely solved equations are used in the tableau part of the proof, thus reducing non-determinism created by applications of MGU replacement rule in Fitting [18]. The equation solution is even more restricted by the use of orderings, basic simplification and subsumption.

A similar idea: combination of proof-search in tableaux and a bottom-up equality saturation of the original formula, is used in Moser, Lynch, and Steinbach [32] for constructing a goal-directed version of model elimination with refined paramodulation.

[3] Recently, we have found that an incomplete procedure for rigid E-unification can be used to give a proof procedure complete for first-order logic [15].

2 Preliminaries

We present here a brief overview of notions and preliminary definitions necessary for understanding the paper. We assume basic knowledge of resolution, rewriting and the tableau method (e.g. Lee and Chang [23], Dershowitz and Jouannaud [17] and Fitting [18]).

Let Σ be a signature, and X be a set of variables. $T(\Sigma, X)$ denotes the set of all terms in the signature Σ with variables from X. The set $T(\Sigma, \emptyset)$ of all *ground terms* in the signature Σ is denoted by $T(\Sigma)$.

A *literal* is either an atomic formula or a negation of an atomic formula. A *clause* is a finite set of literals $\{L_1, \ldots, L_n\}$, denoted $L_1, \ldots, L_n$. If L is a literal and C a clause, then L, C denotes the clause $\{L\} \cup C$. The *empty clause* is denoted by $\Box$. An *equation* is a literal $s = t$, where $s, t \in T(\Sigma, X)$. Literals of the form $\neg(s = t)$ are denoted by $s \neq t$ and called *disequations*. By a *ground expression* (i.e. term, equation, clause etc.) we mean an expression containing no variables. We write $A[s]$ to indicate that an expression A contains s as a subexpression and denote by $A[t]$ the result of replacing a particular occurrence of s in A by t. By $A\sigma$ we denote the result of applying the substitution σ to A. $\sigma \leq \theta$ means that there is a substitution τ such that $\sigma\tau = \theta$. Substitutions θ with the domain $x_1, \ldots, x_n$ are denoted by $\{x_1\theta/x_1, \ldots, x_n\theta/x_n\}$.

From now on we assume that $\succ$ *is a lexicographic path ordering on* $T(\Sigma, X)$ *extending a total precedence relation on* Σ (e.g. Dershowitz and Jouannaud [17]). Nevertheless, our results are valid for *any* reduction ordering total on the set of ground terms $T(\Sigma)$.

Following Bachmair et.al. [2] we distinguish terms occurring in the original formula from terms introduced by substitution by using *closures*, i.e. pairs $C \cdot \sigma$, where C is a clause, σ is a substitution. Such a closure semantically corresponds to the clause $C\sigma$. For any expression E, the set $Var(E)$ is defined as the set of all variables occurring in E. Two closures $C_1 \cdot \sigma_1$ and $C_2 \cdot \sigma_2$ are *variants* iff C_1 is a variant of C_2 and $C_1\sigma_1$ is a variant of $C_2\sigma_2$.

The symbol $\vdash$ stands for the provability in classical first order logic with equality.

A *tableau* is a set $\{C_1, \ldots, C_n\}$ of clauses, denoted $\mid C_1 \mid \ldots \mid C_n \mid$, where $n \neq 0$. The tableau with $n = 0$ is called *the empty tableau* and denoted by $\#$.

There is a correspondence between the traditional representation of tableaux and our representation. For each branch B_i of a tableau represented as a tree (see e.g. Fitting [18]), take the set of all formulas $\varphi_1, \ldots, \varphi_n$ on this branch and replace these formulas on their *names* (literals, defined below), then we obtain the clause C_i.

A formula is in the *Skolem negation normal form* iff it is constructed from literals using the connectives $\wedge, \vee$ and the quantifier $\exists$. There is a provability-preserving translation of formulas without equivalences into formulas in Skolem negation normal form consisting of the standard skolemization and a translation into negation normal form used e.g. in Andrews [1].

Our algorithm uses ***subset unification***[4]. A substitution σ is called a ***subset unifier*** of the clause D_1 against D_2 iff $D_1\sigma \subseteq D_2\sigma$. It is a ***minimal subset unifier*** of D_1 against D_2 iff it is minimal w.r.t. $\leq$ among all subset unifiers of D_1 against D_2. Subset-unifiability is NP-complete. Some other properties of subset unification are discussed in Degtyarev and Voronkov [10].

3 The equality elimination method

For the rest of this section we assume that ξ denotes a closed formula in the Skolem negation normal form to be proved (the "goal"). We assume that all different occurrences of quantifiers in ξ bind different variables. For example, ξ cannot have the form $\exists x A \vee \exists x B$. All formulas in this section are assumed to be subformulas of the goal formula ξ. We shall identify subformulas of ξ and their superformulas with their *occurrences* in ξ. For example, in the formula ξ of the form $A \wedge (A \vee B)$ the second occurrence of A is considered a subformula of $(A \vee B)$, but the first occurrence of A is not.

In order to define a more efficient tableau calculus, we shall only deal with some subformulas of ξ called conjunctive subformulas. The occurrence of a subformula φ of ξ is called ***conjunctive*** iff it is an occurrence in a subformula $\varphi \wedge \psi$ or in $\psi \wedge \varphi$. A ***conjunctive superformula*** of φ is a superformula[5] ψ of φ that is conjunctive. The ***least conjunctive superformula*** of φ is the conjunctive superformula ψ of φ such that any other conjunctive superformula of φ is a superformula of ψ. Let us note that conjunctive superformulas do not necessarily exist. For example, ξ has no conjunctive superformula. Any formula having a conjunctive superformula has the unique least conjunctive superformula.

We can enumerate all conjunctive subformulas $\xi_1, \ldots, \xi_n$ of ξ, for example in the order of their occurrences in ξ. Thus we can unambiguously use "the kth conjunctive (sub)formula" ξ_k of ξ.

Let $A_1, \ldots, A_n$ be predicate symbols not occurring in ξ. We say that the atomic formula $A_k(x_1, \ldots, x_m)$ is ***the ξ-name of a subformula*** φ of ξ iff

1. The least conjunctive superformula of φ is ξ_k;
2. $x_1, \ldots, x_m$ are all free variables of ξ_k in the order of their occurrences in ξ_k.

If a ξ-name of a formula φ exists, then it is unique. Note that different formulas may have the same ξ-names. Also note that some subformulas of ξ do not have ξ-names. We can use ***the set of ξ-names*** of a subformula. The set of ξ-names of a formula φ is either $\emptyset$ or a singleton $\{A_k(x_1, \ldots, x_m)\}$.

Figure 1 illustrates least conjunctive superformulas and ξ-names:

Lemma 1. *Let φ be a subformula of ξ. Then*

[4] A variant of subset unification has been called θ-subsumption in Minker, Rajasekar and Lobo [31].

[5] φ is a superformula of ψ iff ψ is a subformula of φ (not necessarily proper).

Subformula	least conjunctive superformula	
$(\exists x(F(x) \wedge (B(x) \vee \exists y C(x,y))) \wedge \exists z D(z)) \vee E$	no	$\emptyset$
$\exists x(F(x) \wedge (B(x) \vee \exists y C(x,y))) \wedge \exists z D(z)$	no	$\emptyset$
$\exists x(F(x) \wedge (B(x) \vee \exists y C(x,y)))$	$\exists x(F(x) \wedge (B(x) \vee \exists y C(x,y)))$	$\{A_1\}$
$F(x) \wedge (B(x) \vee \exists y C(x,y))$	$\exists x(F(x) \wedge (B(x) \vee \exists y C(x,y)))$	$\{A_1\}$
$F(x)$	$F(x)$	$\{A_2(x)\}$
$B(x) \vee \exists y C(x,y)$	$B(x) \vee \exists y C(x,y)$	$\{A_3(x)\}$
$B(x)$	$B(x) \vee \exists y C(x,y)$	$\{A_3(x)\}$
$\exists y C(x,y)$	$B(x) \vee \exists y C(x,y)$	$\{A_3(x)\}$
$C(x,y)$	$B(x) \vee \exists y C(x,y)$	$\{A_3(x)\}$
$\exists z D(z)$	$\exists z D(z)$	$\{A_4\}$
$D(z)$	$\exists z D(z)$	$\{A_4\}$
E	no	$\emptyset$

Fig. 1. Least conjunctive superformulas and sets of ξ-names of subformulas of the formula $(\exists x(F(x) \wedge (B(x) \vee \exists y C(x,y))) \wedge \exists z D(z)) \vee E$

1. *If ψ is the least conjunctive superformula of φ then $\vdash \forall(\varphi \supset \psi)$.*
2. *If there is no conjunctive superformula of φ then $\vdash \forall(\varphi \supset \xi)$.*

This lemma partially explains the need for introducing least conjunctive superformulas. There are deterministic chains of inferences in sequent systems, where formulas with $\vee$ or $\exists$ occur. For example,

$$\cfrac{\cfrac{\Gamma \to \varphi}{\Gamma \to \varphi \vee \psi}}{\Gamma \to \exists x(\varphi \vee \psi)}$$

By restricting ourselves to conjunctive superformulas only, we eliminate these deterministic chains, making them in one step[6].

Now we can formulate the tableau deductive system we are going to use. The system depends on the goal formula ξ and is denoted T_ξ. The provable objects of the system are of two kinds: closures and tableaux. There are inference rules allowing to derive closures from closures and tableaux from closures and tableaux. The axioms of T_ξ are *initial closures* and the *initial tableau* defined below.

[6] This simple but powerful idea of restricting to conjunctive superformulas (see Voronkov [42]) has been successfully implemented in the theorem prover described in Voronkov [40] and in a theorem prover for intuitionistic logic implemented by T. Tammet (private communications). In the framework of tableau theorem proving, several papers used permutabilities of inference rules in sequent calculi (see e.g. Shankar [38]). In fact, the least conjunctive superformula of φ is a superformula ψ of φ provable from φ and such that all inference rules applied in the proof of ψ from φ are permutable with all other rules. The use of conjunctive superformulas allows us to get rid of non-conjunctive subformulas before the proof-search, unlike the dynamic use of permutabilities as in Shankar [38].

Initial closures of T_ξ are generated according to one of the three rules:

1. Whenever a literal $s \neq t$ occurs in ξ and C is the set of ξ-names of this occurrence of $s \neq t$, the closure $s = t, C \cdot \varepsilon$ is an initial closure.
2. Whenever $s = t$ occurs in ξ and C is the set of ξ-names of this occurrence of $s = t$, the closure $s \neq t, C \cdot \varepsilon$ is an initial closure.
3. Let literals $P(s_1, \ldots, s_n)$ and $\neg P(t_1, \ldots, t_n)$ occur in ξ and C_1, C_2 are their sets of ξ-names. Let the substitution σ rename variables such that variables of $P(s_1, \ldots, s_n)\sigma$ and $P(t_1, \ldots, t_n)$ are disjoint. Then the closure $s_1\sigma \neq t_1, \ldots, s_n\sigma \neq t_n, C_1\sigma, C_2 \cdot \varepsilon$ is an initial closure.

The **initial tableau** of T_ξ is the tableau $\mid \Box \mid$ consisting of the empty clause.

The calculus T_ξ consists of five inference rules: two basic superposition rules, the equality solution rule, the tableau expansion rule and the branch closure rule. The first three inference rules are adapted from basic superposition-based inference systems (Bachmair et.al. [2], Nieuwenhuis and Rubio[33]). As usual, we assume that premises of rules have disjoint variables which can be achieved by renaming variables.

Basic (right and left) superposition

$$\frac{(s = t, C) \cdot \sigma_1 \quad (u[s'] = v, D) \cdot \sigma_2}{(u[t] = v, C, D) \cdot \sigma_1\sigma_2\rho} \qquad \frac{(s = t, C) \cdot \sigma_1 \quad (u[s'] \neq v, D) \cdot \sigma_2}{(u[t] \neq v, C, D) \cdot \sigma_1\sigma_2\rho}$$

where

1. ρ is a most general unifier of $s\sigma_1$ and $s'\sigma_2$;
2. $t\sigma_1\rho \not\succeq s\sigma_1\rho$ and $v\sigma_2\rho \not\succeq u[s']\sigma_2\rho$;
3. s' is not a variable.
4. (for left superposition only) $u[s'] \neq v$ is the leftmost disequation in the second premise.[7]

Equality solution

$$\frac{(s \neq t, C) \cdot \sigma}{C \cdot \sigma\rho}$$

where ρ is a most general unifier of $s\sigma, t\sigma$ and $s \neq t$ is the leftmost disequation in the premise.

Tableau expansion Let D_1, D_2 and D be the sets of ξ-names of φ, ψ and $\varphi \wedge \psi$, where $\varphi \wedge \psi$ is a subformula of ξ. Then the following is a tableau expansion rule:

$$\frac{\mid C_1 \mid C_2 \mid \ldots \mid C_m \mid}{(\mid D_1, C_1 \mid D_2, C_1 \mid C_2 \mid \ldots \mid C_m \mid)\sigma}$$

where σ is a minimal subset unifier of D against C_1 (we assume that the variables of the premise are disjoint from the variables of D, D_1, D_2).

[7] According to our definitions, a clause is a set of literals, so the use of the leftmost disequation is not quite correct, but this restriction can easily be formalized using the selection mechanism (Bachmair et.al. [2]).

Branch closure

$$\frac{| C_1 | C_2 | \ldots | C_m | \quad C \cdot \rho}{(| C_2 | \ldots | C_m |)\sigma}$$

where σ is a minimal subset-unifier of $C\rho$ against C_1. (Note that the closure $C \cdot \rho$ used in this rule cannot contain equality because there is no rule in T_ξ introducing equality in a tableau.)

Let us make some comments on this logical system. Intuitively, tableau expansion rules encode the conjunction rule of sequent calculi or β-rules in the terminology of Smullyan [39] or Fitting [18]:

$$\frac{\Gamma \to \varphi \quad \Gamma \to \psi}{\Gamma \to \varphi \wedge \psi}$$

but turned upside down. The main difference is that instead of formulas φ, ψ and $\varphi \wedge \psi$ we use their ξ-names.

As we can see from the general description of our rules, the branch closure rule is non-local: the substitution produced by it changes the whole tableau, all other rules are local. (The substitution σ used in the tableau expansion rule only renames variables in the premise.)

Theorem 2 (Soundness and completeness). *The formula ξ is provable in first-order logic iff there is a derivation of the empty tableau # in T_ξ.*

Let us consider an example. We shall use a unary function symbol f. In order to avoid formulas having cluttered with parentheses, we shall write ft instead of $f(t)$. The formula to be proven is

$$\exists x((a \neq x \vee \neg G(b) \vee G(ffx)) \wedge (\neg G(fx) \vee x = b))$$

There are two conjunctive subformulas with the following sets of ξ-names:

Formula	Set of ξ-names
$a \neq x \vee \neg G(b) \vee G(ffx)$	$\{A_1(x)\}$
$\neg G(fx) \vee x = b$	$\{A_2(x)\}$

Thus, all tableau expansion rules have the form

$$\frac{| C_1 | C_2 | \ldots | C_m |}{| A_1(y), C_1 | A_2(y), C_1 | C_2 | \ldots | C_m |}$$

where y does not occur in the premise of this rule.

As the order $\succ$ we consider the lexicographic path ordering induced by the precedence relation $f > a > b$. The proof is as follows:

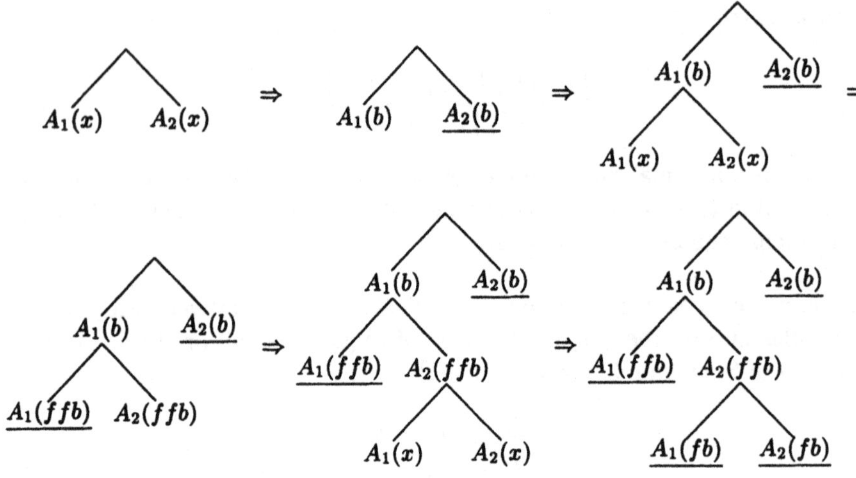

Fig. 2. A tableau proof

1. $\mid \square \mid$ (the initial tableau)
2. $\mid A_1(x) \mid A_2(x) \mid$ (tableau expansion from 1)
3. $x \neq b, A_2(x) \cdot \varepsilon$ (initial closure)
4. $A_2(x) \cdot \{b/x\}$ (equality solution from 3)
5. $\mid A_1(b) \mid$ (branch closure from 2,4)
6. $\mid A_1(x), A_1(b) \mid A_2(x), A_1(b) \mid$ (tableau expansion from 5)
7. $a = x, A_1(x) \cdot \varepsilon$ (initial closure)
8. $b \neq ffx, A_1(x), A_1(y) \cdot \varepsilon$ (initial closure)
9. $b \neq a, A_1(x), A_1(y), A_1(z) \cdot \{ffy/x\}$ (left superposition from 7,8)
10. $b \neq u, A_1(u), A_1(x), A_1(y), A_1(z) \cdot \{ffy/x\}$ (left superposition from 7,9)
11. $A_1(u), A_1(x), A_1(y), A_1(z) \cdot \{ffy/x, b/u\}$ (equality solution from 10)
12. $\mid A_2(ffb), A_1(b) \mid$ (branch closure from 6,11)
13. $\mid A_1(x), A_2(ffb), A_1(b) \mid A_2(x), A_2(ffb), A_1(b) \mid$ (tableau expansion from 12)
14. $fx \neq ffy, A_2(x), A_1(y) \cdot \varepsilon$ (initial closure)
15. $A_2(x), A_1(y)) \cdot \{fy/x\}$ (equality solution from 14)
16. $\mid A_2(fb), A_2(ffb), A_1(b) \mid$ (branch closure from 13,15)
17. # (branch closure from 16,15)

In order to illustrate this proof, we display tableaux in more conventional tree-like form in Figure 2. We omit superposition steps in this proof. In the closed branches, the atom at the bottom of the branch is underlined. The last step here corresponds to two applications of the branch closure rule.

There are several ways to automate proofs in T_ξ. One possibility is to derive closures using a saturation (bottom-up) algorithm in the style described in Lusk [25] and to derive tableaux using the usual top-down search with backtracking. In this case we can apply without losing completeness the following powerful strategies reducing search space:

1. Simplification of tableaux by derived unit closures $s = t \cdot \sigma$;

2. Basic simplification and subsumption for closures (similar to that of Bachmair et.al. [2]).

Another possibility is to use bottom-up (saturation) algorithms for the whole calculus T_ξ. In this case, we shall need powerful strategies for keeping the generated tableaux in a reasonable space. For example, the following subsumption notion for tableaux can be used in the same way as subsumption in resolution, i.e. subsumed tableaux can be removed from the search space.

Definition 3 (Tableau subsumption). A tableau $T = \mid C_1 \mid \ldots \mid C_n \mid$ *subsumes* a tableau $T' = \mid C'_1 \mid \ldots \mid C'_m \mid$, denoted $T \leq T'$, iff there is a substitution θ such that for every $i \in \{1, \ldots, n\}$ there is $j \in \{1, \ldots, m\}$ such that $C'_j \subseteq C_i\theta$.

The ordering $\leq$ on tableaux satisfies all standard properties of subsumption orderings of Voronkov [42] and hence we can remove tableaux subsumed by other tableaux. This ordering has some other properties:

1. The goal tableau # is the least element w.r.t. $\leq$;
2. The initial tableau $\mid \Box \mid$ is the greatest element w.r.t. $\leq$;
3. If the tableau T is obtained from the tableau T' by a tableau expansion, then $T \leq T'$;
4. The tableau $T = \mid C_2 \mid \ldots \mid C_n \mid$ subsumes the tableau $T' = \mid C_1 \mid \ldots \mid C_n \mid$. In particular, it means that for an application of a branch closure rule

$$\frac{\mid C_1 \mid C_2 \mid \ldots \mid C_m \mid \quad C \cdot \rho}{(\mid C_2 \mid \ldots \mid C_m \mid)\sigma}$$

where the subset-unifier σ does not change variables of $C_2, \ldots, C_n$, the conclusion of the rule subsumes the premise of the rule.

When we prove an equation from the set of universally quantified equations, i.e. when the goal formula ξ has the form $\forall s_1 = t_1 \wedge \ldots \wedge \forall s_n = t_n \supset \exists s = t$, there are no conjunctive subformulas in ξ. Hence, the calculus T_ξ becomes the calculus for refutational basic superposition-based (unfailing) completion.

One of the main problems in the tableau method is to decrease the number of applications of γ-rules. A technique for decreasing this number has been proposed in Beckert and Hähnle [4]. This paper introduced an optimization based on the (undecidable) notion of a formula in a tableau branch universal with respect to a variable and proved a sufficient (decidable) condition for universality. A more powerful optimization based on the so-called *usable subformulas* has been used earlier in Voronkov [41]. This optimization is also built-in in T_ξ due to its way of assigning ξ-names to the subformulas of ξ. If a formula φ and a variable x satisfy the sufficient condition of Beckert and Hähnle [4], then x does not occur in the set of ξ-names of φ (the inverse is not true in general). In order to take full advantage of this optimization we have to miniscope existential quantifiers as in Wang [43].

Let us consider an example. The goal is $\forall z \forall v \forall y (v = fy \supset g(v, z) = fv) \supset \exists x (g(g(x, fx), x) = g(fx, g(fx, x)))$. After translation to negation normal form and miniscoping we obtain the goal ξ of the form $\exists v(\exists y(v = fy) \wedge \exists z(g(v,z) \neq fv)) \vee \exists x(g(g(x,fx),x) = g(fx,g(fx,x)))$. We introduce two ξ-names: $A_1(v)$ for $\exists y(v = fy)$ and $A_2(v)$ for $\exists z(g(v,z) \neq fv)$. The tableau expansion rules are the same as in the previous example. As the order $\succ$ we consider the lexicographic path ordering induced by the precedence relation $g > f$. The proof in T_ξ is as follows:

1. $| \Box |$ (the initial tableau)
2. $| A_1(x) | A_2(x) |$ (tableau expansion from 1)
3. $v \neq fy, A_1(v) \cdot \varepsilon$ (initial closure)
4. $A_1(v) \cdot \{fy/v\}$ (equality solution from 3)
5. $| A_2(fy) |$ (branch closure from 2,4)
6. $| A_1(x), A_2(fy) | A_2(x), A_2(fy) |$ (tableau expansion from 5)
7. $| A_2(fy_1), A_2(fy_2) |$ (branch closure from 4,6)
8. $g(v,z) = fv, A_2(v) \cdot \varepsilon$ (initial closure)
9. $g(g(x,fx),x) \neq g(fx,g(fx,x)) \cdot \varepsilon$ (initial closure)
10. $g(fv_1,x) \neq g(fx,g(fx,x)), A_2(v_1) \cdot \{x/v_1\}$ (left superposition from 8,9)
11. $g(fv_1,x) \neq fv_2, A_2(v_1), A_2(v_2) \cdot \{x/v_1, fx/v_2\}$ (left superposition from 8,10)
12. $fv_3 \neq fv_2, A_2(v_1), A_2(v_2), A_2(v_3) \cdot \{x/v_1, fx/v_2, fx/v_3\}$ (left superposition from 8,11)
13. $A_2(v_1), A_2(v_2), A_2(v_3) \cdot \{x/v_1, fx/v_2, fx/v_3\}$ (equality solution from 12)
14. # (branch closure from 13,7)

In the last branch closure rule, we applied subset-unification of $A_2(x), A_2(fx), A_2(fx)$ against $A_2(fy_1), A_2(fy_2)$ using the subset-unifier $\{fy_1/x, fy_1/y_2\}$. If we include the variable z into the ξ-name of $g(v,z) \neq fv$, using $A_2(v,z)$ instead of $A_2(v)$, this subset-unification would not be possible, since $A_2(v_2,z_2)$ and $A_2(v_3,z_3)$ would use different substitutions for z_2 and z_3. In that case, we would have to apply a branch expansion rule once more, that corresponds to additional applications of γ-rules and β-rules.

4 Related work

Semantic tableaux belong to automatic proof methods based on Gentzen sequent calculi. Other methods based on sequent calculi are the inverse method (Maslov [26]), model elimination (Loveland [24]) and the method of matings (Andrews [1]), or the connection method (Bibel [6]). Proof-search in these methods uses the Herbrand theorem and can be considered as the problem of checking (by different methods) that every path through a matrix obtained from the goal formula is inconsistent. For this reason, these methods are sometimes characterized as matrix methods. This idea was originally justified by Prawitz [36] and Kanger [22]. Instead of generating instances of a quantifier-free formula $M(\bar{x})$, they proposed to search for a substitution σ for $\bar{x}_1, \ldots, \bar{x}_n$ such that every path in $(M(\bar{x}_1) \wedge \ldots \wedge M(\bar{x}_n))\sigma$ becomes complementary. In the case without equality, the search for such a *closing* substitution can be carried out using ordinary (simultaneous) unification. Thus, for a given matrix $M(\bar{x}_1) \wedge \ldots \wedge M(\bar{x}_n)$, the problem of the existence of an appropriate substitution is decidable and can be performed by any known unification algorithm. The same idea works for logic with equality, but the closing substitution should make any path E-inconsistent, or E-complementary. The problem of finding such a closing substitution is equivalent to the ***simultaneous rigid E-unifiability problem*** (Gallier, Raatz and Snyder [19]). Simultaneous rigid E-unifiability has recently proved to be undecidable in Degtyarev and Voronkov [13].

As a method for finding a closing substitution, Kanger proposed an algorithm which can now be characterized as an incomplete (but terminating) algorithm for simultaneous rigid E-unifiability. Variables in a matrix (or a tableau) could be consecutively substituted by ground terms already occurring in the matrix. This procedure does not solve simultaneous rigid E-unifiability, but it gives a procedure that is complete for first-order logic with equality. In the terminology of Fitting [18] it means that a closing substitution can be found after a sufficiently high (but not necessarily minimal) number of applications of the γ-rule. The approach to substitutions based on this idea has been characterized as minus-normalization in Matulis [30] and Maslov [27]. Another way of using incomplete but terminating algorithm for rigid E-unification has recently been proposed in Degtyarev and Voronkov [15].

Resolution-like procedures usually handle equality using variants of the paramodulation rule. It is well-known Brand [7] that paramodulation allows one to get rid of all equality axioms (except for $x = x$), and especially of the functional reflexivity axioms $f(\bar{x}) = f(\bar{x})$ for every function symbol f. The functional reflexivity axioms allow one to substitute ***arbitrary terms*** for variables. An analog of paramodulation was introduced in tableaux (Fitting [18]), called MGU replacement rule, ***with*** the functional reflexivity rule. We give an example that functional reflexivity is unavoidable in the method of proving completeness used in Fitting [18]. Consider the formula $\exists xyvu((a = b \supset g(x,u,v) = g(y,fc,fd)) \wedge (c = d \supset g(u,x,y) = g(v,fa,fb)))$. After one quantifier duplication (application of a γ-rule) we obtain the following tableau.

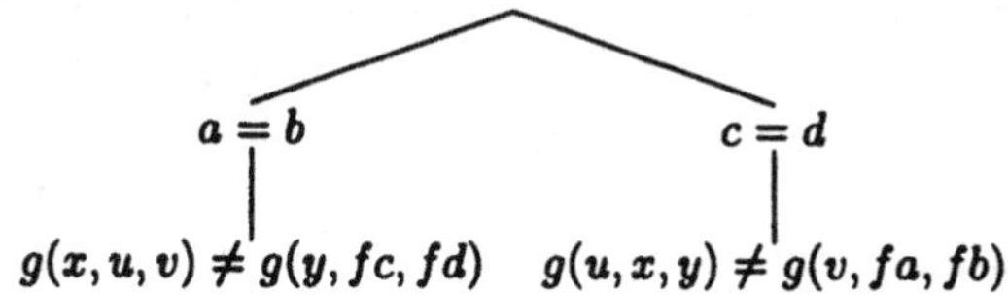

The substitution $\{fc/u, fd/v, fa/x, fb/y\}$ closes this tableau. This substitution can be found by applying the functional reflexivity rule and MGU replacement rule.

The use of the functional reflexivity in the completeness proof of Fitting [18] is dictated by the fact that the substitutions in the tableau method are applied to the whole tableau. In the terminology of Maslov, Mints and Orevkov [29], they are ***global*** unlike the ***local substitutions*** used in resolution. In intuitionistic logic, for which no local methods exist, there are no complete systems without functional reflexivity. In addition, minus-normalization does not hold for the intuitionistic sequent calculus (Maslov [27]).

Consider now approaches based on the simultaneous rigid E-unifiability (Gallier, Raatz and Snyder [19] Gallier et.al. [20]) and related methods. Since simultaneous rigid E-unification is undecidable (Degtyarev and Voronkov [13]), Gallier et.al.'s procedure cannot, in general, find solutions to simultaneous rigid E-unification. However, it is not clear whether this implies incompleteness of this procedure for first-order logic with equality: there are examples when their procedure cannot find a solution for a given amplification in spite that such a solution exists, but can find a solution for a bigger amplification.

In Gallier et.al. [20] it was proved that any (non-simultaneous) rigid E-unification problem has a finite complete set of solutions. According to this result, Goubault [21] proposed to solve simultaneous rigid E-unifiability by using finite complete sets of solutions to the components of the simultaneous problem. The undecidability result of Degtyarev and Voronkov [13] shows that finite complete sets of solutions do not give a solution to the simultaneous problem. The reason for this is that the comparison of solutions to different subproblems is made modulo different congruences.

Petermann [35] introduces a "complete connection calculus with rigid E-unification". Here the completeness is achieved by changing the notion of a complete set of unifiers so that solutions to all subproblems are compared modulo the same congruence (generated by the empty equality theory). In this case, a non-simultaneous problem can have an infinite number of solutions and no finite complete set of solutions. For example, for the rigid E-unification problem $f(a) = a \vdash_{\forall} x = a$ the complete set of solutions in the sense of Gallier et.al. [20] consists of one substitution $\{a/x\}$, but the complete set of solutions in the sense of Petermann [35] is infinite and consists of substitutions $\{f^n(a)/x\}$, for all $n \in \{0, 1, \ldots\}$. This implies that the proof-search by the method of Petermann [35] can be non-terminating even for a limited number of applications of γ-rule (i.e. for a particular tableau), unlike algorithms based on the finite complete sets of unifiers in the sense of Gallier et.al. [20] or based on minus-normalization (Kanger [22], Matulis [30]). The implementation of the method of Petermann [35] uses a completion-based procedure (Beckert [3]) of generation of complete sets of rigid E-unifiers. This procedure is developed with the aim of solving a more general problem — so-called *mixed E-unification* and has been implemented as part of the tableau-based theorem prover ${}_3T^AP$. Complete sets of unifiers both in the sense of Gallier et.al. [20] and in the sense of Petermann [35] can be computed by this procedure. However, the termination is not guaranteed even for complete sets of rigid E-unifiers in the sense of Gallier et.al. [20].

A method not using rigid E-unification is proposed in Beckert and Hähnle [4]. This paper claims the completeness of the method, but this claim is not true. The method expands the tableau using the standard tableau rules, including γ-rules. For finding a closing substitution, an analog of linear paramodulation without function reflexivity has been proposed. As it is well known, linear paramodulation is incomplete without function reflexivity. The same is true for the method of Beckert and Hähnle [4], as the following example shows. Suppose that we prove the formula $\exists x(a = b \wedge g(fa, fb) = h(fa, fb) \supset g(x, x) = h(x, x))$. In order to prove it using paramodulation, we need to paramodulate $a = b$ into $g(fa, fb) = h(fa, fb)$. The method of Beckert and Hähnle [4] only allows for paramodulation into copies of $g(x, x) = h(x, x)$ obtained by the application of γ-rules. Thus, this (provable) formula cannot be proved using the method of Beckert and Hähnle [4].

In several tableau-based theorem provers (see the survey of Schumann [37]) equality is implemented using the modification method of Brand [7]. This method reduces a set of clauses with equality to a set of clauses without equality via a *flattening transformation*. Recently, we proposed a new reduction method based on the so called *basic folding*, demonstrated for Horn clauses in [14].

The equality elimination method was originally introduced in Degtyarev and Voronkov [12] as a method of handling equality in logic programs. Later, it has been applied to the inverse method in Degtyarev and Voronkov [11]. Applications of equality elimination to the tableau method and to the inverse method are based on a common matrix characterization of provability in terms of solution clauses (Degtyarev and Voronkov [11]). This characterization can be considered as a computationally improved version of the Herbrand's theorem. However, the resulting calculi for the inverse method and for the tableau method are quite different because the inverse method is local while the tableau method is global (Maslov, Mints and Orevkov [29]). Among other inference rules, the inverse method uses the *factoring rule* that is absent in the tableau method. The tableau method uses the *branch closure rule* based on subset-unification that has no analogue in the inverse method. Subset-unification has been earlier used in P-deduction (Demolombe [16]) and SLO-resolution (Minker, Rajasekar and Lobo [31]).

References

1. P.B. Andrews. Theorem proving via general matings. *Journal of the Association for Computing Machinery*, 28(2):193–214, 1981.
2. L. Bachmair, H. Ganzinger, C. Lynch, and W. Snyder. Basic paramodulation. *Information and Computation*, 121:172–192, 1995.
3. B. Beckert. A completion-based method for mixed universal and rigid E-unification. In A. Bundy, editor, *Automated Deduction — CADE-12. 12th International Conference on Automated Deduction.*, volume 814 of *Lecture Notes in Artificial Intelligence*, pages 678–692, Nancy, France, June/July 1994.
4. B. Beckert and R. Hähnle. An improved method for adding equality to free variable semantic tableaux. In D. Kapur, editor, *11th International Conference on Automated Deduction (CADE)*, volume 607 of *Lecture Notes in Artificial Intelligence*, pages 678–692, Saratoga Springs, NY, USA, June 1992. Springer Verlag.
5. E.W. Beth. *The Foundations of Mathematics.* North Holland, 1959.
6. W. Bibel. On matrices with connections. *Journal of the Association for Computing Machinery*, 28(4):633–645, 1981.
7. D. Brand. Proving theorems with the modification method. *SIAM Journal of Computing*, 4:412–430, 1975.
8. A. Degtyarev and A. Voronkov. Automated theorem proving I and II. *Cybernetics*, 22(3):290–297, 1986 and 23(4):547–556, 1987.

9. A. Degtyarev and A. Voronkov. Equality control methods in machine theorem proving. *Cybernetics*, 22(3):298–307, 1986.
10. A. Degtyarev and A. Voronkov. Equality elimination for semantic tableaux. UPMAIL Technical Report 90, Uppsala University, Computing Science Department, December 1994.
11. A. Degtyarev and A. Voronkov. Equality elimination for the inverse method and extension procedures. In C.S. Mellish, editor, *Proc. International Joint Conference on Artificial Intelligence (IJCAI)*, volume 1, pages 342–347, Montréal, August 1995.
12. A. Degtyarev and A. Voronkov. A new procedural interpretation of Horn clauses with equality. In Leon Sterling, editor, *Proceedings of the Twelfth International Conference on Logic Programming*, pages 565–579. The MIT Press, 1995.
13. A. Degtyarev and A. Voronkov. Simultaneous rigid E-unification is undecidable. UPMAIL Technical Report 105, Uppsala University, Computing Science Department, May 1995. To appear in Theoretical Computer Science, v.166.
14. A. Degtyarev and A. Voronkov. Handling equality in logic programs via basic folding. In R. Dyckhoff, H. Herre, and P. Schroeder-Heister, editors, *Extensions of Logic Programming (5th International Workshop, ELP'96)*, volume 1050 of *Lecture Notes in Computer Science*, pages 119–136, Leipzig, Germany, March 1996.
15. A. Degtyarev and A. Voronkov. What you always wanted to know about rigid E-unification. In *Submitted to JELIA'96*, page 20, 1996.
16. R. Demolombe. An efficient strategy for non-Horn deductive databases. In G.X. Ritter, editor, *Information Processing 89*, pages 325–330. Elsevier Science, 1989.
17. N. Dershowitz and J.-P. Jouannaud. Rewrite systems. In J. Van Leeuwen, editor, *Handbook of Theoretical Computer Science*, volume B: Formal Methods and Semantics, chapter 6, pages 243–309. North Holland, Amsterdam, 1990.
18. M. Fitting. *First Order Logic and Automated Theorem Proving*. Springer Verlag, New York, 1990.
19. J.H. Gallier, S. Raatz, and W. Snyder. Theorem proving using rigid E-unification: Equational matings. In *Proc. IEEE Conference on Logic in Computer Science (LICS)*, pages 338–346. IEEE Computer Society Press, 1987.
20. J. Gallier, P. Narendran, S. Raatz, and W. Snyder. Theorem proving using equational matings and rigid E-unification. *Journal of the Association for Computing Machinery*, 39(2):377–429, 1992.
21. J. Goubault. Rigid $\overline{E}$-unifiability is DEXPTIME-complete. In *Proc. IEEE Conference on Logic in Computer Science (LICS)*. IEEE Computer Society Press, 1994.
22. S. Kanger. A simplified proof method for elementary logic. In J. Siekmann and G. Wrightson, editors, *Automation of Reasoning. Classical Papers on Computational Logic*, volume 1, pages 364–371. Springer Verlag, 1983. Originally appeared in 1963.
23. R.C.T. Lee and C.L. Chang. *Symbolic Logic and Mechanical Theorem Proving*. Academic Press, 1973.
24. D.W. Loveland. Mechanical theorem proving by model elimination. *Journal of the Association for Computing Machinery*, 15:236–251, 1968.
25. E.L. Lusk. Controlling redundancy in large search spaces: Argonne-style theorem proving through the years. In A. Voronkov, editor, *Logic Programming and Automated Reasoning. International Conference LPAR'92.*, volume 624 of *Lecture Notes in Artificial Intelligence*, pages 96–106, St.Petersburg, Russia, July 1992.
26. S.Yu. Maslov. The inverse method of establishing deducibility in the classical predicate calculus. *Soviet Mathematical Doklady*, 5:1420–1424, 1964.

27. S.Yu. Maslov. An invertible sequential variant of constructive predicate calculus (in Russian). *Zapiski Nauchnyh Seminarov LOMI*, 4, 1967. English Translation in: Seminars in Mathematics: Steklov Math. Inst. 4, Consultants Bureau, NY-London, 1969, p.36–42.
28. S.Yu. Maslov. The generalization of the inverse method to predicate calculus with equality (in Russian). *Zapiski Nauchnyh Seminarov LOMI*, 20:80–96, 1971. English translation in: Journal of Soviet Mathematics 1, no. 1.
29. S.Yu. Maslov, G.E. Mints, and V.P. Orevkov. Mechanical proof-search and the theory of logical deduction in the USSR. In J.Siekmann and G.Wrightson, editors, *Automation of Reasoning (Classical papers on Computational Logic)*, volume 1, pages 29–38. Springer Verlag, 1983.
30. V.A. Matulis. On variants of classical predicate calculus with the unique deduction tree (in Russian). *Soviet Mathematical Doklady*, 148:768–770, 1963.
31. J. Minkér, A. Rajasekar, and J. Lobo. Theory of disjunctive logic programs. In Jean-Louis Lassez and Gordon Plotkin, editors, *Computational Logic. Essays in Honor of Alan Robinson.*, pages 613–639. The MIT Press, Cambridge, MA, 1991.
32. M. Moser, C. Lynch, and J. Steinbach. Model elimination with basic ordered paramodulation. Technical Report AR-95-11, Fakultät für Informatik, Technische Universität München, München, 1995.
33. R. Nieuwenhuis and A. Rubio. Theorem proving with ordering and equality constrained clauses. *Journal of Symbolic Computations*, 19:321–351, 1995.
34. V.P. Orevkov. On nonlengthening rule applications for equality (in Russian). *Zapiski Nauchnyh Seminarov LOMI*, 16:152–156, 1969. English Translation in: Seminars in Mathematics: Steklov Math. Inst. 16, Consultants Bureau, NY-London, 1971, p.77–79.
35. U. Petermann. A complete connection calculus with rigid E-unification. In *JELIA'94*, volume 838 of *Lecture Notes in Computer Science*, pages 152–166, 1994.
36. D. Prawitz. An improved proof procedure. In J. Siekmann and G. Wrightson, editors, *Automation of Reasoning. Classical Papers on Computational Logic*, volume 1, pages 162–201. Springer Verlag, 1983. Originally appeared in 1960.
37. J. Schumann. Tableau-based theorem provers: Systems and implementations. *Journal of Automated Reasoning*, 13(3):409–421, 1994.
38. N. Shankar. Proof search in the intuitionistic sequent calculus. In D. Kapur, editor, *11th International Conference on Automated Deduction*, volume 607 of *Lecture Notes in Artificial Intelligence*, pages 522–536, Saratoga Springs, NY, USA, June 1992. Springer Verlag.
39. R.M. Smullyan. *First-Order Logic*. Springer Verlag, 1968.
40. A. Voronkov. LISS - the logic inference search system. In Mark Stickel, editor, *Proc. 10th Int. Conf. on Automated Deduction*, volume 449 of *Lecture Notes in Computer Science*, pages 677–678, Kaiserslautern, Germany, 1990. Springer Verlag.
41. A. Voronkov. A proof-search method for the first order logic. In P. Martin-Löf and G. Mintz, editors, *COLOG'88*, volume 417 of *Lecture Notes in Computer Science*, pages 327–340. Springer Verlag, 1990.
42. A. Voronkov. Theorem proving in non-standard logics based on the inverse method. In D. Kapur, editor, *11th International Conference on Automated Deduction*, volume 607 of *Lecture Notes in Artificial Intelligence*, pages 648–662, Saratoga Springs, NY, USA, June 1992. Springer Verlag.
43. H. Wang. Towards mechanical mathematics. *IBM J. of Research and Development*, 4:2–22, 1960.

Towards Lean Proof Checking

G. Barthe and H. Elbers*

CWI
PO Box 94079, 1090 GB Amsterdam
The Netherlands
{gilles, elbers}@cwi.nl

Introduction. *Formal mathematics* is the branch of mathematics concerned with the development of mathematical theories within a fixed formal system. Over the last few years, there has been growing interest in doing formal mathematics to verify algorithms, protocols... However, doing formal mathematics can be tedious and time-consuming because complex reasoning steps have to be justified in terms of the rules of the formal system. As a result, the lengths of formal proofs can become disproportionate. This problem can be solved by designing more efficient formal systems and by using computers. *Automated theorem proving* is the branch of computer science concerned with the development of software for formal mathematics. Because they implement *logical formal systems*, theorem provers are especially suited for abstract reasoning but are inefficient at calculations. Indeed, logical formal systems do not have any computational power and calculations have to be *justified* by logical proofs. A natural way to overcome this problem is to consider *formal environments*, i.e. formal systems combining logical and computational power. In such systems, calculations are treated as first-class citizens and do not need to be justified by logical proofs. Formal environments have definite advantages:

- computations do not blow up the size of formal proofs;
- they offer a theoretical model for the combination of theorem provers and symbolic computation systems;
- their properties can be studied independently of any implementation.

The goal of this paper is to study formal environments based on type theory.We feel type theory is a natural basis for formal environments as types play an increasingly important role in symbolic computation systems and theorem provers.

Theoretical Background. We introduce the concept of *oracle type*. An oracle type essentially consists of:

- an algebraic type A, to be thought as a computer algebra system for some mathematical theory T. It is determined by a set of constructors, which determines a set of expressions and a rewriting relation;

* Research supported by the Netherlands Organization for Scientific Research (NWO) under contract SION 612-16-607

- a logical axiom of ***no confusion*** relating computations in A to provability in T, and by a soundness theorem to equality in models of T.

Oracle types allow for a clear separation between the syntax and the semantics (i.e. models) of mathematical theories. For suitable theories, such as the ones considered here, both are related by a soundness theorem, which allows us to prove equalities in a given model of a mathematical theory T by means of logical deductions in T. In turn, logical deductions in T are related to computations in A by the no confusion axiom.

The separation between syntax and semantics is important both for practical and conceptual reasons. On the practical side, it allows us to use oracle types in combination with intensional type theories such as the one of Lego ([4]). On the conceptual side, it gives an account of the interaction between symbolic computation systems and theorem provers: if we think of the algebraic type as a symbolic computation system, the soundness theorem relates computations in the symbolic computation system to proofs of equalities in the theorem prover.

Oracle Types in Theorem Proving. We present an extension of the theorem prover Lego with oracle types. Our extension is both user-friendly (only a few commands are added to Lego) and flexible (rewriting can be implemented using Lego rewrite rules or any symbolic computation system with user-definable rewriting). An oracle type can be defined by specifying its constructors and (possibly conditional) rewrite rules: this results in a representation of the algebraic type of the oracle type in Reduce ([3]) and a representation of its equational theory in Lego. We provide a fully automatic tactic to solve problems of the form $a =_M b$ (where M is a model of the equational theory of an oracle type). In [1], it is estimated that such a tactic will reduce the length of proofs by 10%.

The approach followed in this paper was first suggested in [1] and later developed in [2]. The present work extends the theoretical foundations of [2] (by extending the approach to conditional rewriting) and provides (for the first time) an efficient tool to do computations in Lego.

References

1. G. Barthe. Formalising mathematics in type theory: fundamentals and case studies. Technical Report CSI-R9508, University of Nijmegen, 1995.
2. G. Barthe, M. Ruys, and H. Barendregt. A two-level approach towards lean proof-checking. In S. Berardi and M. Coppo, editors, *Proceedings of TYPES'95*, Lecture Notes in Computer Science. Springer-Verlag, 1996. To appear.
3. Anthony C. Hearn. REDUCE User's Manual Version 3.5. Technical report, RAND, October 1993. RAND Publication CP78.
4. Z. Luo and R. Pollack. LEGO proof development system: User's manual. Technical Report ECS-LFCS-92-211, LFCS, Computer Science Dept., University of Edinburgh, May 1992.

WALDMEISTER: High Performance Equation Theorem Proving

Arnim Buch Thomas Hillenbrand Roland Fettig

Universität Kaiserslautern, FB Informatik, 67653 Kaiserslautern, Germany
email: {buch,hillen,fettig}@informatik.uni-kl.de

Today the application area of deduction systems is yet too small to let them become useful tools e.g. in program verification. In this paper we present a system for equational deduction. Our prover WALDMEISTER avoids the main diseases today's provers suffer from: overindulgence of time and space (see also [HBF96]).

In [KB70], Knuth and Bendix introduced the completion algorithm which tries to derive a set of convergent rewrite rules from a given set of equations. Its extension to unfailing completion in [BDP89] has turned out to be a valuable means of proving theorems in equational theories. Stated as an inference system, completion operates on a set of facts (*rules* and *equations*) which are used to simplify the hypothesis by rewriting. New facts can be derived via superposition of existing ones. Among these possible one–step derivations (*critical pairs* or *cps*) one is chosen to be added to the set of facts. If the hypothesis has not become trivial by now, the next iteration cycle starts.

In [BH96] we elaborate the whole design process starting from a three–level system model: at the bottom the execution of single inference steps, above their aggregation into an inference machine, and at the top the overall control strategy guiding the search for promising derivations. We also give more detailed information on the whole system and a thorough statistical analysis.

Fishing for efficiency

Every cycle of the completion loop starts with the selection of a cp to be processed. As this selection determines success or failure of the proof run, it is the crucial point of completion based theorem provers. It is generally accepted that a valuable analysis of all the cps relies on extracting information from their terms. Our experiments have shown that the extraction is more precise if it is applied to normalized terms. This normalization w.r.t. the actual rewrite relation causes most of the system's time consumption. To avoid expensive recomputation, the selection information already generated should be kept. As millions of cps may arise during a run, economical representations are in demand.

The key to fast rewriting is fast matching from the left–hand sides of the rewrite rules onto the query term. Indexing techniques are one promising solution of this problem. For an overview see [Gra95]. We employ perfect discrimination trees (cf. [McC92]) in a new, refined variant: All branches leading to a single leaf node are shrunk. In the average, this saves about 40% of all tree nodes and therefore quite a lot of backtracking effort. Following McCune, we combine perfect discrimination trees with a linear term representation as in [Chr89], but singly linked only. Using these flatterms in conjunction with free–list based memory management allows us to dispose terms in constant time.

We compared our system with a variant that stores rules and equations in a list with top–symbol hashing, and observed speedups of one order of magnitude.

Lean management. The storage of cps, ordered by some weighting function, has to meet two major requirements: first, to support arbitrary insertion and retrieval of the smallest entry, and second, to minimize memory consumption. We identified three points responsible for the latter problem: the size of the cps themselves, the management overhead, and the number of superfluous cps in the set ("orphans", as their parent rules/eqns. have been discarded). Flatterms are necessary for fast rewriting, but cause high memory consumption when cps are stored in such a manner. For that reason, we developed a string–like term representation which cuts down the size of terms to some 10%. Furthermore, we evolved a minimal representation of constant size using no terms at all but superposition information to recompute them. Topped by a two level data structure, basically a heap of heaps allowing to delete between 60 % and 90 % of all orphans without additional management overhead, this results in a very space–efficient realization of the cp set: as an example, WALDMEISTER kept 734,000 cps within a process size of 16.6 MB when solving mv4 (taken from [LW92]).

A useful theorem prover should allow to solve many problems with its default strategy. Hence, we have statistically approved a parameter setting. Using a simple length–based selection strategy, WALDMEISTER solved two problems claimed as "never proved in a single run unaided" in [LW92]. The table below gives an impression of the high throughput of our system. The power of WALDMEISTER stems from its inference speed and economical stringterm representation, allowing fast intermediate reduction of the cp set.

Problem	Crit.Pairs	Rules +Eqns	Max. Size of CP set	Process Size	Time (Sec.)	Per Second: Crit.Pairs	Reductions	Match Queries
gt6	1,421,317	1,006	856,545	94 MB	968.4	1,467.7	2,344.5	81,740.3
ra4	923,508	1,549	811,044	88 MB	3,759.5	245.7	756.4	59,459.0

References

[BDP89] L. Bachmair, N. Dershowitz, and D.A. Plaisted. Completion without failure. In *Collection on the Resolution of Equations in Algebraic Structures*. Academic Press, Austin, 1989.

[BH96] A. Buch and Th. Hillenbrand. WALDMEISTER: Development of a high performance completion–based theorem prover. SEKI–Report 96–01, Univ. Kaiserslautern, 1996. Available via `ftp://ftp.uni-kl.de/`

[Chr89] J. Christian. Fast Knuth–Bendix completion: A summary. In *Proc. 3rd RTA*, vol. 355 of *LNCS*, 1989.

[Gra95] P. Graf. *Term Indexing*, vol. 1053 of *LNAI*. Springer Verlag, 1995.

[HBF96] Th. Hillenbrand, A. Buch, and R. Fettig. On gaining efficiency in completion-based theorem proving. In *Proc. 7th RTA*, LNCS, 1996.

[KB70] D.E. Knuth and P.B. Bendix. Simple word problems in universal algebra. *Computational Problems in Abstract Algebra*, 1970.

[LW92] E. Lusk and L. Wos. Benchmark problems in which equality plays the major role. In *Proc. 11th CADE*, vol. 607 of *LNCS*, 1992.

[McC92] W. McCune. Experiments with discrimination-tree indexing and path indexing for term retrieval. *Journal of Automated Reasoning, 8(3)*, 1992.

A Reflective Language Based on Conditional Term Rewriting

Masanobu NUMAZAWA[1], Masahito KURIHARA[2] and Azuma OHUCHI[1]

[1] Hokkaido University, Kita 13, Nishi 8, Kita-ku, Sapporo 060, Japan
[2] Hokkaido Institute of Technology, Maeda 7-15, Teine-ku, Sapporo 006, Japan

Abstract. Meta-computation is a computational mechanism that allows computational systems to read and modify meta-objects that represent the current state of its own computation. Implementing meta-computation by high-level language enables us to access meta-level with high-abstract interface. In recent years this notion can be found in several fields of computer science and artificial intelligence. In particular, in the field of intelligent systems meta-computation is often called meta-inference and plays an important role for designing complex systems. We introduce reflective-computation mechanisms into a conditional term rewriting system-based language and discuss about their implementation and application.

First, we introduce a virtual reduction machine equipped with four stacks for storing meta-information during condition evaluation. This machine is used to define the operational semantics of our language in terms of a reduction relation reflecting its state transition. A state of the reduction machine is the tuple $(S, D, X, P, \mathcal{R})$ of four stacks and a program: the *subject term stack* S, the *candidate rewrite rules stack* D, the *context stack* X, the *phase stack* P and the *program* $\mathcal{R}$. Computation in the virtual reduction machine is defined by the reduction relation $\Rightarrow$, on the set of states. $\Rightarrow$ has the fifteen operation steps.

Secondly, we extend the language by adding to our machine two basic meta-computation mechanisms called meta- and base-transformations. The meta-transformation transforms meta-level objects (stacks and programs) into base-level objects (constructor terms), while the base-transformation is just its inverse. The relation $\Rightarrow$ is extended by adding five new operation steps to existing fifteen steps defined before.

In ordinary conditional term rewriting systems, both left- and right-hand sides of the rewrite rules are terms. In our language, however, we introduce *meta-computational expressions* which are not terms ($\notin \mathcal{T}$). A meta-computational expression is not an object to be rewritten, and has a form $t : L$ (with $t \in \mathcal{T}, L \notin \mathcal{T}$) in the left-hand side of the rule, and L (with $L \notin \mathcal{T}$) in the right-hand side of the rule. They play an important role for interfacing with meta-computation of the system. An *expression* is either a term or a meta-computational expression. The symbol L in the meta-computational expressions denote a list, enclosed by $\{\ \}$, of terms of the form $R(s)$, $S(s)$, $D(s)$, $X(s)$, $P(s)$, where $s \in \mathcal{T}$ and R, S, D, X, P are primitive function symbols which present the names of program $\mathcal{R}$ and stacks S, D, X, P, respectively. For example, $L = \{R(r)\ \ S(a)\}$. We assume that at most one term of each form is contained. Conditional rewrite rules may be classified into four types, depending upon whether the left-hand side or the right-hand side is a term or a meta-computational expression. The conditional rewrite rules whose left-hand sides are meta-computational expressions $t : L \rightarrow (\mathit{the\ expression})\ \mathit{if}\ (\mathit{the\ conditions})$ invoke the meta-transformation. Similarly, the rules whose right-hand sides are meta-computational expressions $(\mathit{the\ expression}) \rightarrow L\ \mathit{if}\ (\mathit{the\ conditions})$ invoke base-transformation.

One can think of several applications of our system to the rewrite systems development. For example, it can be applied to the development of the membership-conditional term rewriting systems, in which each rewrite rule can have membership conditions which restrict the substitution values for the variables occurring in the rule. Since membership conditions need meta-level information which cannot be described in ordinary conditional rewrite rules, they have traditionally been described in ordinary programming languages such as Lisp. This has caused serious problems in portability and transparency. In our system, they can be described simply in conditional rewrite rules. As another application, consider an interface with a debugger. Since the computational states are defined in terms of the virtual reduction machine of our system, we can get appropriate computational state even during condition evaluation. Suppose that if the program makes an error then the current program and the contents of the stacks are to be given to a debugger. For example, the following two rules may be added to the program to handle an error caused by some reasons (written by ...).

```
(1)  ...:{ S[?s] D[?d] X[?x] P[?p] R[?r] } = debug[?s, ?d, ?x, ?p, ?r] .
(2)  continue[?s, ?d, ?x, ?p, ?r] = { S[?s] D[?d] X[?x] P[?p] R[?r] } .
```

The meta-transformation is executed by the rule (1). As the result, the meta-representation of the stacks S, D, X, P and the program $\mathcal{R}$ is substituted to `?s`, `?d`, `?x`, `?p`, `?r` , respectively. We assume that when the meta-level function **debug** receives such information via its arguments, it displays and/or updates that information by interacting with the user, and then calls the function **continue** with the updated information in its arguments, in order to continue the base-level computation. Then by the rule (2), the base-transformation is applied to the meta-representation of the information and the computation continues from the state obtained by the transformation.

The transformation could be found useless, if most of the transformed data were not accessed when the meta-transformation has been performed. To avoid such useless transformation, we have implemented our system such that when meta-transformation is invoked, meta-level objects will not be transformed into base-level objects immediately. Actually, meta-level objects are transformed into *delayed-objects* which are partially transformed into base-level objects according to the necessity, i.e., only when those delayed-objects are actually accessed. Since our system is written in CLOS(Common Lisp Object System), the *delayed evaluation* mechanism is implemented in terms of the class and method definitions in CLOS.

Finally, we state further research topics. They include the development of more powerful meta-computation facilities and its efficient implementation.

Term Rewriting Systems: An h-Categorical Semantic

Giulio Balestreri*

Dipartimento di Discipline Scientifiche:
Chimica ed Informatica
Università di Roma Tre
Via della Vasca Navale, 84
00146, Roma-Italy
e-mail:gippo@inf.uniroma3.it

Abstract. In this paper term rewriting is modelled by use of a h-category framework, where 0-cells represent sets of variables, morphisms represent substitutions and contexts and 2-morphisms represent reduction rules. In this context stability and compatibility properties of reduction can be proved. The proposed categorial view allows for a good level of expressivity of the reduction mechanisms, without constraining rewriting to the particular area where it is defined. Moreover, it seems to be a promising framework for the embedding of algebraic reduction.

1 Introduction

Term Rewriting Systems constitute a computation model that is used in several areas of Computer Science, for instance in the semantics of programming languages [Hu80] [Od77] [OH84], automatic theorem provers [La75] [JK89] [HH82], transformation of formulae [Lo74] and, in general, symbolic computation [Bu85] [Wi89].

Since the original work by Knuth and Bendix [KB70], one of the main concerns in the Term Rewriting Systems (TRS) area has been the analysis of their features. The fundamental properties of a TRS are confluence and termination which ensure the system completeness. Therefore, they have been the main subject of research and various approaches to completion procedures have been proposed, with great attention to their efficiency. The completion problem in TRS has been deeply investigated, both in the methodological aspect, for example, studying reduction modulo equivalence classes [JK86] [PS81], and from the standpoint of complexity analysis with the design of more and more powerful and sophisticated techniques [Kl90].

An important area of application of TRS is Computational Algebra. Buchberger [Bu76] pointed out the strong links between completion in TRS and the search of Groebner bases in polynomial rings. In particular, Buchberger and

* This work has been partially supported by the Italian Project MURST ex 40% "Rappresentazione della Conoscenza e Meccanismi di Ragionamento

Loos in [BL83], and Winkler in [Wi89], recognized the close syntactical analogy between rewriting in TRS and reduction in polynomial rings.

Huet [Hu80] and others [Fa79] [HO80], investigated the possibility of establishing a mathematical foundation of deduction with the help of TRS. In particular, equational deduction has been studied in this respect, since the main concepts involved therein are rewriting and substitution. In [RS87], Rydeheard e Steel noted that, since both operations can be composed, they can be suitably framed into a 2-category structure, i.e. a category where morphisms are the objects of a higher level category. Thus rewriting is given a 2-categorial semantics. Recently, 2-categories have been also proposed by Meseguer [Me92], Steel [St92] and Stokkermans [St92a] as an algebraic way to equip TRS with a concurrent semantics.

The above results and the syntactical analogy between rewriting and polynomial reduction, studied by Buchberger and Loos in [BL83], suggest the possibility of defining a categorial semantics of reduction in Computational Algebra. However, a straightforward use of the same semantical structure introduced in [RS87] is not feasible, as shown in [Ba92] and [BM92]. The reason is to be found mainly in the fact that the strong syntactical analogies between TRS and Computational Algebra hide a deep semantical difference.

The distance between rewriting and polynomial reduction appears by observing, for example, that the former is expressed by means of meta-rules, external to the structure we are operating in, that is the free algebra of terms. On the other hand, the second stays internal to the polynomial ring, as it comes out as the rest of division, that is an internal operation. Moreover, while rewriting needs to refer to global concepts, like context and occurrence, polynomial reduction does not. In fact, in the definition of the rewriting relation between terms, in order to refer to the subterm to be reduced, either the notion of context [Kl90] or that of occurence [Hu80] is used. On the other hand, the elements of the polynomial algebra are "flat" structures that are reduced by means of internal operations.

Beyond occurrence, or equivalently context, the second important concept involved in the rewriting relation is substitution. The two notions both correspond to "operations" on terms and they are, in some sense, one the inverse of the other: the localization of an occurrence reduces a term, producing a subterm, while the application of a substitution to a term yelds a new term that is in general deeper and wider than the original one. The two operations are *inverse* in a purely informal and intuitive sense here, and their syntactic translations in Computational Algebra are given by the multiplication with a monomial w.r.t. substitution, and by the individuation of a monomial w.r.t. occurrence; in [BL83] the authors defined this correspondence as:

> *The multiplication with a monomial corresponds to the application of a substitution and the operation of replacement in the present context defined as follows:* $s[u \leftarrow t]$:= *the polynomial that results from the polynomial* s*, by replacing the monomial* u *by the polynomial* t*.*

We believe that a first necessary step for providing a semantic foundations of term and polynomial reduction is making rewriting operations *internal.* To this aim, a h-categorial structure is constructed, that is meant to reflect the algebraic pecularities of a polynomial ring and the complexity of the tree structure of terms.

This paper extends the notion of substitution to a more complex concept, in the attempt of giving a mathematical characterization of rewriting where no reference to either occurence or context is needed. Rewriting is therefore defined as an object-level operation. The aim is not that of simulating rewriting in the Algebraic area through polynomial division, but rather polynomial division is simulated by substitutions and rewriting rules. Term rewriting is finally modeled by means of an *h-category*, that is a more complex structure than a 2-category, and appears to be a promising candidate as the object in which a common semantics to the two structures can be given.

The paper is structured as follows. Section 2 analyzes the notion of substitution and presents new concepts as sets of substitutions and links among them, which are needed in order to define rewriting without any reference to context and occurrence. Section 3 introduces reduction, with the study of some of its properties; a function on substitutions is defined, called *rewriting instance function* (RIF in the following), that is the main tool for the composition of rewriting rules, thus leading to a compositionally clean structure. RIF functions can be composed with substitutions, obtaining new RIF functions, in a sort of action of substitutions on RIF functions. Finally, the h-category of rewriting is constructed, in which substitutions and RIF functions are present at different levels, and can be composed in agreement with the categorial axioms.

We will assume the reader is familiar with tha basic notions of Term Rewriting System, such as in [Hu80a], [Kl90] and [DJ90].

2 Substitutions

A substitution is usually considereds as a function from a set X of variables to the set $T_\Omega(X)$ of terms of a signature Ω with variables in X (Ω-terms). We want to partition the set Σ of substitutions into several disjoint subsets in order to match each substitution with one or more subsets of the partition. To obtain this we make use of terms; in the simplest case, each substitution with codomain $\{t\}$ belongs to the S_t partition. For a given substituion σ, we denote by $D(\sigma)$ the domain of σ, i.e. the subset of X such that $\sigma(x) \neq x$.

Special substitutions are called permutations; θ is a permutation if it is a bijection mapping X in X.

Let σ be a substitution, and U a subset of X. We say that $\sigma|_U$ is the restriction of σ to U if it is the substitution acting only on variables of U.

Classical reduction is defined by substitutions and occurrence, where occurrence identifies the subterm that really matches the left hand side of a rewriting rule. In order to define rewriting only in terms of substitutions, we simulate the role played by the occurrences in a term, by means of substitutions. In the next section we will show how it is possible to replace occurrences by a specified substitution set, playing the role of matching the left hand side of a rule and the term to be reduced.

If t is an Ω-term and θ a permutation over the set $Var(t)$ of the variables occurring in t, then S_t and $S_{\theta(t)}$ are the sets of substitutions with codomain $\{t\}$, and $\{\theta(t)\}$, respectively:

$$S_t = \{\sigma \in \Sigma / \ \exists\theta \in \Theta \ \exists x \in X \ : \ D(\sigma) = \{x\}, \ and \ \sigma(x) = \theta(t)\}$$
$$S_{\theta(t)} = \{\sigma \in S_t / \sigma(x) = \theta(t)\}.$$

Intuitively, for a fixed term t and a permutation θ, $S_{\theta(t)}$ is a subset of Σ, in which the domain of each substitution has cardinality equal to 1, and mapping the only variable to $\theta(t)$.

These definitions are extended to sets of substitutions σ having domain with cardinality equal to n, and such that there exists a variable $x \in D(\sigma)$ such that $\sigma(x) = \theta(t)$.

Definition 1. Let θ be a permutation, σ a substitution, and t an Ω-term.

$$\bar{S}_t = \{\sigma \in \Sigma / \exists\theta \in \Theta, \ \exists x \in D(\sigma) : \ \sigma(x) = \theta(t)\}$$

and

$$\bar{S}_{\theta(t)} = \{\sigma \in \bar{S}_t / \exists x : \sigma(x) = \theta(t)\}.$$

Given two terms t_1 and t_2, we can associate with them, their respective $\bar{S}_{t_i}$. The intersection of these two sets can be defined by:

$$\bar{S}_{t_1} \cap \bar{S}_{t_2} = \{\sigma \in \Sigma : \exists \ x, y \in X \ \exists\theta_1\theta_2 \in \Theta : \sigma(x) = \theta_1(t_1), \ \sigma(y) = \theta_2(t_2)\}$$

Now, for each permutation θ,

$$\cup_{t \in T} \bar{S}_{id(t)} = \cup_{t \in T} \bar{S}_{\theta(t)} = \Sigma.$$

Suppose σ satisfies $card(D(\sigma)) = n$; then we have n different variables $x_1, x_2, ..., x_n$ and n terms $t_1, t_2, ...t_n$ such that $\sigma(x_i) = t_i$; consequently,

$$\sigma \in \bar{S}_{id(t_1)} \cap \bar{S}_{id(t_2)} \cap \bar{S}_{id(t_3)} \cap ... \bar{S}_{id(t_n)}.$$

3 Reduction

We now introduce the definition of reduction between two substitutions, as a generalization of the well known concept of term reduction. Intuitively, reduction of substitutions is linked to term reduction: a substitution σ reduces to a σ' if in its codomain there is a term t, which in turn is reducibile to a term t' belonging to the codomain of σ'.

Definition 2. Let $r = (\lambda, \rho)$ be a rewriting rule and θ a permutation. Let $\sigma_1 \in S_{\theta(t_1)}$ and $\sigma_2 \in S_{\theta(t_2)}$ be two substitutions such that their domain is the same singleton i.e. $D(\sigma_1) = D(\sigma_2) = \{\zeta\}$; we say that σ_1 reduces to σ_2 via r, and we write $\sigma_1 \Rightarrow^r \sigma_2$ if

1. there exist four substitutions σ_c, σ_r, σ_λ, σ_ρ such that $\sigma_\lambda \in S_{\theta(\lambda)}$ and $\sigma_\rho \in S_{\theta(\rho)}$ with $D(\sigma_\lambda) = D(\sigma_\rho)$,
2. $\sigma^r \circ \sigma_\lambda \circ \sigma^c|_{\{\zeta\}} = \sigma_1$ and $\sigma^r \circ \sigma_\rho \circ \sigma^c|_{\{\zeta\}} = \sigma_2$.

In other words, condition 2 requires that

$$\sigma^r \circ \sigma_\lambda \circ \sigma^c \in S_{\theta(t_1)},$$
$$\sigma^r \circ \sigma_\rho \circ \sigma^c \in S_{\theta(t_2)},$$
$$D(\sigma^c) = D(\sigma_1).$$

For example $\sigma_1 : z \mapsto (1 * w^2) + 3$ reduces to $\sigma_2 : z \mapsto w^2 + 3$ via the rule $1 * x \rightarrow x$; in fact, assume $\sigma_c : z \mapsto y + z$, then it is possible to find $\sigma_\lambda : y \mapsto 1 * v$ and $\sigma_\rho : y \mapsto v$, and finally $\sigma_r : \begin{cases} v \mapsto w^2 \\ z \mapsto 3 \end{cases}$

We note that for each pair of substitutions σ_λ and σ_ρ which satisfy the conditions in Definition 2, $\sigma_\lambda \rightarrow \sigma_\rho$ holds; this can be easily proved by taking σ_c and σ_r to be the identity *id*.

The following result extabilishes the equivalence of the above definition with the classical definition of reduction.

Fact 3.1 *Let r be a rewriting rule. $\sigma_1 \Rightarrow^r \sigma_2$ and $D(\sigma_1) = D(\sigma_2) = \{\zeta\}$ iff $\sigma_1(\zeta) \rightarrow_r \sigma_2(\zeta)$*

The following result allows us to define reduction of a set of substitutions; this definition is justified by Definition 2. This will be useful in the sense that we can now replace the concepts of occurrence and context by a most general idea of reduction, which amounts to a sort of substitution rewriting.

Fact 3.2 *If $\sigma_1 \Rightarrow^r \sigma_2$ and $\sigma_i \in S_{\theta(t_i)}$, $i = 1, 2$, then for all pair τ_1, τ_2 and θ' with $\tau_i \in S_{\theta'(t_i)}$ and $D(\tau_i) = \{\xi\}$, $i = 1, 2$, we have $\tau_1 \Rightarrow^r \tau_2$.*

Definition 3. Let r be a rewriting rule. $S_{\theta(t_1)} \Rightarrow^r S_{\theta(t_2)}$ iff there exist σ_1 and σ_2 such that $\sigma_i \in S_{\theta(t_i)}$ and

$$\sigma_1 \Rightarrow^r \sigma_2.$$

Note that if $S_{\theta(t_1)} \Rightarrow^r S_{\theta(t_2)}$, then for all $\sigma_i \in S_{\theta(t_i)}$ with the same domain $\sigma_1 \Rightarrow^r \sigma_2$.

Given a rewriting rule r it is possible to extend the definition of rewriting to substitutions in $\bar{S}_{\theta(t)}$.

Definition 4. Let σ_1 and σ_2 be any two substitutions. We say that $\sigma_1 \Rightarrow^r \sigma_2$ if

- $D(\sigma_1) = D(\sigma_2)$
- $\forall x \in D(\sigma_i)$ we have $\sigma_1(x) = \sigma_2(x)$ or $\sigma_1|_{\{x\}} \Rightarrow^r \sigma_2|_{\{x\}}$.

We recall that $\sigma_i|_{\{x\}}$ is the restriction of σ_i to the domain $\{x\}$.

This definition differs from Definition 2, in that it applies to all σ whereas the previous definition works for those $\sigma \in \bar{S}_t$ for some term t. The following definition is the analogous extension of Definition 3.

Definition 5. $\bar{S}_{\theta(t)} \Rightarrow^r \bar{S}_{\theta(t')}$ iff $\exists \sigma \in \bar{S}_{\theta(t)}$ and $\exists \sigma' \in \bar{S}_{\theta(t')}$ such that $\sigma \Rightarrow^r \sigma'$.

3.1 Rewriting Instance Function

In this section we define a function α^r for each rule r, mapping substitutions to substitutions. Our aim is to model the reduction relation by function reduction. Our goal is to compose reduction rules with objects to be reduced (i.e. substitutions). In this way we not only get closer to the algebraic idea in which polynomials reduce via polynomials remaning inside the structure, but we also get closer to the categorical structure we are looking for, in which morphisms can be composed. We want to prove that if a set $\bar{S}_{\theta(t)}$ is reducibile to a $\bar{S}_{\theta(t')}$, then for each substitution belonging to $\bar{S}_{\theta(t)}$ there is only one substitution in the second set which it reduces to; this fact allows us to define functions α^r on substitutions (one function α^r for each rule r).

Let $r = (\lambda, \rho)$ be a rewriting rule. If $\sigma \Rightarrow^r \sigma'$, $\sigma \in \bar{S}_{\theta(t)}$, and $\sigma' \in \bar{S}_{\theta(t')}$, with $\theta(t) \neq \theta(t')$ then we can conclude that $\bar{S}_{\theta(t)} \Rightarrow^r \bar{S}_{\theta(t')}$.

Definition 6. If $\bar{S}_{\theta(t)} \Rightarrow^r \bar{S}_{\theta(t')}$, then there exists a function $\alpha^r : \bar{S}_{\theta(t)} \rightarrow \bar{S}_{\theta(t')}$, such that

$$\alpha^r(\sigma)(x) = \begin{cases} \theta(t') \text{ if } \sigma(x) = \theta(t) \\ \sigma(x) \text{ otherwise} \end{cases}$$

α^r is called *Rewriting Instance Function*, RIF in the following.

A RIF α^r maps substitutions in $\bar{S}_{\theta(t)}$ into substitutions in $\bar{S}_{\theta(t')}$, if $S_{\theta(t)} \Rightarrow^r S_{\theta(t')}$, and for all substitutions $\sigma \in \bar{S}_{\theta(t)}$ it results $\sigma \Rightarrow^r \alpha^r(\sigma)$.

The function α^r is not injective in general, for example if σ_1 and σ_2 are

$$\sigma_1 : \begin{cases} x \mapsto 1 * x + 4 \\ y \mapsto x + 4 \end{cases} \quad \textit{and } \sigma_2 : \begin{cases} x \mapsto x + 4 \\ y \mapsto 1 * x + 4 \end{cases}$$

then $\sigma_1 \neq \sigma_2$ and $\alpha^r(\sigma_1) = \alpha^r(\sigma_2)$.

However, if σ_1 and σ_2 do not belong to $\bar{S}_{\theta(t)} \bigcap \bar{S}_{\theta(t')}$, then $\forall x \in D(\sigma_1) \cup D(\sigma_2)$ it results either $x \notin D(\sigma_1) \cap D(\sigma_2)$, so that $D(\alpha^r(\sigma_1)) \neq D(\alpha^r(\sigma_2)$; hence $\alpha^r(\sigma_1) \neq \alpha^r(\sigma_2))$, or $\exists x \in D(\sigma_1) \cup D(\sigma_2)$ such that $\sigma_1(x) \neq \sigma_2(x)$ and again $\alpha^r(\sigma_1) \neq \alpha^r(\sigma_2)$.

On the other hand, for each rule r it is easily shown that α^r is a surjection. Stability and compatibility hold for substitution reduction:

Fact 3.3 *If $\sigma_1 \Rightarrow^r \sigma_2$ then for all $\tilde{\sigma} \in \Sigma$ we have*

stability: $\tilde{\sigma} \circ \sigma_1 \Rightarrow^r \tilde{\sigma} \circ \sigma_2$
compatibility: $\sigma_1 \circ \tilde{\sigma} \Rightarrow^r \sigma_2 \circ \tilde{\sigma}$

Let now α^r be a RIF between the two sets:

$$\alpha^r : \bar{S}_{t_1} \rightarrow \bar{S}_{t_2}$$

This means that there exist σ_1 and σ_2 such that

- $\sigma_i \in \bar{S}_{t_i}$ for $i = 1, 2$

- $\alpha^\tau(\sigma_1) = \sigma_2$ i.e. $\sigma_1 \Rightarrow^\tau \alpha^\tau(\sigma_1)$.

For each $\bar{\sigma} \in \Sigma$ it is possible to define a new RIF function $\bar{\sigma} \bullet \alpha^\tau$ of the form:

$$\bar{\sigma} \bullet \alpha^\tau : \bar{S}_{\bar{\sigma}(t_1)} \rightarrow \bar{S}_{\bar{\sigma}(t_2)}$$

by

$$\bar{\sigma} \bullet \alpha^\tau(\sigma)(x) = \begin{cases} \bar{\sigma}(t_2) \text{ if } \sigma(x) = \bar{\sigma}(t_1) \\ \sigma(x) \text{ otherwise} \end{cases}$$

This function has the following properties:

1. $\forall\, \sigma \in \bar{S}_{\bar{\sigma}(t_1)}$ it is $\sigma \Rightarrow^\tau \bar{\sigma} \bullet \alpha^\tau(\sigma)$,
2. $\bar{\sigma} \bullet \alpha^\tau(\bar{\sigma} \cdot \sigma_1) = \bar{\sigma} \cdot \sigma_2$.

Note that property 2 is a consequence of 1, and that in 1 the quantification $\forall\, \sigma \in \bar{S}_{\bar{\sigma}(t_1)}$ could be generalized to $\forall\, \sigma \in \Sigma$.

A diagram is given below, which shows the left composition Substitutions-Instance.

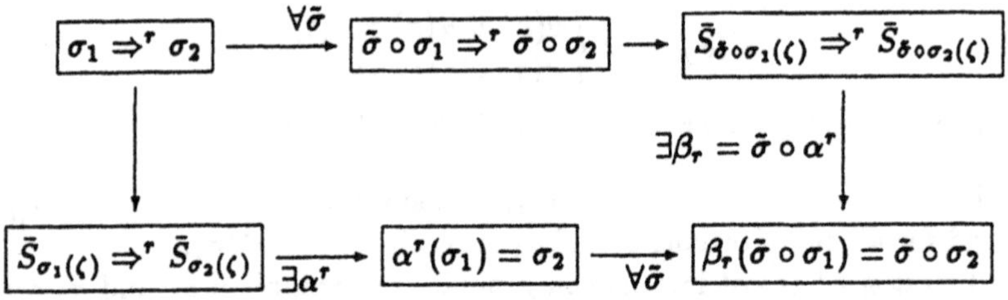

$\sigma_1 \Rightarrow^\tau \sigma_2$ $\xrightarrow{\forall \bar{\sigma}}$ $\bar{\sigma} \circ \sigma_1 \Rightarrow^\tau \bar{\sigma} \circ \sigma_2$ $\longrightarrow$ $\bar{S}_{\bar{\sigma}\circ\sigma_1(\zeta)} \Rightarrow^\tau \bar{S}_{\bar{\sigma}\circ\sigma_2(\zeta)}$

$\downarrow$ $\qquad\qquad$ $\exists \beta_\tau = \bar{\sigma} \circ \alpha^\tau$ $\downarrow$

$\bar{S}_{\sigma_1(\zeta)} \Rightarrow^\tau \bar{S}_{\sigma_2(\zeta)}$ $\xrightarrow[\exists \alpha^\tau]{}$ $\alpha^\tau(\sigma_1) = \sigma_2$ $\xrightarrow[\forall \bar{\sigma}]{}$ $\beta_\tau(\bar{\sigma} \circ \sigma_1) = \bar{\sigma} \circ \sigma_2$

Again, for each $\bar{\sigma} \in \Sigma$ it is possible to define a new function $\alpha^\tau \bullet \bar{\sigma}$ which is a new RIF as follows.

Let $\alpha^\tau : \bar{S}_{t_1} \rightarrow \bar{S}_{t_2}$. It means that there exist σ_1 and σ_2 such that

- $\sigma_i \in \bar{S}_{t_i}$ for $i = 1, 2$
- $\alpha^\tau(\sigma_1) = \sigma_2$ i.e. $\sigma_1 \Rightarrow^\tau \alpha^\tau(\sigma_1)$.

Now for all $\bar{\sigma} \in \Sigma$ we can prove, by compatibility, that

$$\sigma_1 \cdot \bar{\sigma} \Rightarrow^\tau \sigma_2 \cdot \bar{\sigma}$$

and this means that there exists a set of variables $\Xi = \{\zeta_1, \zeta_2, \ldots \zeta_n\}$ which is a subset of $D(\sigma_1 \cdot \bar{\sigma})$ and such that the following holds:

$$\sigma_1 \cdot \bar{\sigma}|_{\{\zeta_i\}} \Rightarrow^\tau \sigma_2 \cdot \bar{\sigma}|_{\{\zeta_i\}} \quad \textit{and} \quad \sigma_1 \cdot \bar{\sigma}(\zeta_i) \neq \sigma_2 \cdot \bar{\sigma}(\zeta_i).$$

We define $\alpha_i^\tau \bullet \bar{\sigma} : \bar{S}_{\sigma_1 \cdot \bar{\sigma}(\zeta_i)} \rightarrow \bar{S}_{\sigma_2 \cdot \bar{\sigma}(\zeta_i)}$ as follows:

$$\alpha_i^\tau \bullet \bar{\sigma}(\sigma)(x) = \begin{cases} \sigma_2 \cdot \bar{\sigma}(\zeta_i) \text{ if } \sigma(x) = \sigma_1 \cdot \bar{\sigma}(\zeta_i) \\ \sigma(x) \qquad \text{otherwise} \end{cases}$$

The functions $\alpha_i^\tau \bullet \bar{\sigma}$ have the properties:

1. $\forall\, \sigma \in \bar{S}_{\sigma_1 \cdot \bar{\sigma}(\zeta_i)}$ it is $\sigma \Rightarrow^\tau \alpha_i^\tau \bullet \bar{\sigma}(\sigma)$

2. $\alpha_i^\tau \bullet \bar{\sigma}(\sigma_1 \cdot \bar{\sigma}) = \sigma_2 \cdot \bar{\sigma}$

Graphically:

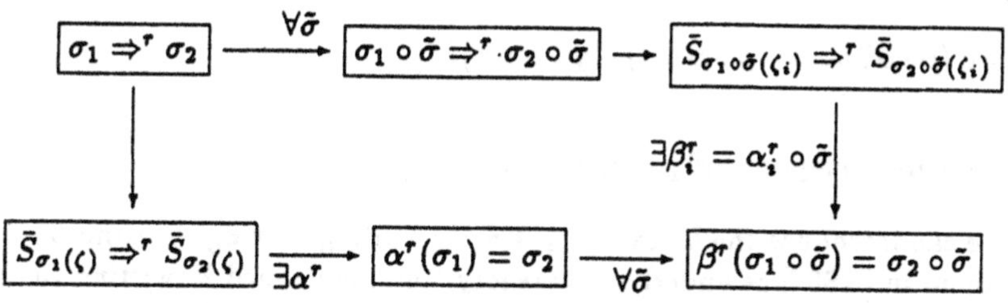

$$\begin{array}{ccccc}
\boxed{\sigma_1 \Rightarrow^\tau \sigma_2} & \xrightarrow{\forall\bar{\sigma}} & \boxed{\sigma_1 \circ \bar{\sigma} \Rightarrow^\tau \cdot \sigma_2 \circ \bar{\sigma}} & \longrightarrow & \boxed{\bar{S}_{\sigma_1 \circ \bar{\sigma}(\zeta_i)} \Rightarrow^\tau \bar{S}_{\sigma_2 \circ \bar{\sigma}(\zeta_i)}} \\
\downarrow & & & & \downarrow \exists \beta_i^\tau = \alpha_i^\tau \circ \bar{\sigma} \\
\boxed{\bar{S}_{\sigma_1(\zeta)} \Rightarrow^\tau \bar{S}_{\sigma_2(\zeta)}} & \xrightarrow[\exists\alpha^\tau]{} & \boxed{\alpha^\tau(\sigma_1) = \sigma_2} & \xrightarrow[\forall\bar{\sigma}]{} & \boxed{\beta^\tau(\sigma_1 \circ \bar{\sigma}) = \sigma_2 \circ \bar{\sigma}}
\end{array}$$

4 The categorical model of substitution rewriting

4.1 h–categories

We briefly sketch here the basic definitions of h–category as in the paper of Grandis [Gr91]; For an introduction to the language of categories we refer to [Ma71] and [Be91]. The idea is to translate rewriting definitions and properties into a categorical framework. The objects and the morphisms of the category will therefore be sets of variables and substitutions, respectively. Furthermore a graph in the h-category provides an interpretation for a RIF function.

Definition 7. An h–category A is a category and a 2–graph, having the same underlying 1–graph and provided with

- a reduced horizontal composition law $y \circ \alpha \circ x$ such that if $x : X \to A$, $\alpha : a_1 \twoheadrightarrow a_2 : A \to B$ and $y : B \to Y$ then

$$y \circ \alpha \circ x : ya_1x \twoheadrightarrow ya_2x : X \to Y$$

  whose horizontal identities are, by definition, the identical morphisms 1_A of the objects.
- a vertical structure consisting just of a cell $1_a : a \twoheadrightarrow a$, called the vertical identity of a for every morphism a, so that this axiom is satified:
  (hc) the horizontal identities are neutral for the horizontal composition, which is associative and distributes with respect to vertical identities:
  1. $1_B \circ \alpha \circ 1_A = \alpha$
  2. $y' \circ (y \circ \alpha \circ x) \circ x' = (y'y) \circ \alpha \circ (xx')$
  3. $y \circ 1_a \circ x = 1_{yax}$

A category can always be thought of as a trivial h–category, whose only cells are vertical identities. Every 2–category has an undelying h–category.

Definition 8. An $h2$–category is an h–category A provided with a vertical composition $\beta \cdot \alpha$, agreeing with vertical domains and codomains and distributive with respect to the reduced horizontal composition, i.e. if

- $\alpha : a_1 \twoheadrightarrow a_2 : A \rightarrow B$
- $\beta : a_2 \twoheadrightarrow a_3 : A \rightarrow B$
- $x : X \rightarrow A$
- $y : B \rightarrow Y$,

then $y \circ (\beta \cdot \alpha) \circ x = (y \circ \beta \circ x) \cdot (y \circ \alpha \circ x)$.

4.2 An h–category of parallel rewriting

Now we can introduce the h–category **TS** of rewriting, having the same T_Ω category found in [RS87] as underlying category, and the defined Rewriting Instance Functions as 2–arcs, in order to make it possible to consider the horizontal composition law as the left-composition (right-composition) of a substitution σ with a RIF α.

The objects of **TS** are subsets of the set of variables X, and the arrows in **TS** are term substitutions. Each rewriting rule r induces a 2–arc structure in **TS**; in fact, if r is the pair (λ, ρ), then for each substitution $\sigma \in \bar{S}_\lambda$ there exists $\sigma' \in \bar{S}_\rho$ such that $\alpha^r(\sigma) = \sigma'$.

For example, in the algebra of groups, the rule $r = (1 \cdot x, x)$, induces a set of 2–arcs of the form

$$X \underset{\sigma'}{\overset{\sigma}{\underset{\Downarrow \alpha^r}{\rightrightarrows}}} Y$$

for each $\sigma \in \bar{S}_{1 \cdot x}$ and $\sigma' \in \bar{S}_x$.

A rewriting rule is not interpreted in **TS** as a schema which, when supplied with variables, generates instances of the rules, but it is interpreted as a function relating the rule of the reduction, the substitution, and the context of reduction, all in a 2–arc. In other words, each rule r does not generate a single 2–cell as in [RS87], with *place holders* for left composition, but it generates a set of 2–arcs.

Fact 4.1 **TS** *is an h–category.*

It is easy to provide the above structure with an horizontal composition law $\sigma_1 \circ \alpha \circ \sigma_2$, where $\circ$ is just the $\bullet$ binary operation to compose substitutions and Rewriting Instance Functions of the previous section. Horizontal identities in **TS** are the restrictions of the identity substitution *id* with respect to subsets of X; i.e. $1_B = id_B$ for all B. Horizontal identities are neutral for the horizontal composition, i.e. if $\alpha : \sigma_1 \twoheadrightarrow \sigma_2$

$$1_B \circ \alpha = \alpha : \sigma_1 \twoheadrightarrow \sigma_2$$

Similarly, $\alpha \circ 1_B = \alpha : \sigma_1 \twoheadrightarrow \sigma_2$.

Each arrow σ has a vertical structure, consisting just of a cell $1_\sigma : \sigma \rightarrow \sigma$, called the vertical identity.

Horizontal composition is associative, and distributes with respect to vertical composition. This follows from the definition of horizontal composition; it is easy to see that, for each σ_1, σ_2, and α,

$$\alpha \circ (\sigma_1 \cdot \sigma_2) = (\alpha \circ \sigma_1) \circ \sigma_2$$

and

$$(\sigma_1 \cdot \sigma_2) \circ \alpha = \sigma_1 \circ (\sigma_2 \circ \alpha).$$

Moreover $1_{\sigma_1} \circ \sigma_2 = 1_{\sigma_1 \cdot \sigma_2}$; in fact $1_{\sigma_1} : \sigma_1 \to \sigma_1$; thus

$$1_{\sigma_1 \circ \sigma_2} : \sigma_1 \circ \sigma_2 \to \sigma_1 \circ \sigma_2$$

and, on the other hand,

$$1_{\sigma_1 \circ \sigma_2} : \sigma_1 \circ \sigma_2 \to \sigma_1 \circ \sigma_2.$$

Similarly

$$\sigma_1 \circ 1_{\sigma_2} = 1_{\sigma_1 \cdot \sigma_2}$$

We have proved that **TS** is an h–category, but we may say that the above defined h–category of rewriting is an $h2$–category. To enrich our category **TS** with vertical composition we can define vertical composition between two 2–graphs as the usual function composition.

Given two RIF $\alpha : \sigma_1 \twoheadrightarrow \sigma_2$, and $\beta : \sigma_2 \twoheadrightarrow \sigma_3$, there exists a new Rewriting Instance Function named $\beta \cdot \alpha : \sigma_1 \twoheadrightarrow \sigma_3$.

Again by definition of $\cdot$, we have that vertical composition is distributive with respect to horizontal composition. Indeed we can prove that for all RIF α and β, and substitutions τ_1 and τ_2, it results:

$$\tau_1 \circ (\beta \cdot \alpha) \circ \tau_2 = (\tau_1 \circ \beta \circ \tau_2) \cdot (\tau_1 \circ \alpha \circ \tau_2).$$

5 Concluding Remarks

This paper contains a categorical approach to rewriting, in the spirit of adapting the main ideas of algebraic reduction to term rewriting. The approach led to a deeper study of substitutions and to a natural categorical view of reduction which takes into account this new point of view. Future research will deal with the attempt to reach h-category from polynomial reduction, with the aim of providing both rewriting of terms and algebraic reduction with a common semantics.

ACKNOLEDGEMENTS. I wish to thank Gianfranco Mascari, who provided much of the initial inspiration of this work, Marta Cialdea Mayer and the anonymous referees for helpful comments and suggestions.

References

[Ba92] Balestreri G. (1992) *External and Internal Rewriting* Internal Report n. 13/1992 I.A.C. CNR Italy.

[BM92] Balestreri G., Mascari G. (1992) *Concurrent Rewriting and Knuth Bendix Completion Theorem in a 2−Category* Internal Report n. 15/1992 I.A.C. CNR Italy.

[Be91] Benabou J., *Lectures held in Rome at the Mathematics Departement of University La Sapienza*, 1991.

[Bu76] Buchberger B. (1976) *A Theoretical Basis for the reduction of polynomials to canonical form* ACM SIGSAM Bull. 10/4 (19-24).

[Bu85] Buchberger B. (1985) *Grobner Bases: An Algorithmic Method in Polynomial Ideal Theory.* Multidimensional Systems Theory, (ed. N.K. Bose), D. Reidel Publ. Comp.(184-232).

[BL83] Buchberger B., Loos R. (1983) *Algebraic Simplification.* Computer Algebra-Symbolic and Algebraic Computation, 2nd ed. Springer, (11-43).

[DJ90] Dershowitz N., Jouannaud J.P. (1990) *Rewrite Systems* Handbook of Theoretical Computer Science, chapter 15 (243-320).

[Fa79] Fay M. (1979) *First Order unification in an equational Theory* Proc. fourth workshop on automated deduction Austin Texas (161-167).

[Gr91] Grandis M. (1991) *Homotopical Algebra: a two-dimensional categorical setting* Internal Report n. 191/1991 Dept. Mat. University of Genova-Italy.

[HH82] Huet G., Hullot J.M. (1982) *Proof by Induction in Equational Theories With Constructors* J.A.C.M. 25/2 (239-266).

[HO80] Huet G., Oppen D.C. (1980) *Equations and Rewrite Rules - A Survay.* Formal Language Theory, (ed. R.V. Book), Academic Press, (349-405).

[Hu80] Huet G. (1980) *Deduction and Computation.* L.N.C.S. 232 (39-74).

[Hu80a] Huet G. (1980) *Confluent Reductions: Abstract Properties and Applications to Term Rewriting Systems.* J.ACM 27/4, (797-821).

[JK86] Jouannaud J.P., Kirchner H. (1986) *Completion of a set of rules modulo a set of equations* SIAM J. Comp. Vol 15/4.

[JK89] Jouannaud J.P., Kounalis E. (1989) *Automatic proofs by induction in equational theories without constructors* Information and Computation 82 (1-33).

[Kl90] Klop J.W. (1990) *Term Rewriting Systems.* C.W.I. Amsterdam rep.CS-R9073.

[KB70] Knuth D.E., Bendix P.B. (1970) *Simple word problems in universal algebras.* Computational Problems in Abstract Algebra (ed. J. Leech), Pergamon Press, 1970, (263-279).

[La75] Lankford D. (1975) *Canonical Inference* Memo ATP-32 Aut. Theor. Prov. Project. Univ. Texas.

[Lo74] Loos R. (1974) *Toward a formal implementation of Computer Algebra* EUROSAM 1974 (9-16).

[Ma71] MacLane S. (1971) *Categories for the Working Mathematician* Vol. 5 Graduate Texts in Mathematics. Springer Verlag.

[Me92] Meseguer J. (1992) *Conditional rewriting as a unified model of concurrency* T.C.S. 96 (73-155).

[Od77] O'Donnell M.J. (1977) *Computing in systems described by equations* L.N.C.S. 58.

[OH84] O'Donnell M.J., Hoffman C. (1984) *Implementation of an Interpreter of abstract equations* Proc. 12th POPL.

[PS81] Peterson G.E., Stickel M.E. (1981) *Complete sets of reductions for some equational theories* J. ACM 28/2 (233-264).

[RS87] Rydeheard D.E., Stell J.G. (1987) *Foundation of Equational Deduction: A Categorical Treatment of Equational Proofs and Unification Algorithms* L.N.C.S. 283 (114-139).

[St92] Steel J.G. (1992) *Categorical Aspect of Unification and Rewriting* Ph.D. Thesis, University of Manchester.

[St92a] Stokkermans K. (1992) *A Categorical Formulation for Critical Pair Completion Procedures* L.N.C.S. 656 (328-342).

[Wi89] Winkler F. (1989) *Knuth-Bendix Procedure and Buchberger Algorithm: A Synthesis* ISAAC-89 (1-8).

Generative Geometric Modeling in a Functional Environment

Alberto Paoluzzi

Dip. di Discipline Scientifiche – Sez. Informatica
Università degli Studi di Roma Tre
Via della Vasca Navale, 84 – 00146 Roma, Italy

Abstract

Some aspects of geometric design programming using a functional language and a dimension-independent approach to geometric data structures are discussed in this paper. In particular it is shown that such an environment allows for a very easy implementation of geometric transformations, hierarchical assemblies and parametric curves, surfaces and solids. Since geometric shapes are associated to generating functions, and geometric expressions can be passed to functions as actual parameters, this approach allows for a very powerful programming approach to variational geometry. The paper also aims to show that this language can accommodate both the description of methods for generating geometric shapes (see e.g. the definition of either the Coons surfaces or the Bezier curves) as well as the use of such methods to generate specific shape instances. Finally, the language allows for both bottom-up and top-down development of the designed shape, as it is shown in the appendix, where the generation of the model of a parametric umbrella by successive refinements is discussed.

1 Introduction

Functional programming enjoys several good properties. In general, in a functional language the set of rules is very small; each rule is very simple; program code is concise and clear; the meaning of a program is well understood (no state); functions are used both as programs and as data; programs are easily connected by concatenation and nesting.

Conversely, a complex geometric shape is an assembly of components, highly dependent from each other. In geometric modeling of complex objects each part results from computations involving other parts. The main idea of geometric programming with a functional language is to associate a generating function to each part and in using geometric expressions as actual parameters of function invocations. So, a functional programming approach results in a natural environment for geometric computations.

`PLASM` [7, 8] is a language for geometric design based on a subset of the functional language FL developed by Backus, Williams and others [1, 2, 3] at IBM research. The basic geometric object in `PLASM` is the polyhedral complex [10].

This is defined as a quasi-disjoint decomposition of a pointset into polyhedral cells. An algebraic calculus over polyhedral complexes is defined, and embedded within a modern functional language.

Since polyhedral-valued expressions may appear as parameters of functions, PLASM implements a programming approach to variational geometry. In fact the geometric object resulting from the evaluation of a function will often depend on the values of other geometric expressions. In some sense a PLASM function can be seen as a shape prototype, or "generating form", which is able to produce infinitely many *different* geometric objects, all with some common structure. When a polyhedral-typed form is evaluated using actual values of parameters, often obtained by evaluation of other forms, the evaluation mechanism produces (and sometime stores) a geometric model associated to the form.

The language, besides the regularized Boolean operations of union, intersection, difference and the complementation (A+B, A&B, A/B, -A), provides some additional dimension-independent geometric operators, like the k-skeletons @k ($k = 0, 1, 2, \ldots$) and the boundary @ of a polyhedral complex, as well the product *, offset ++ and intersection of extrusions && defined in [7]. A generalized product of cell-complexes which contains as special cases the standard intersection of cell-decomposed polyhedra, the finite and non-finite extrusion, the intersection of extrusions and the cartesian product is discussed in [9]. The language allows for overloading of operators. For instance, + and * are used both for addition and multiplication of numbers and for union and product of polyhedra, respectively.

The paper aims to show the amazing descriptive power of the PLASM language. Section 2 discusses two main topics defined and implemented in a dimension-independent way, i.e. hierarchical polyhedral complexes and simplicial maps over polyhedral complexes. Section 3 introduces to the use of affine tensors, both pre-defined and user-definable. This section also discusses the PLASM approach to hierarchical assemblies by animating a simple robot with respect to any path in configuration space. Section 4 then introduces to the PLASM appproach to parametric geometry, by showing: how to implement a Bezier curve of any degree; the Coons patch defined by any four boundary curves; the thin narrow solid defined by a Bezier curve. In Appendix A a complete example of development of a quite complex parametric solid is presented. A prerequisite for full comprehension of the PLASM code given in this paper (which is completely self-contained) is a preliminary reading of Reference [8]. The author hopes that some taste of the language flavor is given anyway.

2 Dimension-independence

2.1 Hierarchical Polyhedral Complexes

PLASM is characterized by the use, as primitive objects, of dimension-independent [11] complexes of polyhedra, where each polyhedron may be non-convex and even unconnected. Each primitive polyhedron in a complex is represented as a decomposition in convex cells, where each convex cell is described as an intersection of half-spaces. The set of polyhedral cells in a complex is represented as a directed acyclic multigraph, where nodes of outdegree zero (leaves) are associated to elementary polyhedra. Each leave is represented as a pair (F, C) of sets of faces and convex cells. For a discussion of such a representation see the paper [10].

The representation scheme used within the language is so defined on the domain of multilevel ***hierarchical collections of polyhedral complexes.*** The range of the representation scheme is the set of ***direct acyclic multigraphs.*** We call ***dimension*** (d, n) of a polyhedron the ordered pair of "intrinsic" and "embedding" dimensions. For instance, a ***plane polygon*** has dimension $(2, 2)$, whereas a ***space polygon*** has dimension $(2, 3)$.

2.2 Simplicial maps

A very important primitive operator in `PLASM` allows for generating curves and surfaces via parametric maps from a simplicial decomposition of a polyhedral complex. The predefined operator MAP is used as

```
MAP:f:domain
```

where `f` ≡ `[fx1, fx2, ..., fxn]` is the **cons** of a number of coordinate functions which equate the dimension n of the space where the `domain` has to be embedded and incurved. The operational semantics of the map function can be shortly described as follows: (a) compute a simplicial decomposition of the polyhedral domain; (b) apply to each vertex of such a decomposition the coordinate functions `fx1, fx2, ..., fxn`, in order to generate its image in the target space. For instance, consider the generation of a polyhedral approximation (with m segments) of the unit circle in $\Re^2$:

```
DEF Circle (m::IsInt) = MAP:[fx,fy]:domain
WHERE
  fx = cos~S1,   fy = sin~S1,
  domain = QUOTE:(#:m:(2*PI/m))
END;
```

3 Geometric modeling with functions

3.1 Pre-defined affine tensors

According to [5], we use *tensor* $\mathbf{T}$ as a synonym for "linear transformation from a vector space $\mathcal{V}$ to $\mathcal{V}$". In other words a tensor is a linear map $\mathbf{T}$ which maps each vector $\mathbf{u}$ in a vector $\mathbf{v} = \mathbf{T}\mathbf{u}$. The *product* $\mathbf{ST}$ of tensors $\mathbf{S}, \mathbf{T} \in Lin\mathcal{V}$ is defined as a function composition

$$\mathbf{ST} = \mathbf{S} \circ \mathbf{T},$$

where $(\mathbf{ST})\mathbf{v} = \mathbf{S}(\mathbf{T}\mathbf{v})$ for all $\mathbf{v} \in \mathcal{V}$.

In **PLASM** a set of pre-defined operators **S, T, R** and **H** is given, corresponding to elementary affine transformations of scaling, translation, rotation and shearing. They are defined in a dimension-independent way, so that the user may define the set of coordinates which are affected by the tensor, and give the corresponding parameters. For example, a scaling in the first and fourth coordinates and a rotation around the z axis are respectively given as:

```
S:<1,4>:<s1,s4>;  R:<1,2>:alpha
```

According to the standard meaning of tensors, the product of tensors is given as function composition:

```
S:<1,4>:<s1,s4> ~ R:<1,2>:alpha
```

The pre-defined affine tensors may either directly apply to polyhedral objects:

```
(S:<1,4>:<s1,s4> ~ R:<1,2>:alpha): pol
```

or may be accumulated within structures, with a semantics similar to that of the structures of the PHIGS ISO graphics standard:

```
STRUCT:< pol1, S:<1,4>:<s1,s4>, pol2, R:<1,2>:alpha, pol3>
```

so that the evaluation of the expression produces the following polyhedra in world coordinates:

```
pol1,
S:<1,4>:<s1,s4>:pol2,
(S:<1,4>:<s1,s4>~R:<1,2>:alpha):pol3
```

3.2 User-definable affine tensors

It may be sometimes useful to apply tensors not to polyhedra but to single points, i.e. directly to a set of coordinates. This may be done by implementing the desired tensors as standard **PLASM** functions. In the following the elementary rotations, translations and scalings in $\Re^3$ depending on just one real parameter are respectively given:

```
DEF Rx (a::IsReal)(x,y,z::IsReal) =
  <x, cos:a * y - sin:a * z, sin:a * y + cos:a * z>;
DEF Ry (a::IsReal)(x,y,z::IsReal) =
  <cos:a * x + sin:a * z, y, (-~sin):a * x + cos:a * z>;
DEF Rz (a::IsReal)(x,y,z::IsReal) =
  <cos:a * x - sin:a * y, sin:a * x + cos:a * y, z>;

DEF Tx (a::IsReal)(x,y,z::IsReal) = <x + a, y, z>;
DEF Ty (a::IsReal)(x,y,z::IsReal) = <x, y + a, z>;
DEF Tz (a::IsReal)(x,y,z::IsReal) = <x, y, z + a>;

DEF Sx (a::IsReal)(x,y,z::IsReal) = <x * a, y, z>;
DEF Sy (a::IsReal)(x,y,z::IsReal) = <x, y * a, z>;
DEF Sz (a::IsReal)(x,y,z::IsReal) = <x, y, z * a>;
```

It is interesting to note that using the standard composition of the language such functions may be partially specified, freely composed and later they can be applied to the target point. For example, a rotation of α degrees of the point (p_x, p_y, p_z) around the y axis with fixed point $(0, 0, h)$ can be given as:

```
(Tz:h ~ Ry:(PI/180 * alpha) ~ Tz:(-:h)):<px,py,pz>
```

where **PI** is the **PLASM** denotation for π.

3.3 Hierarchical assemblies

PLASM allows for easy specification of hierarchical assemblies by using the **STRUCT** function, which is applied to sequences of polyhedra, tensors and invocations of **STRUCT** functions. Each polyhedral complex in a structure is defined in local coordinates and is transformed in the coordinates of the first object of the structure at traversal time, when the application of the **STRUCT** function is evaluated.

For example, a plane robotic arm with four rods and three rotational joints (depending on rotations $\alpha1, \alpha2, \alpha3$) can be specified as

```
DEF arm (alpha1,alpha2,alpha3::IsReal) = STRUCT:
  <rod, T:2:-18 ~ (R:<1,2>:(PI/180 * alpha1)),
   rod, T:2:-18 ~ (R:<1,2>:(PI/180 * alpha2)),
   rod, T:2:-18 ~ (R:<1,2>:(PI/180 * alpha3)), rod >
WHERE rod = T:<1,2>:<-1,-19>:(CUBOID:<2,20>) END;
```

So, the evaluation of the expressions **arm:<30,45,60>** and **arm:<60,45,30>** produces the two arm configurations shown in Figure 1.

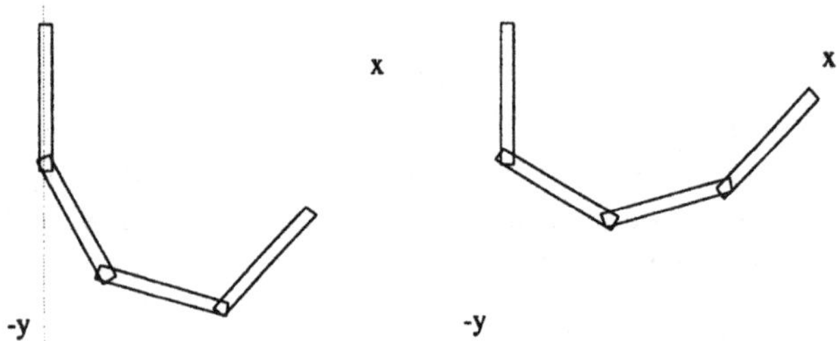

Figure 1: The two robot configurations generated by evaluating the expressions **arm:<30,45,60>** and **arm:<60,45,30>**, respectively.

Notice that dimension-independence implies that it is sufficient to add a third parameter to the **CUBOID** function, i.e. to write **CUBOID:<2,20,1>**, to get a 3D articulated model.

The following definitions generate the Bezier mapping which maps the unit interval $[0, 1]$ into a cubic Bezier curve in $\Re^3$ and a uniform sampling of such unit interval into n samples. The **PLASM** code for the **Bezier** function is given in the next section.

```
DEF CSpath = Bezier:<<0,0,0>,<90,0,0>,<90,90,0>,<90,90,90>>;
DEF Sampling (n::IsIntPos) = (AA:LIST~AA:/~DISTR):<0..n,n>;
```

The evaluation of **Sampling:10** generates, e.g.:

```
<<0>,<1/10>,<1/5>,<3/10>,<2/5>,<1/2>,<3/5>,<7/10>,<4/5>,<9/10>,<1>>
```

The configuration space curve of Figure 2a is generated by the **PLASM** expression

```
MAP:CSpath:(Intervals:18)
```

The sampling of the arm configurations given in Figure 2b is instead generated by the expression

```
(STRUCT~AA:arm~AA:CSpath):(Sampling:18)
```

4 Parametric geometry

4.1 Bezier curves

One interesting feature of **PLASM** is its ability to specify parametric geometry, i.e. parametric representation of curves, surfaces and multivariate varieties. In such an approach a curve is specified as the image set of a mapping over a parametric domain. The **PLASM** construct to be used is:

```
MAP:Fun:Domain;
```

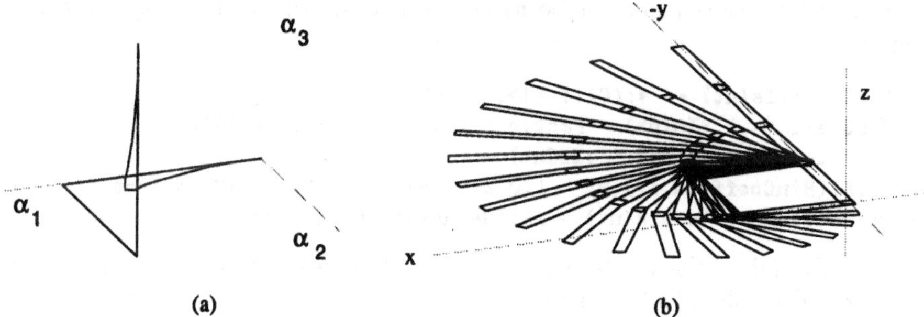

Figure 2: (a) The configuration space (CS) path between the start and the goal positions is defined as a Bezier cubic. (b) The set of model configurations in work space (WS) corresponding to a sampling of the CS curve.

where Fun is the **CONS** of the coordinate functions, i.e. **FUN = [Fx,Fy,Fz]** if the curve is embedded in $\Re^3$. Also, **Domain** is a polyhedral complex which partitionate the parametric domain. E.g., for a parametric curve, it is a 1D complex, for a surface it is a 2D complex, and so on.

The following definition is given to generate a polyhedral partition of the interval $[0,1]$ with n segments of size $1/n$:

```
DEF Domain (n::IsIntPos) = QUOTE:(#:n:(1/n));
```

Remembering that a Bezier curve of degree n can be computed as

$$\mathbf{c}(u) = \sum_{i=0}^{n} B_i^n(u)\, \mathbf{q}_i, \quad \text{where} \quad B_i^n(u) = \binom{n}{i}(1-u)^i u^{n-i}$$

we can write, to define a quadratic Bezier function:

```
DEF Bezier2 (q0,q1,q2::IsSeq) = [Bx, By, Bz]
WHERE
  Bx = B0 * x:q0 + B1 * x:q1 + B2 * x:q2,
  By = B0 * y:q0 + B1 * y:q1 + B2 * y:q2,
  Bz = B0 * z:q0 + B1 * z:q1 + B2 * z:q2,
  B2 = u*u,
  B1 = K:2*u*(K:1 - u),
  B0 = (K:1 - u)*(K:1 - u),
  u  = S1, x = (K~S1), y = (K~S2), z = (K~S3)
END;
```

where the parametric variable is substituted by the FL selector function **S1** and where all the algebraic operators work directly between functions. In particular, scalar coefficients (numbers in $\Re$) are transformed into constant functions.

A more general approach to the computation of Bezier functions for generating parametric curves of *any degree* can be obtained by firstly defining a small toolbox of utility functions. The top-level function **BernsteinBase** allows to generate the whole sequence of Bernstein's functions $B_i^n(u)$, $0 \leq i \leq n$, for any given n.

```
DEF Fact (n::IsInt) =  *:(CAT:<<1>, 2..n>);
DEF BinCoeff (n,i::IsInt) = Fact:n / (Fact:i * Fact:(n-i));
DEF Bernstein (n::IsInt)(i::IsInt) =
  * ~ [K:(BinCoeff:<n,i>),** ~ [ID,K:i],** ~ [- ~ [K:1,ID],K:(n-i)]] ~ S1;
DEF BernsteinBase (n::IsInt) = AA:(Bernstein:n):(0..n);
```

So, the graph of the whole Bernstein base of degree 4 shown in Figure 3 can be generated by the **PLASM** expression:

```
STRUCT:<
  MAP:[S1, Bernstein:4:4]: (Domain:22),
  MAP:[S1, Bernstein:4:3]: (Domain:22),
  MAP:[S1, Bernstein:4:2]: (Domain:22),
  MAP:[S1, Bernstein:4:1]: (Domain:22),
  MAP:[S1, Bernstein:4:0]: (Domain:22)
>;
```

or, more compactly, by the equivalent expression

```
STRUCT:(
  (CONS ~ AA:MAP ~ AA:CONS~DISTL):<S1, BernsteinBase:4>:
  (Domain:22) );
```

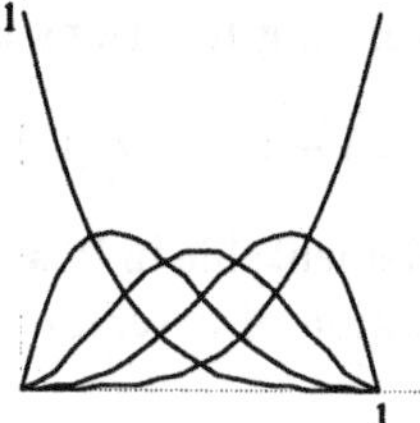

Figure 3: The Bernstein-Bezier base functions of degree 4.

The full power of the "combinatorial engine" of the FL language is exploited in the sequel to define a Bezier map of any (positive integer) degree, which may be applied to a sequence of (**degree+1**) control points in $\Re^n$, for any finite n.

```
DEF Bezier (ControlPoints::IsSeq) = (CONS~AA:(+~AA:*~TRANS)~DISTL):
     <BernsteinBase:degree,(AA:(AA:K)~TRANS):ControlPoints>
WHERE
  degree = LEN:ControlPoints - 1
END;
```

For example, the cubic Bezier curve in configuration space $\Re^3$ shown in Figure 2a is generated as follows:

```
DEF CSpath = Bezier:<<0,0,0>,<90,0,0>,<90,90,0>,<90,90,90>>;
MAP:CSpath:(Domain:22)
```

4.2 Coons surfaces

A well know approach in geometric design to the interpolation of curves are the Coons surfaces [4]. Let suppose to assign four boundary curves:

$$\mathbf{r}(u,0), \quad \mathbf{r}(u,1), \quad \mathbf{r}(0,v), \quad \mathbf{r}(1,v).$$

A surface patch which interpolates the given curves is obtained by using two blending functions $\alpha_0, \alpha_1 : \Re \to \Re$, which are monotone in $[0,1]$ and satisfy:

$$\begin{aligned} \alpha_0(0) = 1, \qquad & \alpha_0(1) = 0, \\ \alpha_1(0) = 0, \qquad & \alpha_1(1) = 1, \\ \alpha_0(t) + \alpha_1(t) \quad = \quad & 1, \qquad t \in [0,1]. \end{aligned}$$

A well-know solution [6] to this problem is:

$$\begin{aligned} \mathbf{r}(u,v) = \quad & \begin{bmatrix} \alpha_0(u) & \alpha_1(u) \end{bmatrix} \begin{bmatrix} \mathbf{r}(0,v) \\ \mathbf{r}(1,v) \end{bmatrix} \\ & + \begin{bmatrix} \mathbf{r}(u,0) & \mathbf{r}(u,1) \end{bmatrix} \begin{bmatrix} \alpha_0(v) \\ \alpha_1(v) \end{bmatrix} \\ & - \begin{bmatrix} \alpha_0(u) & \alpha_1(u) \end{bmatrix} \begin{bmatrix} \mathbf{r}(0,0) & \mathbf{r}(0,1) \\ \mathbf{r}(1,0) & \mathbf{r}(1,1) \end{bmatrix} \begin{bmatrix} \alpha_0(v) \\ \alpha_1(v) \end{bmatrix}, \qquad u,v \in [0,1]. \end{aligned}$$

The obvious choice of the blending functions is linear:

$$\alpha_0(t) = 1 - t, \qquad \alpha_1(t) = t.$$

This method is directly implementable in **PLASM** . First define two operators to perform the scalar multiplication of a vector times a scalar and the sum of two compatible vectors:

```
DEF times = AA:*~DISTL;
DEF plus  = AA:+~TRANS;
```

The **PLASM** operator which generate a bivariate Coons mapping for any 4-tuple of boundary curves (univariate vector functions) is a direct translation of its mathematical definition:

```
DEF CoonsPatch (ru0,ru1,r0v,r1v::IsFun) =
  plus~aa:times~[[a0u,r0v],[a1u,r1v],[a0v,ru0],
                 [a1v,ru1],[-~a0v,b0u],[-~a1v,b1u]]
WHERE
  b0u = plus~aa:times~[[a0u,r00],[a1u,r10]],
  b1u = plus~aa:times~[[a0u,r01],[a1u,r11]],
  r00 = (K~ru0):<0,0>,  r01 = (K~r0v):<0,1>,
  r10 = (K~ru0):<1,0>,  r11 = (K~r1v):<1,1>,
  a0u = K:1 - u,    a1u = u,
  a0v = K:1 - v,    a1v = v,
  u = S1,           v = S2
END;
```

In order to generate a polyhedral approximation the Coons function has to be mapped over a simplicial decomposition of a plane interval. This can be obtained as a polyhedral product [9] of the 1D complexes generated by the evaluation of the expressions **Domain:n** and **Domain:m**.

```
DEF Domain2D (n,m::IsIntPos) = Domain:n * Domain:m;
```

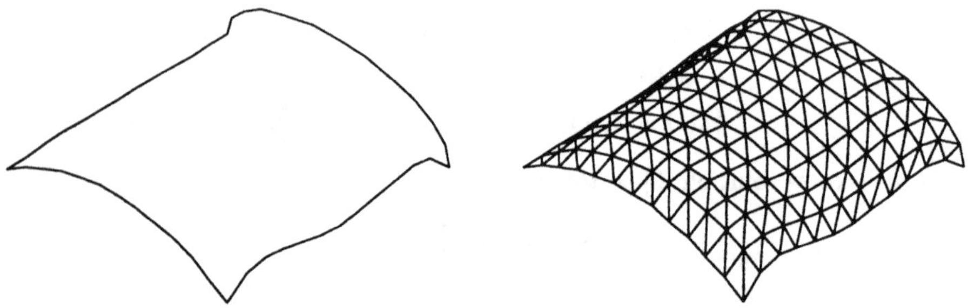

Figure 4: Four boundary curves and the Coons patch which bilinearly interpolates them.

So, the surface shown in Figure 4, which interpolates the four boundary curves defined as Bezier maps of various degrees, is generated as

```
MAP:(CoonsPatch:<ru0,ru1,r0v,r1v>):(Domain2D:<12,12>);
```

where

```
DEF ru0 = Bezier:<<0,0,0>,<10,0,0>>~[S1];
DEF ru1 = Bezier:<<0,10,0>,<2.5,10,3>,<5,10,-3>,<7.5,10,3>,<10,10,0>>~[S1];
DEF r0v = Bezier:<<0,0,0>,<0,0,3>,<0,10,3>,<0,10,0>>~[S2];
DEF r1v = Bezier:<<10,0,0>,<10,5,3>,<10,10,0>>~[S2];
```

The composition with the functions [S1] and [S2] is introduced in order to consider ru0,ru1,r0v,r1v as isovalued coordinate curves mappable from a 2D parametric domain.

It is interesting to note that no constraints exist in PLASM about the polyhedral domain of a parametric mapping. So, polyhedral complexes of any topology and even of non full dimensionality can be embedded in parametric space and used as the argument of a mapping.

E.g., a spiral curve in parametric space can be defined as follows and mapped using the previous Coons mapping. The resulting 3D curve is shown in Figure 5.

```
DEF SpiralFun = [radius*COS~S1, radius*SIN~S1]
  WHERE radius = K:1 - (K:(0.2 / (6 * PI)) * S1) END;

DEF Spiral = (T:<1,2>:<0.5,0.5>~S:<1,2>:<0.5,0.5>):
  (MAP:SpiralFun:(QUOTE:(#:100:(6 * PI/100))));

MAP:(CoonsPatch:<ru0,ru1,r0v,r1v>):Spiral;
```

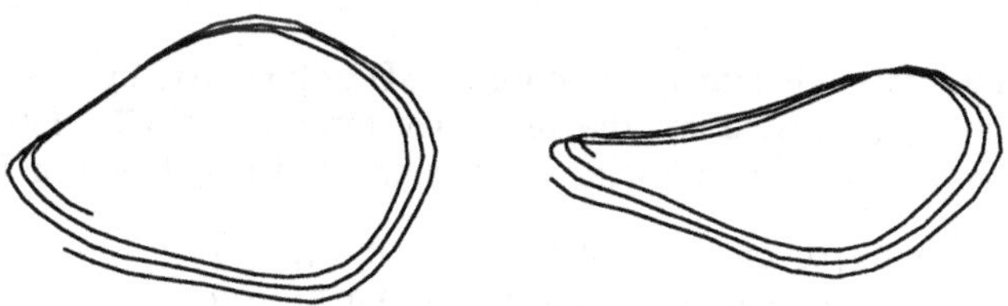

Figure 5: Two different projections of a spiral curve in parametric space $[0,1] \times [0,1]$ mapped over the Coons surface defined by four boundary curves.

4.3 Thin solid rods

In this section we derive thin and narrow solids starting from Bezier curves. In particular, we define a narrow surface adjacent to a plane quadratic Bezier curve, and then we get a 3D solid by polyhedral product of such a surface times a 1D interval. First, the normal unit vector to a quadratic Bezier curve is given as a polynomial function of the 3 control points. The basis functions are

$$\frac{d}{du}B_0^2(u) = 2u - 2, \qquad \frac{d}{du}B_1^2(u) = 1 - 2u, \qquad \frac{d}{du}B_2^2(u) = 2u.$$

```
DEF NormOrthBezier2 (q0,q1,q2::IsSeq) = [-~yu/den, xu/den]
WHERE
  den = MySQRT~+~AA:sqr~[-~yu, xu],
  xu = b0 * x:q0 + b1 * x:q1 + b2 * x:q2,
  yu = b0 * y:q0 + b1 * y:q1 + b2 * y:q2,
  b2 = K:2*u,
  b1 = K:1 - K:2*u,
  b0 = K:-2 + K:2*u,
  u  = S1, x = (K~S1), y = (K~S2)
END;
```

Two non primitive functions are given to compute the square root and the square of any argument (both function or number):

```
DEF MySQRT = IF:<EQ~[K:0,ID], K:0, SQRT>;
DEF sqr = ID * ID;
```

At this point a stripe of given `width`, polyhedrally approximated with `n` quadrilaterals, is generated as a 2D mapping depending on the 3 control points of a quadratic Bezier curve.

```
DEF StripBezier2 (q0,q1,q2::IsSeq; width::IsReal;n::IsIntPos) =
  MAP:[Xuv,Yuv]:(S:2:width:(Domain2D:<n,1>))
WHERE
  Xuv = S1~Bezier:<q0,q1,q2> + (v * S1~NormBezier2:<q0,q1,q2>),
  Yuv = S2~Bezier:<q0,q1,q2> + (v * S2~NormBezier2:<q0,q1,q2>),
  v = S2
END;
```

Finally, a 2D stripe is transformed into a 3D polyhedron by a product times a 1D polyhedron generated by the primitive function **QUOTE**. Two such thin solids are generated by the following expressions and are shown in Figure 6.

```
StripBezier2:<<1,0>,<0,0>,<0,1>,0.2,12> * QUOTE:<0.2> ;
StripBezier2:<<2,0>,<0,0>,<0,1>,0.2,12> * QUOTE:<0.2> ;
```

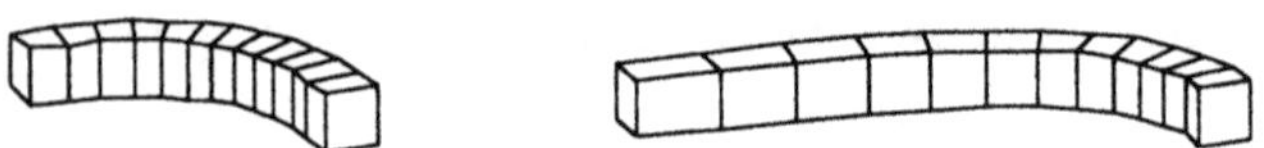

Figure 6: Two thin solids associated to quadratic Bezier curves.

5 Conclusion

This paper has introduced some geometric programming ideas by using the functional design language **PLASM** . It has been shown that the **PLASM** environment is characterized by the use of a modern functional language like FL and by a dimension independent approach to geometric modeling based on (both simplicial and convex) cell decompositions. A quite unusual point of view was adopted in the geometric kernel underlying the language by relaxing the recording of topological relations and by using pointset coverings more than pointset partitions.

The paper has also shown that such a functional language allows to manipulate functions as we do with numbers; this results in a very natural approach to parametric geometry, where some basis of polynomials in one or more variables is combined with control points or vectors to generate curves or surfaces.

Also, the **PLASM** approach seems to give a powerful alternative to variational geometry based on the simultaneous resolution of geometric constraints. Actually, the author believes that two such approaches can be usefully combined. So, one goal of the next language revision is to empower it with: (a) a type system, which is currently not supported; (b) a mechanism of constraint satisfaction; (c) some kind of logic, in order to support automatic reasoning over geometric models and environments.

Acknowledgments

This language would not exist without the contribution of ideas and the implementation work of Valerio Pascucci and Michele Vicentino. The first one developed most part of the dimension-independent modeler underlying the language; the second one was responsible for the language interpreter and the design environment. Other significant contributions came from Vincenzo Ferrucci and Fausto Bernardini. More recently, Claudio Baldazzi extended the modeler with dimension-independent regularized Booleans making use of BSP-trees. The preparation of graphics courses at "La Sapienza" and "Rome III" universities gave the author a chance to explore some **PLASM** application areas which were not foreseen at language design time. Several stimulating discussions with Antonio Di Carlo are also gratefully acknowledged.

References

1. BACKUS, J. Can Programming Be Liberated from the Von Neumanns Style? A Functional Style and its Algebra of Programs. *Communications of the ACM*, **21**(8):613–641, (ACM Turing Award Lecture). 1978.
2. BACKUS, J., WILLIAMS, J.H. AND WIMMERS, E.L. An Introduction to the Programming Language FL. In *Research Topics in Functional Programming*, D.A. Turner (Ed.), Addison-Wesley, Reading, MA, 1990.
3. BACKUS, J., WILLIAMS, J.H., WIMMERS, E.L., LUCAS, P., AND AIKEN, A. FL Language Manual, Parts 1 and 2. IBM Research Report RJ 7100 (67163), 1989.
4. COONS, S.A. Surfaces for Computer Aided Design of Space Forms. Report MAC-TR-41, Project MAC, MIT, 1967.
5. GURTIN, G. An Introduction to Continuum Mechanics. Academic Press, Boston, MA, 1981.

6. GREGORY, J.A. (ED.) The mathematics of surfaces. The Institute of Mathematics, Oxford University Press, New York, NY, 1986.

7. PAOLUZZI, A., AND SANSONI, C. Programming language for solid variational geometry. *Computer Aided Design 24*, 7 (1992), 349–366.

8. PAOLUZZI, A., PASCUCCI, V., AND VICENTINO, M. Geometric programming: A programming approach to geometric design. *ACM Transactions on Graphics 14*, 4 (Oct. 1995).

9. BERNARDINI, F., FERRUCCI, V., PAOLUZZI, A. AND PASCUCCI, V. Product operator on cell complexes. In *Proc. of the Second ACM/IEEE Symp. on Solid Modeling and Application*, Montreal, Canada, ACM Press, (May 1993), 43–52.

10. PASCUCCI, V., FERRUCCI, V., AND PAOLUZZI, A. Dimension-independent convex-cell based HPC: Skeletons and Product. *Int. Journal of Shape Modeling 2*, 1 (1996), 37–67.

11. PAOLUZZI, A., BERNARDINI, F., CATTANI, C., AND FERRUCCI, V. Dimension-independent modeling with simplicial complexes. *ACM Transactions on Graphics 12*, 1 (Jan. 1993), 56–102.

12. TAKALA, T. Geometric Boundary Modelling Without Topological Data Structures. In *Eurographics '86* (Amsterdam, 1986), A.A.G. Rquicha, Editor, North-Holland.

APPENDIX

A Umbrella modeling: a top-down work session

In this appendix it is shown a work session with the **PLASM** functional environment. The goal of the geometric programming example is the generative modeling of a parametric umbrella, parameterized on the opening angle. In a first subsection it is presented the modeling from scratch of the cinematic mechanism, using wire frame parts. Later on, the model is refined first by imposing that some rods may be incurved (as quadratic Bezier curves) depending on the opening angle; then by defining the umbrella canvas as Coons patches delimited by polynomial curves of degree 2, 1 and 0. Finally, the various rods are modeled as 3D solid parts, so substituting their first definitions as wire frames. This example aims to demonstrate that the **PLASM** language, conversely than current development environments for graphics programming and virtual reality, actually allows for progressive refinement of working models and top-dow development. This seems should be a distinctive feature of a design language.

A.1 Starting from scratch

A 1D rod of length `len` is defined along the z-axis in $\Re^3$ by using the primitive constructor **MKPOL**.

```
DEF Rod (len::IsReal) = MKPOL:<<<0,0,0>,<0,0,len>>,<<1,2>>,<<1>>>;
DEF Axis (len::IsReal) = Rod:len;
```

The heart of the moving mechanism (shown in Figure 7) is then described as a structure depending on the height **h** and on the opening angle **alpha**. At this purpose two instances of the **Rod** primitive are properly combined with affine transformations. According to the semantics of standard ISO PHIGS structures, both **Rod1** and **Rod2** are defined in local coordinates and jointly transformed into world coordinates.

```
DEF RodPair (h, alpha::Isreal) = STRUCT:<
  T:3:h, R:<3,1>:(-:alphaRad), Rod1,
  T:3:(-:AB), R:<3,1>:(2*alphaRad), Rod2 >
WHERE
  alphaRad = alpha*PI/180,
  Rod1 = S:3:-1:(Rod:(2*AB)),
  Rod2 = S:3:-1:(Rod:AB),
  AB = h*4/10
END;
```

The whole parametric **Umbrella** is then defined as a structure by catenating a sequence with one **Axis** and **Handle** and 12 pairs, each containing a rotation of $\pi/6$ around the z-axis and one instance of the **RodPair** model previously defined.

```
DEF Umbrella (h, alpha::Isreal) = (STRUCT~CAT):
  <[Axis, Handle]:h,
   ##:12:< RodPair:<9/10*h,alpha>, R:<1,2>:(PI/6) > >;
```

In particular, the **Handle** is defined as a properly positioned halfcircle linearly approximated with 12 segments. The whole umbrella model at this stage is shown in Figure 8.

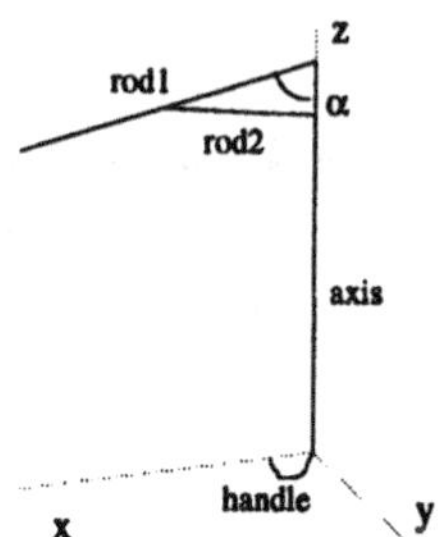

Figure 7: The model generated by the evaluation of the expression **(STRUCT~[Axis~S1,Handle~S1,RodPair]):<10,80>;**

```
DEF Handle (h::Isreal) = (T:1:Radius ~ S:<1,3>:<Radius,Radius>):
  (MAP:[COS~S1, K:0, SIN~S1]:dom )
WHERE
  dom = T:1:PI:(QUOTE:(#:12:(PI/12))),
  Radius = h/18
END;
```

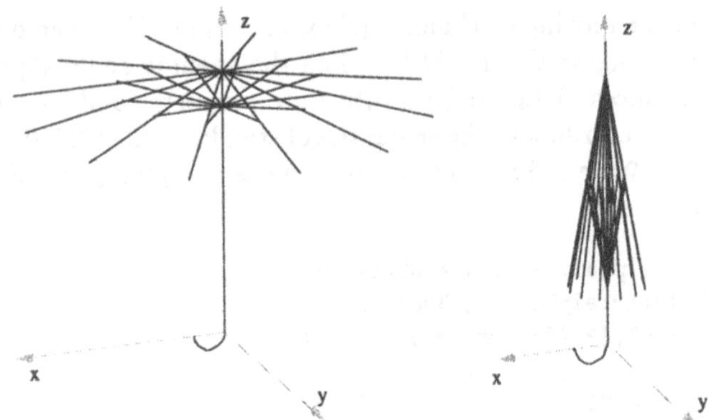

Figure 8: The 1D models embedded in $\Re^3$ generated by evaluating the expressions **Umbrella:<10,80>** and **Umbrella:<10,15>**.

A.2 First refinement: 1D curved rod specification

The function **RodPair** is redefined so that the previous **Rod1** component is now split into two components **Rod10** and **Rod11**, where **Rod11** is generated as a 1D polyhedral approximation with 4 segments of a quadratic Bezier curve. The resulting **Umbrella** values are shown in Figure 9.

```
DEF RodCurve (len,beta,n::IsReal) =
  MAP:(Bezier:<<0,0,0>,<0,0,len/2>,<len/2 * sin:beta,0,len>>):(Domain:n);
  DEF RodPair (h, alpha::Isreal) = STRUCT:<
    T:3:h, R:<3,1>:(-:alphaRad), Rod10,
    STRUCT:<T:3:(-:AB), Rod11>,
    T:3:(-:AB), R:<3,1>:(2*alphaRad), Rod2 >
  WHERE
    alphaRad = alpha*PI/180,
    Rod10 = S:3:-1:(Rod:(AB)),
    Rod11 = S:3:-1:(RodCurve:<AB,-:alphaRad/4,4>),
    Rod2 = S:3:-1:(Rod:AB),
    AB = h*4/10
  END;
```

Figure 9: The new values of **Umbrella:<10,80>** and **Umbrella:<10,30>** after redefinition of **RodPair**.

A.3 Second refinement: design of canvas

A pair of canvas is now defined for the umbrella sector associated to the single instance of **RodPair**. The two canvas, shown in Figure 10a and Figure 10b, are defined as Coons surfaces delimited by four Bezier curves of various degrees.

In the case of **Canvas1**, two quadratic Bezier limiting curves (**ru0** and **ru1**) are generated by the same control points of **Rod11** and by their rotated instance (around the z axis); the other two curves are obtained by taking respectively the first and the last points (**p1** and **p3**) of such curves, and by computing two more points (**p12** and **p22**) on the horizontal plane. The position of such control points is a proper function of the opening angle of the umbrella.

A function **MoveToWC** is needed to reproduce the pipeline of 3D transformations (discussed in Section 3.2) which must be applied to the extreme points of **rod11**. Notice that within the **Umbrella** generating function the same transformations are applied to polyhedra by using primitive **PLASM** affine operators.

```
DEF Canvas1 (h, alpha::Isreal) =
  MAP:(CoonsPatch:<ru0,ru1,r0v,r1v>):(Domain2D:<4,4>)
WHERE
  handles1 = MoveToWC:<<0,0,0>,<0,0,AB/2>,<AB/2 * sin:beta,0,AB>>,
  MoveToWC = AA:(Tz:h ~ Ry:(-:alphaRad) ~ Tz:(-:AB) ~ Sz:-1),
  handles2 = AA:(Rz:(PI/6)):handles1,
  ru0 = Bezier:handles1,
  ru1 = Bezier:handles2,
  r0v = Bezier:<p1, p12, Rz:(PI/6):p1>~[S2],
  r1v = Bezier:<p2, p22, Rz:(PI/6):p2>~[S2],
  p1  = FIRST:handles1,
  p2  = LAST:handles1,
  p12 = Rz:(PI/12):<S1:p1 - AB/2*COS:alphaRad, S2:p1, S3:p1>,
  p22 = Rz:(PI/12):<S1:p2 - AB*COS:alphaRad, S2:p2, S3:p2>,
  beta = -:alphaRad/4,
  alphaRad = alpha*PI/180,
  AB = h*4/10
END;
```

The second canvas just differs from the first one for the choice of the limiting curves. In particular, the functions **ru0** and **ru1** are now given as linear Bezier with two control points, and **r0v** is given as a Bezier function of degree 0 depending on a single control point.

```
DEF Canvas2 (h, alpha::Isreal) =
  MAP:(CoonsPatch:<ru0,ru1,r0v,r1v>):(Domain2D:<4,4>)
WHERE
  handles1 = MoveToWC:<<0,0,0>,<0,0,AB>>,
  MoveToWC = AA:(Tz:h ~ Ry:(-:alphaRad) ~ Sz:-1),
  handles2 = AA:(Rz:(PI/6)):handles1,
  ru0 = Bezier:handles1,
  ru1 = Bezier:handles2,
  r0v = Bezier:<p1>~[S2],
```

```
  r1v = Bezier:<p2, p22, Rz:(PI/6):p2>~[S2],
  p1  = FIRST:handles1,
  p2  = LAST:handles1,
  p22 = Rz:(PI/12):<S1:p2 - AB/2*COS:alphaRad, S2:p2, S3:p2>,
  beta = -:alphaRad/4,
  alphaRad = alpha*PI/180,
  AB = h*4/10
END;
```

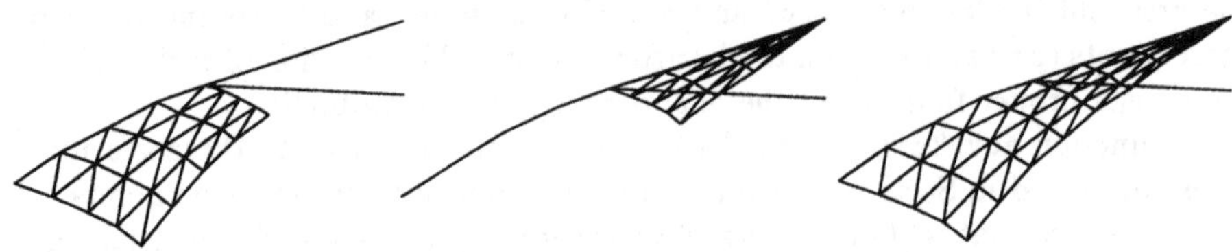

Figure 10: The picture shows the first, the second and both canvas mounted together with the complex `RodPair:<10,80>`

The function `RodPair` is so defined again, by inserting in it a proper pair of canvas, shown in Figure 10. The new configurations of the whole umbrella for different opening angles are shown in Figure 11.

```
DEF RodPair (h, alpha::Isreal) = STRUCT:<
  Canvas1:<h, alpha>,
  Canvas2:<h, alpha>,
  T:3:h, R:<3,1>:(-:alphaRad), Rod10,
  STRUCT:<T:3:(-:AB), Rod11 >,
  T:3:(-:AB), R:<3,1>:(2*alphaRad), Rod2 >
WHERE
  alphaRad = alpha*PI/180,
  Rod10 = S:3:-1:(Rod:(AB)),
  Rod11 = S:3:-1:(RodCurve:<AB,-:alphaRad/4,4>),
  Rod2 = S:3:-1:(Rod:AB),
  AB = h*4/10
END;
```

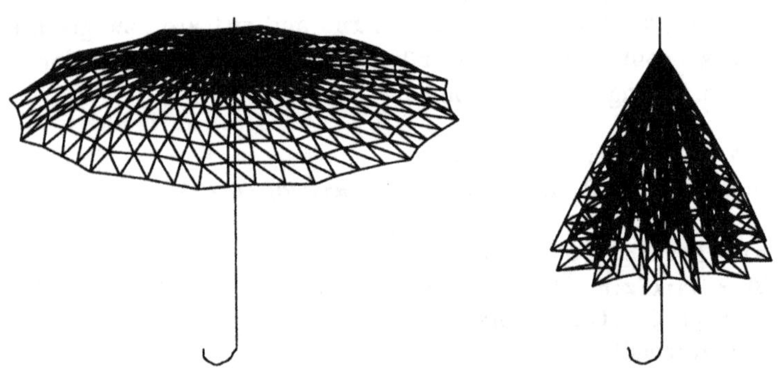

Figure 11: The new values of `Umbrella:<10,80>` and `Umbrella:<10,30>` after the second redefinition of `RodPair`.

A.4 Third refinement: solid parts

The 1D subcomplex with handle and rods defined in the first umbrella version is now substituted by 3D solid parts. In particular, the new handle is generated as a 3-variate mapping of the boundary Dom of a translated parallelopided D1 * D2 * D3 in parametric space, where D1 is a partition in 8 parts of the interval $[\pi, 2\pi]$, D2 is a partition in 1 parts of the interval $[0, r/3]$, where r is the handle radius, and D3 = D2.

```
DEF Handle (h::Isreal) = T:<1,2>:<Radius*7/6,-:Radius/6>:
  (MAP:[S2*COS~S1, S3, S2*SIN~S1]:dom )
WHERE
  D1 = (T:1:PI ~ S:1:PI ~ Domain):8,
  D2 = (S:1:(1/3*Radius) ~ Domain):1,
  D3 = D2,
  Dom = @2:(T:2:Radius:(D1 * D2 * D3)),
  Radius = h/18
END;
```

```
DEF Axis (h::IsReal) = (@2 ~ STRUCT):<
  Handle2, T:3:(h/10), MetalRing, Rod, T:3:(2*AB), Tip >
WHERE
  Handle2 = T:<1,2>:<-1/6*Radius,-1/6*Radius>:
            (CUBOID:<1/3*Radius,1/3*Radius,h/10>),
  MetalRing = CYLINDER:<h/50, 0.5*h/10, 12>,
  Rod = CYLINDER:<h/90, 2*AB, 12>,
  Tip = Handle2,
  Radius = h/18,
  AB = h*4/10
END;
```

```
DEF Rod (len::IsReal) = T:<1,2>:<-2*a,-1/2*a>:((@2~CUBOID):<2*a,a,len>)
WHERE  a = len/100  END;
```

Finally, the curved tiny rods of the umbrella are generated from their initial wire frame definition, by using the StripBezier2 discussed in Section ??. Such a function will generate a plane surface stripe depending on the 3 control points of a quadratic Bezier curve. The 3D solid rods are so generated by the polyhedral product of the surface stripe times a 1D segment of size len/100.

```
DEF RodCurve (len,beta,n::IsReal) = (R:<2,3>:<PI/2> ~ @2):
  (StripBezier2:<<0,0>,<0,len/2>,<len/2 * sin:beta,len>,2*len/100,n>
  * QUOTE:<len/100>);
```

Some projections of the refined model, for different values of the opening angle, are given in Figure 12.

Figure 12: Four perspective projections of the model generated as **Umbrella:<10,80>**, and **Umbrella:<10,30>**, respectively, where 80 and 30 (degrees) are the opening angles.

Exploiting SML for Experimenting with Algebraic Algorithms: The Example of p-adic Lifting

Wolfgang Gehrke and Carla Limongelli *

Università degli Studi di Roma Tre, Dipartimento di Informatica
Via della Vasca Navale 84, I - 00146 Roma, Italy
{wgehrke, limongel}@inf.uniroma3.it

Abstract. This paper shows the expressive power of the functional programming language Standard ML (SML) in the context of computer algebra. It is focused on a special application of the p-adic lifting technique, the Hensel algorithm, that is utilized in a symbolic but also numeric context. This experiment demonstrates that SML provides a suitable frame for the implementation of abstract algebraic notions together with the possibility to code related algorithms in a generic way on the corresponding level of abstraction.

1 Introduction

The functional programming language Standard ML (SML) originally evolved as a "meta-language" in the context of logic deductions. Meanwhile it was turned into a general purpose high-level programming language [MTH90, MT91]. It provides imperative features, an exception mechanism, and a powerful parametric module system (examples can be found in [Pau91]).

As the benefits for computer algebra resulting from the application of SML we see the following:

1. The SML notation of algebraic notions comes very close to algebraic specifications.
2. The clear typing of all operations makes the understanding for a newcomer to the field of computer algebra easier.
3. Since SML is widely applied for theorem proving systems this allows the integration of both components: a computational one and a deductive one.

Recently in [San95] SML was used to implement an expressive type system suitable for computer algebra. Furthermore also Extended ML [San89] provides a promising language since it additionally allows logical axioms inside signatures but it currently lacks an implementation.

There are several reasons that carried us to apply SML in the field of computer algebra:

* This work has been partially supported by the Italian project MURST ex 40% "Rappresentazione della conoscenza e meccanismi di ragionamento"

its strong polymorphic type system which enforces a considerable discipline in coding,
its advanced module systems which allows generic programming of algebraic notions on a corresponding level of abstraction,
its formalized semantics which provides a mean to reason about SML programs.

We can also take advantage of further extensions of the language like Concurrent ML [Rep91] or the higher-order module system [Tof94].

In order to verify our claim we have chosen the Hensel algorithm as a suitable candidate. On the one hand this algorithm is defined at a high level of abstraction. On the other hand it is also suitable to perform numeric computations.

In this first experiment we have restricted the implementation to univariate polynomial root finding. The general case of multivariate polynomials has been studied in detail in the literature [Lau83, Yun74]. A special case of this algorithm is the exact representation of algebraic numbers in a p-adic domain.

Our implementation differs from existing computer algebra systems since there the algorithms presented in this paper are mostly built-in. In contrast we make the implementation techniques visible and provide different instantiations for the same abstract notion. Therefore the final algorithm can be easily customized.

From the functional point of view we contribute:

- a set of sufficiently general operations for polynomials in the context of the Hensel algorithm,
- an interesting example of the usefulness of higher-order functors,
- a first step towards a library for specialized computations with algebraic numbers.

Nevertheless the efficiency of such generic algorithms still has to be investigated.

In Sec. 2 we give a brief overview of the Hensel algorithm and in Sec. 3 we summarize main features of SML. The Sec. 4 provides details of the implementation and a discussion of design choices. Furthermore we describe how to exploit the functional style and SML modules. We demonstrate how algebraic notions can be implemented on an appropriate level of abstraction. Examples and tests are shown in Sec. 5. Finally we conclude and suggest future work.

2 Some Background

Given a polynomial equation and a suitable initial approximation *mod* p^t of its solution, with p being prime, lifting algorithms compute a solution *mod* p^{t+1}, where p belongs to the domain where the polynomial is defined. They are based on Newton's method for root finding, translated into an appropriate algebraic domain, that is in the most general case a commutative ring. The following theorem states the convergence of the lifting algorithm.

Theorem 1 (Abstract Linear Lifting). *Let I be a finitely generated ideal in a commutative ring R and $f_1, \ldots, f_n \in R[x_1, \ldots, x_r]$, $r \geq 1$, $a_1, \ldots, a_r \in R$ with*

$$f_i(a_1, \ldots, a_r) \equiv 0 \ mod\ I, \quad \text{with } i = 1, \ldots, n.$$

Further let $U = (u_{i,j}), i = 1, \ldots, n,\ j = 1, \ldots, r,$ with $u_{i,j} = \frac{\partial f_i}{\partial f_j}(a_1, \ldots, a_r) \in R$ (U is the Jacobian matrix of $f_1, \ldots, f_n$, evaluated at $a_1, \ldots, a_r$). Assume that U is invertible mod I. Then for each positive integer t, there exist $a_1^{(t)}, \ldots, a_r^{(t)} \in R$, such that

$$f_i(a_1^{(t)}, \ldots, a_r^{(t)}) \equiv 0 \ mod\ I^t, \quad i = 1, \ldots, n \ \ and\ \ a_j^{(t)} \equiv a_j \ mod\ I, \quad j = 1, \ldots, r.$$

Proof. The proof is given by induction on t. See [Lau83]. □

The approximation methods for p-adic construction are based on the following computational steps:

1. start from an appropriate initial approximation,
2. compute the first order Taylor series expansion,
3. solve the obtained equation,
4. find an update of the solution.

The Hensel algorithm is based on this theorem and its constructive proof. We will consider the restriction to univariate polynomials and algebraic numbers. In particular we will assume $R = \mathbb{Z}[x]$, $I = (p)$ the ideal generated by a prime number, $a_1 = G^1, a_2 = H^1$ both in $R = \mathbb{Z}[x]$ and $\Phi(x, G, H)$ a polynomial function. According to the previous theorem, for any positive integer t, there exists $G^{(t)}, H^{(t)} \in \mathbb{Z}_p^{(t)}[x]$, such that

$$G^{(t)} \equiv G \ mod\ p, \quad \text{and} \quad H^{(t)} \equiv H \ mod\ p.$$

Given $F \in \mathbb{Z}[x], n \in \mathbb{N}$ we want to find the solution of the equation $\Phi(x, G, H) = 0$ where $G, H, \in \mathbb{Z}[x]$ and $x \in \mathbb{Z}$.

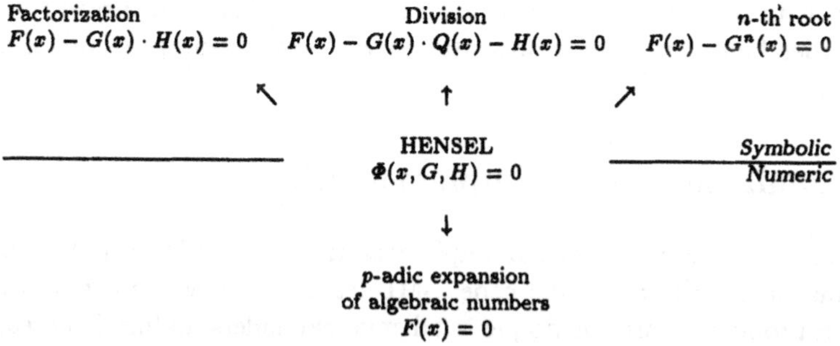

Fig. 1. Applications of Hensel Method

Already at this less general level it is possible to appreciate the intrinsic abstraction of this method that can solve different computing problems. These are either symbolic (factorization, n-th root of a polynomial, polynomial division) or numeric (zero of a polynomial, p-adic expansion of an algebraic number), according to different instantiations of its input parameters as the Fig. 1 shows.

3 Functional Programming with SML

SML is a strict and impure functional programming language, i.e. also functions are first-class objects. It is strict since the evaluation mechanism works based on the rule "call-by-value". It is impure since it contains reference types, exceptions, and an imperative I/O mechanism.

SML is a statically typed language using a sophisticated polymorphic type system where types can be inferred at compile time. The type system is accompanied by a highly advanced module system. In particular these modules make the language very attractive since they are a powerful tool to structure large programs.

Another important feature of SML is that it has a formal semantics, actually its definition [MTH90, MT91] proceeds in a completely formal style. This is an important point when properties of SML programs have to be proved. A general introduction and examples of SML programs can be found in [Pau91].

The module system as a mean to structure complex programs consists of:

structures as a way to bundle various declarations together,
signatures as the type checking information of structures,
functors as mappings from structures to structures.

The descriptive power of signatures comes close to algebraic specifications. Functors can describe generic constructions of new structures in terms of given ones.

This module discipline allows additionally separate compilation. Finally it should be mentioned that meanwhile the module concept has been generalized to the higher-order case [Tof94]. This means that now also functors can be the input but also output of other functors and they have their own signatures.

For our implementation we have been using the SML/NJ compiler [2]. It comes together with several useful tools, like a make facility, and already supports higher-order functors. For our presentation we took advantage of this generalization.

4 Organization of the Implementation

In the following we comment our implementation. It should be noted that this is not just a description of the code. SML leads the programmer to structure the given problem clearly. A deeper mathematical understanding is necessary to achieve an appropriate implementation.

[2] Copyright 1989, 1990, 1991, 1992, 1993, 1994, 1995, 1996 by AT&T Bell Laboratories

4.1 Signatures for Algebraic Notions

The initial algebraic notion for our purpose is a *ring* and a *commutative ring* with *unit*. These notions can be represented in a straightforward fashion. Here we present their signatures:

```
signature Ring =
    sig
        type $
        val eq : $ * $ -> bool
        val r2s : $ -> string
        val zero : $
        val neg : $ -> $
        val plus : $ * $ -> $
        val times : $ * $ -> $
    end
```

```
signature UnitCommutativeRing =
    sig
        include Ring
        val one : $
    end
```

In the signature `Ring` we make the equality between elements explicit in form of the function `eq` together with a function `r2s` which allows to print ring elements. A commutative ring with unit has just one more constructor `one` for the type `$`. Properties like commutativity are not represented in this coding.

This last point is controversial. Of course it would also be possible to maintain a set of properties for every structure together with some rules how to compute new properties from given ones. Nevertheless this merely means a hard coding of mathematical theorems.

Another approach to this problem could be the use of Extended ML as demonstrated in [San89]. In Extended ML signatures can contain logical axioms which describe further constraints. Unfortunately there does not exist a system to support Extended ML.

We proceed to the next mathematical notion, an *Euclidean domain*. This notion can be enriched generically by the computation of the *greatest common divisor* and the *extended Euclidean algorithm*. The other two functions are special calls to the extended Euclidean algorithm.

```
signature EuclideanDomain =
    sig
        structure ucr : UnitCommutativeRing
        exception DivMod
        val div : ucr.$ * ucr.$ -> ucr.$
        val mod : ucr.$ * ucr.$ -> ucr.$
    end

signature EnrichedEuclideanDomain =
    sig
        include EuclideanDomain
        val gcd : ucr.$ * ucr.$ -> ucr.$
        val eea : ucr.$ * ucr.$ -> ucr.$ * ucr.$ * ucr.$
        exception DiophFail
        val dioph : ucr.$ * ucr.$ * ucr.$ -> ucr.$ * ucr.$
        exception NoEinv
        val einv : ucr.$ * ucr.$ -> ucr.$
    end
```

This representation of an Euclidean domain is slightly different from the literature as in [Lip81]. We do not provide a *degree* function since we do not need it explicitly. **div** and **mod** are implemented as functions and they are not just existential statements.

The integers will be an example of a commutative ring with unit and also of an Euclidean domain. Furthermore we need modular arithmetic. In case of prime numbers with these ingredients we can construct a finite commutative field.

```
signature CommutativeField =
    sig
        structure ucr : UnitCommutativeRing
        exception DivisionByZero
        val inv : ucr.$ -> ucr.$
        type integer
        exception NotPrime
        val char : integer
    end
```

As already done for the Euclidean domain we make use of SML exceptions to code inadmissible computations. Intentionally we have an own type **integer** in order to be flexible in its implementation which here is needed for coding the type for the *characteristic*. The exception **NotPrime** will be used to check a semantical requirement during the creation of a finite field.

Finally we present the signature for polynomials. For some operations we make use of the built-in type **int** for integers since these support other operations like comparison. The substructures **base** and **ucr** describe the rings of coefficients and the univariate polynomials, resp.

```
signature UniPolynomial =
    sig
        structure base : UnitCommutativeRing
        structure ucr : UnitCommutativeRing
        val simplify : ucr.$ -> ucr.$
        val map : (base.$ -> base.$) -> ucr.$ -> ucr.$
        val p2l : ucr.$ -> base.$ list
        (* use dense list of coefficients beginning with the smallest *)
        val l2p : base.$ list -> ucr.$
        val embed : base.$ -> ucr.$
        exception NoProjection
        val project : ucr.$ -> base.$
        val degree : ucr.$ -> int
        val lcf : ucr.$ -> base.$
        val subst : ucr.$ * ucr.$ -> ucr.$
        val shift : ucr.$ * int -> ucr.$
        val derive : ucr.$ -> ucr.$
        val power: ucr.$ * int -> ucr.$
    end

signature PolynomialEuclideanDomain =
    sig
        include UniPolynomial
        structure base' : CommutativeField
        exception DivMod
        val div : ucr.$ * ucr.$ -> ucr.$
        val mod : ucr.$ * ucr.$ -> ucr.$
    end
```

Several times we made use of substructures with the **UnitCommutativeRing** signature in order to express better sharing constraints in the following code. For the polynomials it was necessary to have conversion functions between the polynomials and its coefficients (**embed** and **project**) and polynomials and a generally available type (**p21** and **12p**). Note also the generality of the substitution **subst** which does not just take an element but an entire polynomial.

4.2 Example of a Functor

To illustrate the module concept of SML we present an interesting functor in detail. This is the enrichment of an Euclidean domain by further functions. The enrichment is coded here once and for all. Later it can be applied in different situations for example for integers but also for polynomials over a commutative ring.

```
functor EnrichED(structure ED: EuclideanDomain): EnrichedEuclideanDomain=
    struct
        local open ED.ucr
        in
            fun gcd(a, b) = if eq(b, zero) then a
                            else gcd(b, ED.mod(a, b))
            (* eea(a,b) = (x,y,z) s.t. a*y + b*z = x = gcd(a,b) *)
            fun eea(a, b) =
                let fun aux_loop((a0, a1), (s0, s1), (t0, t1)) =
                    if eq(a1, zero) then (a0, s0, t0)
                    else let val q = ED.div(a0, a1)
                         in
                            aux_loop((a1, plus(a0, neg(times(a1, q)))),
                                     (s1, plus(s0, neg(times(s1, q)))),
                                     (t1, plus(t0, neg(times(t1, q)))))
                         end
                in
                    aux_loop((a, b), (one, zero), (zero, one))
                end
            (* solve the equation f*u + g*v = h if gcd(f,g) divides h *)
            (* returns (u, v)                                          *)
            exception DiophFail
            fun dioph(f, g, h) =
                let val (gcd, s, t) = eea(f, g)
                    val c = ED.div(h, gcd)
                    val b = ED.div(g, gcd)
                    val a = times(c, s)
                    val q = ED.div(a, b)
                    val r = ED.mod(a, b)
                in
                    (r, plus(times(c, t),times(q, ED.div(f, gcd))))
                end handle DivMod => raise DiophFail
            (* computes a^(-1) mod m if this is possible *)
            exception NoEinv
            fun einv(a, m) = let val (gcd, s, t) = eea(m, a)
                             in if eq(gcd, one) then ED.mod(t, m)
                                else if eq(gcd, neg(one))
                                       then ED.mod(neg(t), m)
                                     else raise NoEinv
                             end
        open ED
    end
```

The entire program contains some more auxiliary functors which we will not present in all detail here. These concern the construction of polynomials from rings or fields and the creation of a finite field. The latter is parameterized in an implementation of integers for more flexibility.

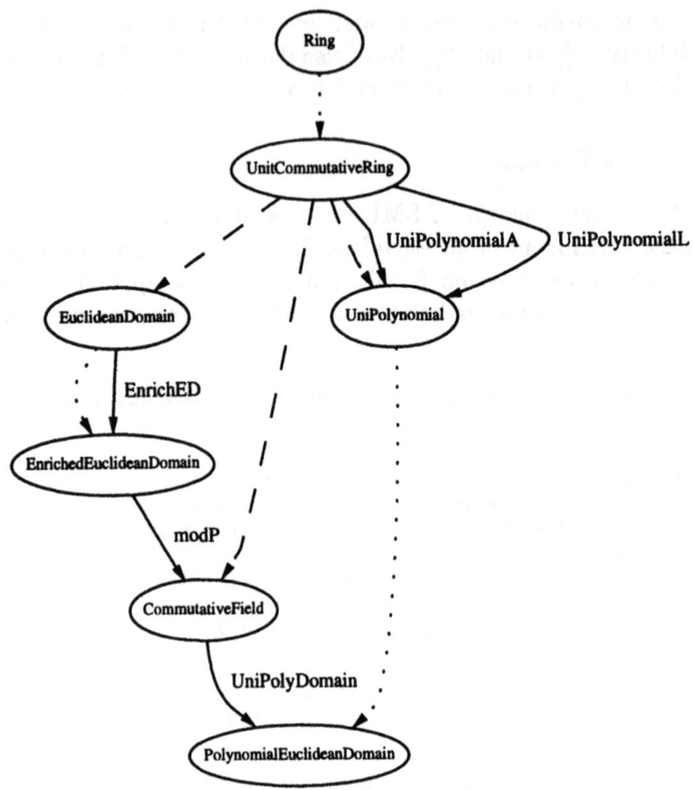

Fig. 2. Main Signatures and Functors

Although we do not present the code we give the *functor signatures*. We follow the style of higher-order modules as suggested in [Tof94]. SML/NJ implements this module discipline, too.

```
funsig Enrich (structure ED : EuclideanDomain) =
sig
    include EnrichedEuclideanDomain
    sharing ED.ucr = ucr
end
funsig UniPolys (structure UCR : UnitCommutativeRing)=
sig
    include UniPolynomial
    sharing UCR = base
end
funsig UniPolyd (structure CF : CommutativeField
                 functor UP : UniPolys) =
sig
    include PolynomialEuclideanDomain
    sharing CF.ucr = base and CF = base'
end
funsig ModP (structure IEED : EnrichedEuclideanDomain
             val prime : IEED.ucr.$) =
sig
    include CommutativeField
    sharing type IEED.ucr.$ = ucr.$ and type IEED.ucr.$ = integer
end
```

When writing the functor with the signature `UniPolys` we have to decide the way of implementing polynomials. It has been done in two different ways: with lists and with arrays. Since the functor with the signature `UniPolyd` makes use of the previous functor it is turned into a parameter.

This functor `UniPolydomain` is a non-trivial example of an implementation using the higher-order style. On the one hand we want the flexibility of allowing different implementations of polynomials, on the other hand we do not want to code polynomials again. This situation is characterized in that `UniPolydomain` is an extension of a functor implementing polynomials.

The relationship between the signatures and functors can be seen in the Fig. 2. For the `modP` functor there is a further parameter, a prime number. As said before the `UniPolyDomain` functor is also parameterized over a functor with the signature `UniPolys`.

The dotted arrows indicate the inclusion relation between signatures. The dashed arrows mean that there is a substructure of the given signature. The functors are shown by normal arrows.

4.3 The Central Algorithm

The main algorithm is implemented in form of a single functor `Hensel`. Its main input are a representation of the integers and a prime number. The interface of this functor is described in Fig. 3.

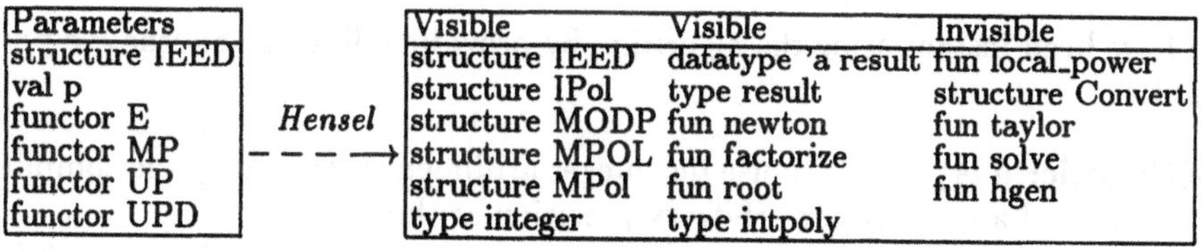

Parameters		Visible	Visible	Invisible
structure IEED		structure IEED	datatype 'a result	fun local_power
val p		structure IPol	type result	structure Convert
functor E	*Hensel*	structure MODP	fun newton	fun taylor
functor MP	- - - -→	structure MPOL	fun factorize	fun solve
functor UP		structure MPol	fun root	fun hgen
functor UPD		type integer	type intpoly	

Fig. 3. Hensel Functor

All the other auxiliary structures are created internally with the help of the provided functors:

`IEED` a copy of the parameter,
`IPol` the integer polynomials,
`MODP` the finite commutative field of characteristic p,
`MPOL` the polynomials over this field,
`MPol` these polynomials enriched by further operations.

The type definitions `integer` and `intpoly` make use of these structures. The signature hides a further structure `Convert` which is needed for conversion between integers and modular numbers and for the conversion of the corresponding polynomials.

The main internal functions which are also hidden are:

taylor gives the Taylor series expansion,
solve computes the solution of the equation obtained by the Taylor series expansion and finds a next approximation,
hgen triggers the iteration of this process.

The input of **hgen** is **Phi**, its derivation, the approximations for **x, g, h** and the number of iterations. This coding closely follows the Maple code as in [Lim93].

The functions visible in the interface, **newton**, **factorize**, **divide** and **root** are all specialized calls to **hgen**. Since **Phi** is a function in the variables **x,g,h** the code uses a lambda abstraction. The same applies for the partial derivatives of **Phi**.

This functional representation allows an easy way of implementing all the 4 special cases. Furthermore we can completely avoid multivariate polynomials which would make the code considerably more complex. The provided operations for univariate polynomials suffice to code an arbitrary polynomial function.

Parts of the code for the functor **Hensel** as the heart of the implementation can be found in the Appendix. The use of external functors is made explicit in the parameter list. The other functors are mainly routine implementations of the corresponding notion.

5 Example Instantiations and Tests

We have been instantiating the algorithm by providing different alternative possibilities.

Arithmetic: it is possible to use the integer arithmetic provided by the compiler or instead a package for arbitrary precision arithmetic,
Modular Arithmetic: here it is possible to experiment with different modular representation for numbers. Once p is fixed we can decide to represent the numbers either as $[0, p-1]$ (structure IntED0), or as $[-\lfloor \frac{p}{2} \rfloor, \lfloor \frac{p}{2} \rfloor]$ (structure IntED1),
Polynomials we can decide for different implementation of polynomials. In our case we decided to implement them in two different ways, by lists and arrays.

The structures IntEE0 and IntEE1 result from applying the functor EnrichED to IntED0 and IntED1, resp.

We show how the different versions of modular arithmetic are used. We are going to instantiate the functor Hensel and in this case the arithmetic we will use is given by the structure IntEE0.

```
structure E5 = Hensel(structure IEED = IntEE0
                      val p = 5
                      functor E = EnrichED
                      functor MP = modP
                      functor UP = UniPolynomialL
                      functor UPD = UniPolyDomain)
```

When we want to modify the number representation, the only necessary change is to alter this parameter. In this case **IntEE1** can be used.

If we want to work with arrays instead of lists we only have to provide the functor **UniPolynomialA** instead of **UniPolynomialL**. This will instantiate the polynomial structures implemented as arrays.

```
structure E5' = Hensel(structure IEED = IntEE1
                       val p = 5
                       functor E = EnrichED
                       functor MP = modP
                       functor UP = UniPolynomialA
                       functor UPD = UniPolyDomain)
```

Let us show now some examples of the application of the Hensel algorithm itself. We present three main examples, related to the zero of a polynomial, polynomial factorization and a root of a polynomial.

Example 1. We want to find the complex root of the equation $x^2 + 1 = 0$.

```
- val x41 = E5.newton(E5.IPol.l2p([1,0,1]), 2, 3);
val x41 = Approx [(2,-,-),(1,-,-),(2,-,-),(1,-,-)] : E5.results
```

Here the structure **E5** is the one described in the previous example. The function **newton** calls the function **hgen** that is the core of the algorithm. The function **newton** shown below, describes the shape of the polynomial function Φ and its derivative by means of a lambda abstraction.

```
fun newton(f, x0, r) =
    hgen((fn (x,g,h)=>IPol.subst(f,IPol.embed(x))),
         ((fn (x,g,h)=>IPol.subst(IPol.derive(f),IPol.embed(x))),
          (fn (x,g,h)=>IPol.ucr.zero),      (* SINCE NOT NEEDED *)
          (fn (x,g,h)=>IPol.ucr.zero)),     (* SINCE NOT NEEDED *)
         x0, IPol.ucr.zero, IPol.ucr.zero, r)
```

Since in this case the partial derivatives with respect to **g** and **h** are not used in the computation, we set them to zero.

```
- val x42 = E5.newton(E5.IPol.l2p([1,0,1]), 3, 3);
val x42 = Approx [(3,-,-),(3,-,-),(2,-,-),(3,-,-)] : E5.results
```

The results are the p-adic expansions of $-i$ and $+i$.

Example 2. For the following factorization we get an exact result.

```
val x5' = E5'.factorize(E5'.IPol.l2p([~119,309,~163,~22,12,1]),
                (E5'.IPol.l2p([2,0,0,1]),E5'.IPol.l2p([~2,2,1])), 5)
step 0
g = +1 *x^3 +2
h = +1 *x^2 +2*x -2
step 1
g = +2*x -2
h = +2*x -1
step 2
g = -1*x +1
h = +0
```

From the output of the result we can see that

$$\Phi(x, G, H) = x^5 + 12x^4 - 22x^3 - 163x^2 + 309x - 119 =$$

$$\left((x^3+2)\cdot 5^0 + (2x-2)\cdot 5^1 + (-x+1)\cdot 5^2\right)\cdot\left((x^2+2x-2)\cdot 5^0 + (2x-1)\cdot 5^1 + 0\cdot 5^2\right)$$

Example 3. The example of an exact root is shown below:

```
- val x14 = E5.root((E5.IPol.l2p([49,126,95,18,1]), 2),
                                       E5.IPol.l2p([2,4,1]), 3)
exact result
step 0
g = +1 *x^2 +4*x +2
h = +1
step 1
g = +1*x +1
h = +0
```

Again we can verify that:

$$\Phi(x, G, H) = x^4 + 18x^3 + 95x^2 + 126x + 49 = \left((x^2 + 4x + 2) \cdot 5^0 + (x + 1) \cdot 5^1\right)^2$$

Note that in the last two examples the polynomials have been represented by a list of coefficients, starting with the least significant one, where this is the external form of coding polynomials and it does not need to coincide with the internal representation.

6 Conclusions and Future Work

We have demonstrated the useful application of a high-level functional programming language to an abstract algebraic algorithm, the p-adic lifting. The implementation benefits from the functional style as well as from the parametric modularity of the program. SML provides means to describe the procedures on the right level of abstraction.

Furthermore this flexibility allows different instantiations of the overall program. Different implementations of integers can be used (built-in or arbitrary precision) as well as different forms of modular arithmetic (different normal forms). Also the concrete implementation of polynomials can be altered using lists or arrays for example.

Because of its clear typing this implementation can also be used for teaching the Hensel algorithm. Different implementation techniques for polynomials are another possible example. The availability of various instantiations allows a variety of practical tests.

After a detailed performance analysis we plan to extend this kernel to a SML library for specialized computations with algebraic numbers. This performance analysis should not only investigate the use of different implementation methods. Also the influence of generic programming on the performance has to be considered.

As an alternative such an implementation could be contrasted with a realization of computable complex numbers in the style of [Vui87]. Again both approaches have to be analyzed. In form of libraries both are highly desirable for performing exact scientific computations.

References

[Lau83] M. Lauer. Computing by Homomorphic Images. In R. Loos, B. Buchberger, and Collins E.G., editors, *Computer Algebra - Symbolic and Algebraic Computation*. Springer-Verlag, 1983.

[Lim93] C. Limongelli. *The Integration of Symbolic and Numeric Computation by p-adic Construction Methods*. PhD thesis, Università degli Studi di Roma "La Sapienza", 1993.

[Lip81] J.D. Lipson. *Elements of Algebra and Algebraic Computing*. Addison-Wesley Publishing Company, 1981.

[MT91] R. Milner and M. Tofte. *Commentary on Standard ML*. MIT Press, 1991.

[MTH90] R. Milner, M. Tofte, and R. Harper. *The Definition of Standard ML*. MIT Press, 1990.

[Pau91] L.C. Paulson. *ML for the Working Programmer*. Cambridge University Press, 1991.

[Rep91] J.H. Reppy. CML: A Higher-Order Concurrent Language. In *SIGPLAN'91 Conference on Programming Language Design and Implementation*, pages 293–305, June 1991.

[San89] D. Sannella. Formal Program Development in Extended ML for the Working Programmer. Technical Report ECS-LFCS-89-102, LFCS, Department of Computer Science, University of Edinburgh, December 1989.

[San95] P.S. Santas. A Type System for Computer Algebra. *Journal of Symbolic Computation*, 19:79–109, 1995.

[Tof94] M. Tofte. Principal Signatures for Higher-Order Program Modules. *Journal of Functional Programming*, 4(3):285–335, July 1994.

[Vui87] J. Vuillemin. Exact Real Computer Arithmetic with Continued Fractions. Technical report, INRIA, Rocquencourt, France, 1987.

[Yun74] D.Y.Y. Yun. *The Hensel Lemma in Algebraic Manipulation*. PhD thesis, Massachusetts Institute of Technology, November 1974.

A Partial Code of the Functor Hensel

```
functor Hensel(structure IEED : EnrichedEuclideanDomain
               val p : IEED.ucr.$
               functor E : Enrich
               functor MP : ModP
               functor UP : UniPolys
               functor UPD : UniPolyd) : Hensel =
    struct
        structure IEED = IEED
        structure IPol = UP(structure UCR = IEED.ucr)
        structure MODP = MP(structure IEED = IEED
                            val prime = p)
        structure MPOL = UPD(structure CF = MODP
                             functor UP = UP)
        structure MPol = E(structure ED = MPOL)
        type integer = IEED.ucr.$
        type intpoly = IPol.ucr.$
        (* ... *)
        datatype 'a result =
            (* distinguish the type of the result *)
            Exact of 'a
          | Approx of 'a
        type results = ((integer * intpoly * intpoly) list) result
        fun taylor (x, g, h) Phi Pder pp =
            let val (pdx, pdg, pdh) = Pder
            in
                ((IPol.map (fn xx => IEED.div(xx,pp)) (Phi(x, g, h))),
                 (pdx(x, g, h)), (pdg(x, g, h)), (pdh(x, g, h)))
            end
```

```
    fun solve (x, g, h) (phi, dx, dg, dh) =
        if (IPol.ucr.eq(g,IPol.ucr.zero) andalso
            IPol.ucr.eq(h,IPol.ucr.zero)) then
            (Convert.i2m(IEED.ucr.times
                         (IEED.einv(IPol.project(dx), p),
                          IEED.ucr.neg(IPol.project(phi)))),
             IPol.ucr.zero, IPol.ucr.zero)
        else
            if IPol.ucr.eq(h, IPol.ucr.one) then
                (x,
                 Convert.mp2ip(MPol.div(MPol.ucr.neg
                                        (Convert.ip2mp(phi)),
                                        Convert.ip2mp(dg))),
                 IPol.ucr.zero)
            else
                let val (g´, h´) =
                    MPol.dioph(Convert.ip2mp(dg),
                               Convert.ip2mp(dh),
                               MPol.ucr.neg(Convert.ip2mp(phi)))
                in
                    (x, Convert.mp2ip(g´), Convert.mp2ip(h´))
                end

    (* global entrance point for all calls *)
    fun hgen(Phi, Pder, x0, g0, h0, r) =
        (* Phi as a function in x, g, and h
         F and n are implicitly coded by Phi
         Pder as a triple of functions in x, g, and h
         representing the partial derivatives wrt x, g, h, resp.
         *)
        let fun aux (x, g, h) r Res pp =
            let val exact = IPol.ucr.eq((Phi(x,g,h)), IPol.ucr.zero)
            in
                if (r = 0) then
                    if exact then Exact(Res)
                    else Approx(Res)
                else
                    if exact then Exact(Res)
                    else
                        let val phi_exp = taylor (x, g, h) Phi Pder pp
                            val (xs, gs, hs) = solve (x, g, h) phi_exp
                            val nx = IEED.ucr.plus(x,
                                     IEED.ucr.times(xs, pp))
                            val ng = IPol.ucr.plus(g,
                                     IPol.ucr.times(gs,IPol.embed(pp)))
                            val nh = IPol.ucr.plus(h,
                                     IPol.ucr.times(hs,IPol.embed(pp)))
                        in
                            if (IPol.ucr.eq((Phi(nx,ng,nh)),
                                            IPol.ucr.zero)) then
                                Exact(Res @ [(xs, gs, hs)])
                            else
                                aux (nx, ng, nh) (r - 1)
                                (Res @ [(xs, gs, hs)])
                                (IEED.ucr.times(pp, p))
                        end
            end
        in
            aux (x0, g0, h0) (Integer.abs(r)) [(x0, g0, h0)] p
        end
     (* ... *)
end
```

Conditional Categories and Domains

Philip S. Santas

Institute for Scientific Computation, ETH Zurich, 8050 Switzerland

Abstract. We extend the Type System defined in [San95] with Axiom-like Conditional Categories with the additional property of Static Typing and Checking. Categories and Domains may contain conditionals in their bodies, which are elaborated by our compiler by techniques used in standard typing. We define an appropriate calculus and discuss its properties. Examples inspired by the Axiom library illustrate the power of our approach and its application in constructing algebraic concepts. The full calculus has been implemented and tested with our LA compiler which generates executable files.

1 Introduction

Various relations and constructions in computer algebra involve the combination of several algebraic domains. For instance a univariate polynomial domain in some indeterminate over a commutative ring is a commutative ring with additional functions for computation of the leading coefficient, etc. Such domain is generally implemented by defining a functor *Poly* which takes a commutative ring D as argument, and returns another ring (with some additional operations such as constructors and selectors) of type PolyCat(D) (where $PolyCat(D)$ is an extension of the category $CRing$). When the functor is defined, it can be applied to any desired ring resulting in the intended domain. If a domain is an instance of a category σ, it is also an instance of the categories which σ extends (i.e. of all the supertypes of σ)

Since addition, multiplication, 0 and 1 on polynomials are defined in terms of the corresponding operations in the coefficient ring D, the properties of the resulting univariate polynomial domain $D[x]$ depend on the properties of D ([GCL92]). Thus if D is a unique factorization domain, then $D[x]$ is also a unique factorization domain. This property can be expressed by constructing another functor from UFD to $UFDPolyCat(D)$, such that every instance of the type $PolyUFDCat(D)$ is also an instance of the type $PolyCat(D)$.

Unfortunately, additional functors and types have to be defined for constructing polynomial domains over various algebras which assume the axioms[1] for rings. Even worse, if D assumes some additional specifications, such as ordering, etc. it is useful to express this in the resulting polynomial domain: then the number of functor definitions increases dramatically, for something that algebraists use only one definition.

---

[1] When we examine types in the rest of the paper, we refer to axioms as specifications

1.1 Conditional Categories and Domains in Axiom

Systems like Axiom, *A♯* [WBDI+, JS92] and Gauss [Mon94] allow the user to introduce all the desired properties and assumptions in a single type (category) and functor. Certain assumptions about the functor arguments can be expressed with the use of conditional category and conditional domain definitions. The previous two types and functors can be merged into one type and one functor:

```
PolyCat(r:CRing):Category = CRing with {
     if r has UFD then UFD,
     monomial: r -> N -> %,
     lCoeff: % -> r };
Poly(r:CRing):PolyCat(r) = {
     Rep = List(N, r),
     0 : % = [],
     + : % -> % -> % = ...,
     * : % -> % -> % = ...,
     monomial(r)(n) = ...,
     lCoeff(p) = ...,
     if r has UFD then { gcd: % -> % -> % = ...,
                         factor: % -> List(%) = ... } }
```

The test *r has UFD* checks if R is actually an instance of the category UFD. If true, the type $PolyCat(r)$ assumes the specifications of UFD, otherwise it doesn't. Although r is specified as $CRing$, it is possible to infer some assumptions which allow it to be of a more specific type: consider the application of *PolyCat* to a UFD. This implies that specifications are not the only source of information for an identifier; various program dependencies can enrich its type. The primitive *has* can be viewed as an infix operator with the type $has : Domain \to Category \to Boolean$ in some higher order system that allows the passing of categories and domains in functions that return values. The same test exists in the body of the functor *Poly*: if true, the implementations of the functions *gcd* and *factor* can be added.

The most intriguing part of this approach is that conditional expressions of the form $if - then$ statements (eventually with an *else* component) are allowed in type expressions. From a type theoretic point of view it would be nonsense to allow types based on conditionals, for they could be one type in one line of a program, and another in some other below. If conditionals remain inside types, then typing is not guaranteed to be sound, and arbitrary tests are not always decidable. In fact, Axiom treats conditional categories in an unsound way and *A♯* relies on run-time tests.

1.2 Desired Features

No systematic account for conditional types and categories has been given in the literature. The only description of conditional categories is given in the book of Axiom [JS92]; however, type theoretic issues have not been handled. On the other hand, the majority of the papers on type-checks concentrate on dynamic typing, which is a different problem: here we examine conditional type definitions and not just tests whether some variable has some ordinary type. Additionally, conditional types are important for constructions in computer algebra as the previous example (which is a miniature of the tremendous application of conditional categories in the Axiom library and Gauss) shows.

In this paper we develop a calculus for conditional types (conditional dependent sums and products) with the hope to introduce a reference for designers and users of computer algebra systems. This calculus is an extension of the system described in [San95] with the following features:

1. Conditional types (conditional sums) which extend the notion of dependent sums and products. The tests involve membership of a domain in a category or sharing between two domains and can be decided at compile time (testing the value of constants can be viewed as testing for sharings). This makes necessary the introduction of rules for removing the dependencies (and conditionals) from types, in order to ensure sound typing.
2. Definition of a subtyping relation among conditionals so that abstract definitions can be properly typed, even if there is not enough information for the elimination of dependencies.
3. Stratification of the type system in two universes so that types are elaborated and decided at compile time. Although domains are first class values, types remain strictly second class.
4. Allowence for abstraction and separate compilation of modules.

The syntax of the abstract language (simplified for the purposes of this paper) is the following (optional syntactic components are enclosed in <>):

expressions values/types:	e	$::= x \mid (x : \tau) \mapsto e \mid e\, e' \mid \omega.x \mid \tau$
(modules as values)		$\mid \mathsf{add}\{< dec >\}$
access:	ω	$::= x \mid \omega.x$
types and type constructors:	τ	$::= t \mid T \mid \sigma \mid arr \mid \omega.t \mid (x : \tau_1) \mapsto \tau_2 \mid \tau\, e$
function types	arr	$::= \tau_1 \to \tau_2 \mid \Pi(x : \tau_1).\tau_2$
categories/signatures	σ	$::= \mathsf{with}\{< sdec >\}$
value/type declarations	dec	$::= id = exp \mid id(x : \tau) = e \mid dec_1, dec_2$
(conditional)		$\mid \mathsf{if}\ access\ \mathsf{has}\ \tau\ \mathsf{then}\ strdec$
signature declarations:	$sdec$	$::= id <: tau > \mid sdec_1, sdec_2$
(conditional)		$\mid \mathsf{if}\ access\ \mathsf{has}\ \tau\ \mathsf{then}\ sdec$
(type/value sharing)		$\mid t = \tau \mid x = \omega$
identifiers	id	$::= x \mid t$

The calculus of the presented type system is given as usually by a set of selected rules. Each rule has a number of antecedent judgements above a horizontal line (optional) and a conclusion judgement below the line. Each judgement has the form $E \vdash \mathcal{A}$, for an environment E and an assertion $\mathcal{A}$, depending on the judgement. Environments contain typing assumptions for variables, type variable declarations and subtyping assumptions. Typical judgements are $E \vdash x$ asserting that x is definable in E, $E \vdash x : \sigma$ meaning that x has type σ under assumptions E and $E \vdash \sigma' \sqsubseteq \sigma$ asserting that σ' is a subtype of σ (or equivalently, σ is a supertype of σ') in E; we demand from all the environments to be well formed. An environment can be extended with new assumptions as in $E, x : \tau; E' \vdash x : \tau$; this environment is well formed provided that $E \vdash \tau$ if there is no other assumption for x in E, E'; variables can be renamed in case of conflicts; The overline on dependent sum types is a striping operator which removes the enclosing brackets. The verbose keywords *add* and *with* are not used in the calculus. It must be clear from the context which expressions refer to types and which to modules.

2 Dependencies and Conditionals

[ΣTHIN]	$\frac{\forall i{\in}I.\,\exists j{\in}j.\ E,\overline{\sigma_j}_{\forall j \in J} \vdash \sigma_j \sqsubseteq \sigma_i}{E \vdash \{\forall j{\in}J.\overline{\sigma_j}\} \sqsubseteq \{\forall i{\in}I.\overline{\sigma_i}\}}$	[ΣVAL]	$\frac{E \vdash \tau' \sqsubseteq \tau}{E \vdash x{:}\tau' \sqsubseteq x{:}\tau}$
[ΣNEST]	$\frac{E \vdash \sigma_1' \sqsubseteq \sigma_1 \quad E,x{:}\sigma_1' \vdash \sigma_2' \sqsubseteq \sigma_2}{E \vdash \{x{:}\sigma_1', \sigma_2'\} \sqsubseteq \{x{:}\sigma_1, \overline{\sigma_2}\}}$	[ΣINST]	$\frac{E \vdash e{:}\tau \quad E,x{:}\tau \vdash e'{:}\sigma}{E \vdash \{x{=}e, e'\}{:}\{x{:}\tau, \overline{\sigma}\}}$
[ΣMER]	$\frac{E \vdash \{\overline{\sigma_1}, \overline{\sigma_2}, \overline{\sigma_3}\},\ \{\overline{\sigma_1}, \overline{\sigma_3}\} \sqsubseteq \sigma_2}{E \vdash \{\overline{\sigma_1}, \overline{\sigma_2}, \overline{\sigma_3}\} \equiv \{\overline{\sigma_1}, \overline{\sigma_3}\}}$	[ΣACC]	$\frac{E \vdash e{:}\{x{:}\tau\}}{E \vdash e.x{:}\tau}$
[ΠINST]	$\frac{E;x{:}\sigma \vdash e{:}\sigma'}{E \vdash (x{:}\sigma) \mapsto e : \Pi.(x{:}\sigma).\sigma'}$	[ΣSEQ]	$\frac{E \vdash \tau \equiv \tau'}{E \vdash \tau \sqsubseteq \tau'}$
[ΠSUB]	$\frac{E \vdash \sigma' \sqsubseteq \sigma \quad E;x{:}\sigma' \vdash \tau \sqsubseteq \tau'}{E \vdash \Pi(x{:}\sigma).\,\tau \sqsubseteq \Pi(x{:}\sigma').\,\tau'}$	[ΣFOR]	$\frac{E \vdash \tau \quad E;t{=}\tau \vdash \diamond}{E \vdash t{=}\tau \sqsubseteq t}$
[ΠAPP]	$\frac{E \vdash x_1{:}\Pi(x{:}\sigma_1).\sigma_2,\ x_2{:}\sigma_1,\ [x_2/x]\sigma_2 \sqsubseteq \sigma}{E \vdash x_1(x_2){:}\sigma}$	[SUB]	$\frac{E \vdash x{:}\tau' \quad E \vdash \tau' \sqsubseteq \tau}{E \vdash x{:}\tau}$
[τAPP]	$\frac{E \vdash x'{:}\tau' \quad E,x{=}x' \vdash \tau \equiv \tau''}{E \vdash ((x{:}\tau') \mapsto \tau)\ x' \equiv \tau''}$	[SUBTR]	$\frac{E \vdash \tau_1, \tau_3, \tau_1 \sqsubseteq \tau_2, \tau_2 \sqsubseteq \tau_3}{E \vdash \tau_1 \sqsubseteq \tau_3}$
[EQTRE]	$\frac{E \vdash \tau}{E \vdash \tau \equiv \tau} \quad \frac{E \vdash \tau_1, \tau_3, \tau_1 \equiv \tau_2, \tau_3 \equiv \tau_2}{E \vdash \tau_1 \equiv \tau_3}$	[FORMΣ]	$\frac{E,\overline{\sigma} \vdash \tau \equiv \tau' \quad E \vdash \{\overline{\sigma}, x{:}\tau, \overline{\sigma'}\}}{E \vdash \{\overline{\sigma}, x{:}\tau, \sigma', x{:}\tau'\}}$
[FORM]	$\frac{E \vdash e{:}\sigma \quad E,\overline{\sigma} \vdash e{:}\tau \quad x{:}\tau' \notin \overline{\sigma}}{E \vdash \{\overline{e}, x{=}e\}{:}\{\overline{\sigma}, x{:}\tau\}}$	[FORMx]	$\frac{x{:}\tau \in E}{E \vdash x{:}\tau}$

Fig. 1. General Subtyping and Formation rules for Σ and Π-types

Our system supports dependent sum (**translucent sums**) and product types (known as Σ and Π types) with sharing information, which are extensions of the record and function (arrow) types respectively. Their definition and properties have been discussed extensively in the literature (for some recent results with many references to previous research, read [HM93, MQ86, HL94, Ler94, San95]). Dependent sum types correspond to **categories**, and dependent product types to **category constructors** in Axiom [San95]. For example, a monoid M can be defined by a dependent sum as:

$Monoid = \{Rep : Type, 0 : Rep, + : Rep \to Rep \to Rep\}$
$M : Monoid = \{Rep = (Int, Int), 0 = (0,0), + = \lambda(x_1, y_1).\lambda(x_2, y_2).(x_1 + x_2, y_1 + y_2)\}$

A sum type σ_1 is subtype of σ_2 if σ_1 declares at least as many fields as its supertype σ_2; concerning the common fields, the fields declaring products and sums in σ_1 must assume types which are subtypes of those in the corresponding fields in σ_2. The typing rules for dependent sums are given in Fig. 1. Most of them are standard, with the exception of the normalization rule ΣMER which merges dependent sums with multiple occurrences of the same component, by normalizing the sum type $\{t : Monoid, t : Ring, x : t\}$ to $\{t : Ring, x : t\}$[2]

For dependent product types we assume the additional properties mentioned in [San95]. Thus a product type can be viewed as a type constructor which can be applied to other sums or products. The functor *Poly* from the introduction forms a dependent product while the category constructor *PolyCat* can be used as the dependent product type of the functor *Poly*:

$$PolyCat = \Pi(D : CRing).\{\overline{CRing},\ monomial : D.Rep \to N \to Rep,\ ...\}$$
$$Poly : PolyCat = ...$$

Application of *PolyCat* results in a sum type.

Subtyping in product types involves contravariance in the type of the argument and covariance in the type of the result (rule ΠSUB). A Π-type π_1 is subtype of π_2, if the type of its argument is supertype of the type of the argument of π_2, and the type of module it returns is subtype of the type of the return module of π_2.

Conditional types include a condition and one branch which defines another type based on the conditional assumptions. We refer to conditional types as $?(x : \sigma_1).\sigma_2$ for the statement *if x has* σ_1 *then* σ_2. Thus the category *PolyCat* from the second part of the introduction is defined as:

$$\Pi(r : CRing).\{\overline{CRing}, ?(r : UFD).UFD, monomial : r.Rep \to N \to Rep, ...\}$$

Additionally, we introduce a conditional for testing identity of two domains and we notate it by $?(x = x').\sigma$; this stands for the type *if x is x' then* σ and is the conditional equivalent of the notion of sharings in the sense that *has*-conditionals correspond to declarations. The above notation for conditional categories resembles that of dependent products $(\Pi(x : \sigma_1).\sigma_2)$; this resemblance involves type theoretic issues too.

In order to avoid meaningless tests and arbitrary substitutions (as we see later on), conditional tests can be performed only on names (access ω in the grammar), otherwise there is nothing that can be shared. If a name has been declared (ie. it has a concrete implementation), then it is represented with upper case characters and we refer to it is concrete name while names which have been only specified are indicated with lower case characters (variables). In the rest of tha paper we substitute ω with x or X.

The introduction of names in our calculus can increase the flexibility of the system. Given the definition $X : \sigma = X'$, where X' is concrete, and $X' : \sigma'$ with $\sigma' \sqsubseteq \sigma$, the type of X in the rest of the program does not have to be restricted to σ, but it can be σ' as well. While this is true within a module, where we can infer all the dependencies, it is not true out of it, where the only information available is that provided by the declarations. In this way we ensure separate compilation while we handle declarations and conditionals in a uniform way.

[2] Axiom's new language A♯ considers the initial sum type as normalized, resulting in different behavior.

3 The System of Conditional Types

The rules for formation and normalization of conditional types are given in Fig. 2. In order to give a sensible account for the type $?(x : \sigma_1).\sigma_2$, it is important that the type of the domain x is expected to be an enrichment of its assumed type as the rule ?-HAS-Σ illustrates. If x hasn't been specified before the conditional, then there is nothing to test and consequently nothing to enrich, resulting in errors. Another assumption for this rule is that the types of x before and after the assumptions introduced by the conditional are dependent sum types, that is, categories; in general, the types are of the same kind.

The formation rule for conditionals involving product types is ?-HAS-Π. Here, a conditional is allowed to enrich the product type $T = \Pi(x' : \sigma'_1).\sigma'_2$ only if it assumes another product type whose argument type σ_1 is a supertype of σ'_1, and the return type σ_2 is a sum (both σ_2 and σ'_2 are sum types, which can include fields with other product types, resulting in higher order product types). In such case, x is given the type $\Pi(x' : \sigma_1).\{\overline{\sigma'_2}, \overline{\sigma_2}\}$ which, according to the rule ΠSUB in Fig. 1, is a subtype of T. This is the only rule we provide for conditional types with product type tests; the rest are similar to the numerous rules given for tests involving dependent sums which we do not repeat in this paper.

[?-HAS-Σ] $$\frac{E \vdash x{:}\sigma \quad E, x'{:}\{\overline{E(x)}, \overline{\sigma_1}\} \vdash [x'/x]\sigma_2}{E \vdash \{?(x{:}\sigma_1).\sigma_2\}}$$

[?-IS] $$\frac{E, x'_2{:}\{[x_2/x_1]\overline{E(x_1)}, \overline{E(x_2)}\} \vdash [x'_2/x_1]\sigma}{E \vdash \{?(x_1{=}x_2).\sigma\}} \quad \frac{E \vdash E(X) \sqsubseteq (x) \quad E \vdash [X/x]\sigma}{E \vdash \{?(x{=}X).\sigma\}}$$

[?-N-HAS] $$\frac{E \vdash x{:}\sigma_1 \quad E \vdash \sigma \equiv \sigma'}{E \vdash \{?(x{:}\sigma_1).\sigma\} \equiv \sigma'} \quad \frac{E \vdash E(X) \sqsubseteq \sigma_1 \quad E \vdash \sigma \equiv \sigma'}{E \vdash \{?(X{:}\sigma_1).\sigma\} \equiv \sigma'}$$

[?-N-NHAS-X] $$\frac{E \vdash E(X) \not\sqsubseteq \sigma_1}{E \vdash \{?(X{:}\sigma_1).\sigma\} \equiv \{\}}$$

[?-N-NIS-X] $$\frac{E \vdash E(X) \not\sqsubseteq (x)}{E \vdash \{?(x{=}X).\sigma\} \equiv \{\}} \quad \frac{E \vdash X_1 \not\equiv X_2 \quad E \vdash \sigma}{E \vdash \{?(X_1{=}X_2).\sigma\} \equiv \{\}}$$

[?-N-IS] $$\frac{E \vdash x \equiv x' \quad E \vdash \sigma}{E \vdash \{?(x{=}x').\sigma\} \equiv \sigma} \quad \frac{E \vdash X_1 \equiv X_2 \quad E \vdash \sigma}{E \vdash \{?(X_1{=}X_2).\sigma\} \equiv \sigma}$$

Fig. 2. Formation and Normalization of ?-Types (modulo Π-Types)

If we test the membership of an already defined (concrete) domain such as *Integer*, then there is no place for any assumption, and the test performed results in the elimination of the conditional. Rules ?-EQ-HAS and ?-NEQ-HAS perform this normalization. The test $(X : \sigma)$ succeeds if the type of X is subtype of σ, otherwise the test fails. The rules for subtyping of conditional types are discussed in Sect. 3.2.

Domain sharing tests are treated similarly (rule ?-IS): when a declared but undefined domain x is tested for equality with another undefined domain the elaboration of the conditional category proceeds with the assumption that the two domains are equal in that scope and have the same kind of types (here dependent sum types). Testing sharing with a concrete domain X treats the conditional in a similar way. If the type of X is not subtype of the type of x (rule ?-N-NIS-X), then the conditional type is discarded, since there is no possibility that x can ever share with X. The subtyping relation in the antecedent judgement, provides the additional information that the type of x can be further enriched. In this sense, $?(x = X).\sigma \ \equiv ?(x : E(X)).?(x = X).\sigma$. Testing if two already defined domains share in ?-N-IS-X and ?-N-NIS-X removes the conditional, while ?-N-IS eliminates the conditional when the system can infer that two declared domains share.

3.1 Scope of Conditional Declarations

Since the assumptions from the conditional $?(x : \sigma_1).\sigma_2$ influence only the resulting type σ_2, the declarations in σ_2 when the conditional is present, are not visible from an outer scope. The type

$$T = \Pi(x : Type).\{?(x : Ring).\{t : Type\}, x' : t\}$$

is ill defined, because a type t is introduced only if x is a *Ring*, while the field x' is introduced independently of the condition. On the other hand, the type

$$\Pi(x : Type).\{?(x : Ring).\{t : Type\}, ?(x : UFD).\{x' : t\}\}$$

should be valid, because it can be transformed to the equivalent

$$\Pi(x : Type).\{?(x : Ring).\{t : Type, ?(x : UFD).\{x' : t\}\}\}$$

However such automatic transformations are context dependent (x cannot be simply assumed as UFD without influencing the type field t if the latter depends on x) and out of the scope of this paper.

Another approach is to introduce conditionals in the dependent components. For instance, the type T can be transformed to

$$T = \Pi(x : Type).\{?(x : Ring).\{t : Type\}, x' :?(x : Ring).t\}$$

[?-DEC-HAS-Σ]
$$\frac{E, x'' : \{\overline{E(x)}, \overline{\sigma}\} \vdash [x''/x]\{dec\} : [x''/x]\sigma'}{E \vdash \{?(x{:}\sigma).\{dec\}\} : \{?(x{:}\sigma).\sigma'\}}$$

[?-IS-x]
$$\frac{E, x_2' : \{[x_2/x_1]E(x_1), E(x_2)\} \vdash \{dec\} : [x_2'/x_1]\sigma}{E \vdash \{?(x_1{=}x_2).\{dec\}\} : \{?(x_1{=}x_2).\sigma\}}$$

[?-DEC]
$$\frac{E, x'' : \{\overline{E(x)}, \overline{\sigma}\} \vdash [x''/x]\{dec\} : [x''/x]\sigma'}{E \vdash \{?(x{:}\sigma).\{dec\}\} : \{?(x{:}\sigma).\sigma'\}}$$

[?-SUB-HAS-Σ]
$$\frac{E \vdash \sigma_1 \sqsubseteq \sigma_1' \quad E, x'' : \{\overline{\sigma_1}, E(x)\} \vdash [x''/x]\sigma_2' \sqsubseteq [x''/x]\sigma_2}{E \vdash \{?(x{:}\sigma_1').\sigma_2'\} \sqsubseteq \{?(x{:}\sigma_1).\sigma_2\}}$$

[?-SUB-IS-x]
$$\frac{E, x'' : \{[x/x']E(x'), E(x)\} \vdash [x''/x']\sigma' \sqsubseteq [x''/x']\sigma}{E \vdash \{?(x'{=}x).\sigma'\} \sqsubseteq \{?(x'{=}x).\sigma\}}$$

[?-SUB-IS-X]
$$\frac{E \vdash X \equiv X' \quad E \vdash E(X) \sqsubseteq (x) \quad E \vdash [X/x]\sigma' \sqsubseteq [X/x]\sigma}{E \vdash \{?(x{=}X').\sigma'\} \sqsubseteq \{?(x{=}X).\sigma\}}$$

[?-SUB-COND-T]
$$\frac{E \vdash \sigma_2 \quad E \vdash \{?(x{:}\sigma).\sigma_2\} \sqsubseteq \{?(x{:}\sigma).\sigma_1\}}{E \vdash \sigma_2 \sqsubseteq \{?(x{:}\sigma).\sigma_1\}}$$

Fig. 3. Selected Typing and Subsumption Rules for ?-Types

If x happens to be a ring then x' has the type t. otherwise it has the type $\{\}$ according to the rule ?-N-HAS. Although this approach is sound, it demands an additional treatment of dependencies with respect to sharings, something which complicates the calculus considerably, without adding any substantial flexibility to the type system. In the rest of the paper we do not assume any such technique.

An important consequence of the above is that conditional components cannot be accessed. If they were accessed, then we violate scoping issues. Thus only components which have no conditional type can be accessed.

3.2 Typing of Conditional Expressions

Conditional types are used mainly for the declaration of conditional domains and functors. Referring to the example in Sect. 1.1 we conclude that the functor *Poly* has the type *PolyCat*. In order to support this in our calculus, we need rules for typing of domains and functors, as in Fig. 3. In this paper we provide only two rules for typing. The rest can be easily constructed from the introduction and normalization rules defined in Fig. 2.

The rules for subtyping of conditional types are based on the constraints specified in the conditional. Conditional types which test for membership are contravariant in their conditional and covariant in their result type (rule ?-SUB-HAS) Intuitively the type of $DF_1 : \Pi(x).?(x : CRing).CRing$ is a subtype of $DF_2 : \Pi(x).?(x : UFD).CRing$ because the former adds *CRing* components only if x is a ring while the latter having a stronger condition, introduces additional tests for checking if x is a *UFD*. On the same spirit, if a conditional category is subtype of another, the resulting type of the former should be a subtype of the latter.

From the above rule, various other rules are derivable. Comparison of a conditional type and a well formed non-conditional type reduces to comparison of two conditional types (rule ?-SUB-COND-T). The treatment of conditional sharings is similar (rules ?-SUB-IS-x, ?-SUB-IS-X and ?-SUB-COND-T).

More typings are allowed with the equivalence rules we give in Sects. 3.3.

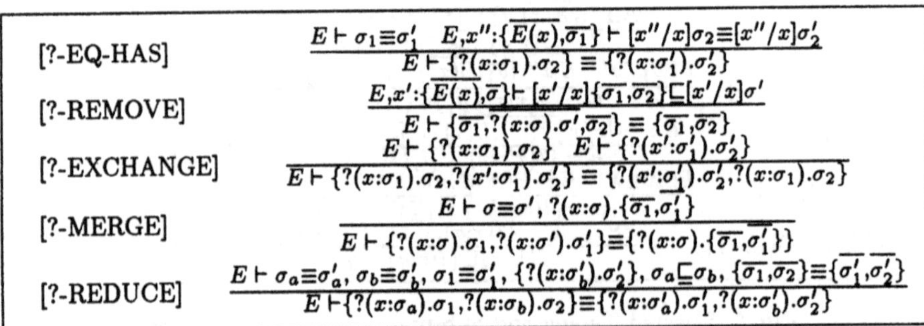

[?-EQ-HAS]	$\dfrac{E \vdash \sigma_1 \equiv \sigma_1' \quad E, x'':\{\overline{E(x)}, \overline{\sigma_1}\} \vdash [x''/x]\sigma_2 \equiv [x''/x]\sigma_2'}{E \vdash \{?(x{:}\sigma_1).\sigma_2\} \equiv \{?(x{:}\sigma_1').\sigma_2'\}}$
[?-REMOVE]	$\dfrac{E, x':\{\overline{E(x)}, \overline{\sigma}\} \vdash [x'/x]\{\overline{\sigma_1}, \overline{\sigma_2}\} \sqsubseteq [x'/x]\sigma'}{E \vdash \{\overline{\sigma_1}, ?(x{:}\sigma).\sigma', \overline{\sigma_2}\} \equiv \{\overline{\sigma_1}, \overline{\sigma_2}\}}$
[?-EXCHANGE]	$\dfrac{E \vdash \{?(x{:}\sigma_1).\sigma_2\} \quad E \vdash \{?(x'{:}\sigma_1').\sigma_2'\}}{E \vdash \{?(x{:}\sigma_1).\sigma_2, ?(x'{:}\sigma_1').\sigma_2'\} \equiv \{?(x'{:}\sigma_1').\sigma_2', ?(x{:}\sigma_1).\sigma_2\}}$
[?-MERGE]	$\dfrac{E \vdash \sigma \equiv \sigma', \ ?(x{:}\sigma).\{\overline{\sigma_1}, \sigma_1'\}}{E \vdash \{?(x{:}\sigma).\sigma_1, ?(x{:}\sigma').\sigma_1'\} \equiv \{?(x{:}\sigma).\{\overline{\sigma_1}, \sigma_1'\}\}}$
[?-REDUCE]	$\dfrac{E \vdash \sigma_a \equiv \sigma_a', \ \sigma_b \equiv \sigma_b', \ \sigma_1 \equiv \sigma_1', \ \{?(x{:}\sigma_b').\sigma_2'\}, \ \sigma_a \sqsubseteq \sigma_b, \ \{\overline{\sigma_1}, \overline{\sigma_2}\} \equiv \{\overline{\sigma_1'}, \overline{\sigma_2'}\}}{E \vdash \{?(x{:}\sigma_a).\sigma_1, ?(x{:}\sigma_b).\sigma_2\} \equiv \{?(x{:}\sigma_a').\sigma_1', ?(x{:}\sigma_b').\sigma_2'\}}$

Fig. 4. Equality rules for ?-types

3.3 Equivalence

Except for the normalization rules of Fig. 2, we introduce more rules in Fig. 4 for equality of conditional types, in order to accept the typing of more expressions.

Syntactic equality of conditional types is defined in ?-EQ-HAS and ?-EQ-IS.

A rule similar to ΣMER is ?-REMOVE: conditional types can be removed from a sum type if they do not offer any additional fields. Notice that the condition $E; x : \sigma \vdash \{\overline{\sigma_1}, \overline{\sigma_2}\} \sqsubseteq \sigma_2$ does not add any extra information in the bodies of σ_1 and σ_2. If σ_1 and σ_2 make tests such as $x : \sigma$, the extra information is welcome, because we would have to add $x : \sigma$ anyway; in the absence of such tests, the enrichment of the type of x with σ is ignored completely. Tests involving subtypes of σ succeed, but the final type $\{\overline{\sigma_1}, \overline{\sigma_2}\}$ does not include the condition $x : \sigma$.

If conditionals define local scope, they can be interchanged when they have adjoining positions in a dependent sum type (rule ?-EXCHANGE). If both conditionals are the same, then their branches can be merged (rule ?-MERGE) provided they are well formed (this rule calls the FORM rules).

The rule ?-REDUCE can eliminate fields from a conditional type, if these fields are defined in an adjoining conditional type involving a weaker condition. Here we have to examine if the type $?(x : \sigma_b').\sigma_2'$ is well-formed, for σ_2 can be a dependent sum: if some of its components are removed resulting in σ_2', we have to establish that this transformation does not violate the dependencies. Consider the case:

$F : \Pi(x : \{\}).\{?(x : Ring).\{t : Type, x : t\}, \{?(x : UFD).\{t : Type, x : t, y : t\}\}$

Here $t : Type$ cannot be removed from the second conditional. Finally we remark that the rule ?-REMOVE can be derived from the rule ?-REDUCE if we introduce the (dummy) test $?(x : \sigma_a)$ to the adjoining types.

3.4 Removing Conditionals

A conditional category involves a dependency between a type and a domain. In order to provide a concrete account for the property of an inclusion in a category, the dependency has to be removed. Assume that the type of the functor *Poly* is $Poly : \Pi(x : CRing).PolyCat(x)$, denoting a dependent product type. In order to apply *Poly* to the domain R (say *Integer*) of type *CRing*, all the dependencies have to be removed/discharged, otherwise we are left with a variable x the scope of which is only within the functor *Poly*. The steps followed involving the coercion of the dependent type of *Poly* to an arrow type are:

$Poly : \Pi(r : CRing).Poly_{Cat}(r)$
$Poly : \Pi(r : CRing = Integer).Poly_{Cat}(r)$
$Poly : (r : CRing = Integer) \rightarrow Poly_{Cat}(Integer)$
$Poly : CRing \rightarrow Poly_{Cat}(Integer)$

PolyInt has the category $Poly_{Cat}(Integer)$. The same happens for Σ-types:

```
M: with {r:CRing, n:PolyCat(r)} = ...;  PolyInt = n$M
```

Due to the manipulation of names in our system, the type of M is coerced to a non-dependent sum:

$M : \{r : CRing, n : Poly_{Cat}(r)\}$
$M : \{r : CRing = r\$M, n : Poly_{Cat}(r)\}$
$M : \{r : CRing = r\$M, n : Poly_{Cat}(r\$M)\}$

PolyInt has the category $Poly_{Cat}(r\$M)$.

Similar transformations can be applied on conditional categories. Expanding the definition of $Poly_{Cat}$ we discharge the type of the functor *Poly*:

$Poly : CRing \rightarrow Poly_{Cat}(Integer)$
$Poly : CRing \rightarrow (\Pi(r : CRing).\{\overline{CRing}, ?(r : UFD).\overline{UFD}, ...\})(Integer)$
$Poly : CRing \rightarrow \{\overline{CRing}, ?(Integer : UFD).\overline{UFD}, ...\})(Integer)$
$Poly : CRing \rightarrow \{\overline{CRing}, \overline{UFD}, ...\}$

assuring that the resulting polynomial domain forms a unique factorization domain. If *Poly* was applied to a domain R not satisfying UFD, then its type would be

$Poly : CRing \rightarrow \{\overline{CRing}, monomial : R \rightarrow N \rightarrow \%,\ lCoeff : \% \rightarrow R\}$

resulting in a commutative ring.

At this point we examine an example of detected unsoundness involving product dependencies and removal of conditionals, so that a type is coerced to its subtype! Consider the functions:

```
f1(d1:{},d2:{if d1 has Ring then Ring}):{if d1 has UFD then Ring} = d2;
f2(d1:{},d2:{if d1 has UFD then Ring}):{if d1 has {} then Ring} = d2;
f3(d1:{},d2:{if d1 has UFD then Ring}):Ring = d2
```

The definition of f_1 involves the subtyping rules we have discussed, and consequently it is accepted. However, the definition of f_2 has the reverse typing: d_2 seems to be coerced to a subtype since $?(d_1 : Type).Ring \sqsubseteq ?(d_1 : UFD).Ring$. One can argue that in the declaration of d_2, d_1 is tested for membership in UFD. Since d_2 is the returned domain, the condition on d_1 still holds for the output. Thus, d_1 is an instance of *Ring*, if it is an instance of UFD, and the test $?(d_1 : Ring)$ returns true. If we assume that d_1 satisfies UFD, then we can also assume that d_1 satisfies *Type*, allowing the conditional in the output to be removed entirely as in f_3: allowing this, permits the definition of a casting operator on domains!

3.5 Maintaining Soundness

The rules provided so far handle dependent and conditional sums and products. Our language does not allow explicit declarations of conditional types (although declarations of conditional components are allowed), while the rule ΣACC and the handling of scoping on conditionals assure that a conditional component cannot be accessed unless the condition is removed by any of the normalization rules. The rule FORMΣ ensures that repeated declarations of the same component will have the same type. Thus all the accessed components have well defined non-conditional types, and the system reduces to standard typing.

However, conditionals have effects similar to introduction of overloading in the system. Thus constants, variables, and module components have to be named

$$[\Sigma\text{THIN}]\quad \frac{\forall i\in I.\ \exists j\in J.\ E;([\overline{\sigma_j}])_{\forall j\in J\vee J'}\vdash\overline{\sigma_j}\sqsubseteq\overline{\sigma'_i}\quad \forall i\in I'.\ \exists j\in J\vee J'.\ E;([\overline{\sigma_j}])_{\forall j\in J\vee J'}\vdash\overline{\sigma_j}\sqsubseteq\overline{\sigma_i}}{E\vdash(\{\overline{\sigma_j}\}_{\forall j\in J},\{\sigma'_{j'}\}_{\forall j'\in J'})\sqsubseteq(\{\overline{\sigma_i}\}_{\forall i\in I},\{\overline{\sigma'_{i'}}\}_{\forall i'\in I'})}$$

$$[\Sigma\text{NEST}]\quad \frac{E\vdash\sigma'_1\sqsubseteq\sigma_1\quad E;x{:}\sigma'_1\vdash\overline{\sigma'_2}\sqsubseteq\overline{\sigma_2}\quad E;x{:}\sigma'_1\vdash\overline{\sigma'_2,\sigma'}\sqsubseteq\overline{\sigma_2,\sigma}}{E\vdash(\{x{:}\sigma'_1,\sigma'_2\},\sigma')\sqsubseteq(\{x{:}\sigma_1,\overline{\sigma_2}\},\sigma)}$$

Fig. 5. Modified rules for support of Hidden Components

not only with their syntactic name, but also a signature of their type. Implementations which ignore this fact tend to be unsound. An example will clarify the case:

```
f1(d:Type) = add{if d has with{f:Int->Int} then x=(f$d)(3)}
```

Here we test if the argument d has a function component f. Assuming that a domain is implemented as a list (or any equivalent structure) of pairs of the form $(name, value)$, then an implementation will have to find a pair whose first component is f (or any other transformation of f), in order to retrieve its value. However if d includes a f component whose type is not the required (it could be $f : Int$ for example), then errors may occur at run-time. In such case, static compilation is no help, unless it enriches names with a type signature. However such a restriction poses difficulties on the naming of variables whose type is generic, ie. with a type including type variables.

In order to avoid decoration of names with a type signature, we include the required type for f in the type of f_1, and then test if the values to which f_1 is applied include any component f and if so, whether the type of this component is the required one.

Only one change is needed. Dependent sum types in the static analysis were one list with the types of their components. By adding a second list with the required types for eventually **hidden components**, type checking can be performed statically in case the type of a module includes such components. Propagation of the otherwise hidden type information is crucial. Therefore the second list shouldn't be thinned as it happens in the presence of subtyping. If the first list is thinned, then the removed components have to be passed to the second list.

Subtyping is extended so that the additional components of the subtype have to be subtypes of the hidden components of the supertype, in case these are specified. Similar for the hidden components of the subtype, which have to be subtypes of the corresponding hidden components of the supertype. The second list of the subtype is enriched with the hidden components of the supertype, in order to propagate all the information. This enrichment is a form of updating a type in the presence of type inference, not easily expressible in a calculus: therefore, in the modification of ΣTHIN we accept that the subtype has all the hidden components of the supertype (and possibly more) which is the case after the enrichment. There is nothing unsound which such effect, because if a type did not have these hidden components, it means that there was nothing to introduce them up to this point. The modified typing rules ΣTHIN and ΣNEST are given in Fig. 5.

Some brief examples which have been compiled with the LA compiler developed by the author in ETH Zurich follow. They demonstrate the benefits of hidden components and the typing rules we provide specially in complex cases involving higher order modules. The types mentioned in the comments are inferred at compile time.

```
c1 = with{x:int}; c2 = with{x:int; y:int->int}; c3 = with{y:int};
h(f,m) = {mm = f m;
          add{ if mm has with{y:int->int} then a=(mm.y)(m.x))} }
f1(m:c1):c2 = add{ x = x$m; y = x +> x };
f2(m:c1):c1 = m;
f3(m:c1):c1 = add{ x = m.x; if m has with{y:int->int} then y = (m.y) x};
g m = add{ if m has c1 then y = m.x;
           if m has c3 then x = m.y;
           if m has with{x:int; z:int} then z = m.z + m.x };
m0 = add{ x = 3; z = 4 };
m1:c1 = add{ x = 3 };
m2:c2 = add{ x = 3; y = x +> x+1 };
m3:c1 = add{ x = 3; y = 4 };
mg1 = g(m0); mg2 = g(mg1); -- mg1: with{y:int; z:int}; mg2: with{x:int}
mh1 = h(f1,m1);            -- mh1: with{}
mh2 = h(f2,m2);            -- mh2: with{a:int}
mh3 = h(f2,m3); mh4 = h(f3,m2);  -- both rejected
```

3.6 A word on Separate Compilation

Our system supports separate compilation. Once a module has been compiled, any operation on it, or access of its components is based entirely on its type.

The *hidden components* however are also part of a module's type. Thus, if the type of non exported variables changes, so the type of the module changes. Removal of hidden components from the original code does not influence the type of the module, in the sense that the system can work in a sound way even if the type of the module includes more hidden components than otherwise specified. These are only additional constraints, on top of the existing type information.

In practice, the majority of changes in modules and modifications or extensions of algorithms implemented in a function do not involve change of their type. Thus, except for a few cases (when types of even hidden components change), separate compilation works in a conventional fashion.

4 Alternative Approaches, Further Work

There is a lack of references on conditional types in the literature of type theory and computer algebra systems. There is a brief introduction on conditionals in the reference book of Axiom [JS92, Sects. 11.8 and 12.11]. The solution of *A♯* in which conditionals are tested at run-time is unsatisfactory, not because they slow down the evaluation of the code, but because errors tend to increase exponentially. With run-time tests on types of components, one can write code in which all modules are checked at run-time: this is actually the opposite of what we want to accomplish in a typed system.

Computation of type information in dynamically typed programs which involves conditional types is discussed in [AWL94] for a *Scheme*-like language with a type system. Given the expression *if* e_1 *then* e_2 *else* e_3 the authors use conditional types to express that e_2 is evaluated only in environments where e_1 is true, and by using a well elaborated control flow analysis they can derive the best type for e_2 (and e_3). Although we examine conditional definitions and *static* properties, the results of the above paper can be reused in any extension of our system with dynamic type checks.

There has been discussion on two other concepts involving dynamic typing [ACPPR92, ACPP91, LM93] and run-time type tests and soft typing [CF91] or partial types [Tha88] The main idea here is the introduction of a type *Top* or *Dynamic* with the property that any type can be coerced to it. This means that objects can loose type information, which however, can be recovered at run time with type checks. A similar scenario has been proposed for *B*♮ with the type *User* [JT94].

Currently we implement a computer algebra library based on Axiom's library, therefore we have urgent need for proper typing with conditionals. Constructs from category algebra fit well into this framework.

A future plan is to test conditionals in proving properties of modules. Modules as first class objects can be used as terms or expressions, while conditionals propagate information statically; the run-time system can provide results depending on the values stored in each module. In this way theorems can be proved.

I thank Manuel Bronstein, Richard Jenks, Barry Trager and Stephen Watt for support and discussions on the Axiom system.

References

[ACPP91] M. Abadi, L. Cardelli, B. Pierce, G. Plotkin. Dynamic Typing in a Statically Typed Language. ACM Transactions on Programming Languages and Systems. **13(2)**. April 1991.

[ACPPR92] M. Abadi, L. Cardelli, B. Pierce, G. Plotkin, D. Remy. Dynamic Typing in Polymorphic Languages. In Proc. the ACM Workshop on ML. ACM. June 1992.

[AWL94] A. Aiken, E.L.Wimmers, T.K. Lakshman. Soft Typing with Conditional Types. In Proc. of the 21st Sumposium on Principles of Programming Languages (POPL'94). ACM Press. 1994.

[CF93] R. Cartwright, M. Fagan. Soft Typing. POPL'93. ACM Press. 1993.

[GCL92] K. O. Geddes, S. R. Czapor, G. Labahn. Algorithms for Computer Algebra. Kluwier Academic Publishers. 1992.

[HL94] R. W. Harper, M. D. Lillibridge. A Type-Theoretic Approach to Higher-Order Modules with Sharing. POPL'94. ACM Press. 1994.

[HM93] R.W. Harper, J.C. Mitchell. On the Type Structure of SML. ACM Transactions on Programming Languages and Systems. **15(2)**. pp. 211-252. 1993.

[JS92] R. D. Jenks, R. S. Sutor. Axiom: The Scientific Computation System: NAG, Springer Verlag. 1992.

[JS92] R. D. Jenks, B. M. Trager. How to make Axiom into a scratchpad. In Proc. of ISSAC'94. ACM Press. 1994.

[Ler94] X. Leroy. Manifest Types, Modules and Separate Compilation. POPL'94. pp. 109-122. ACM Press. 1994.

[LM93] X. Leroy, M. Mauny. Dynamics in ML. J. of Functional Progr., 3(4). 1993

[MQ86] D. B. MacQueen. Using Dependent Types to express Modular Structure. Proc. of POPL'86. pp. 277-286. ACM Press, Fla. 1986.

[Mon94] M. Monagan. Gauss: a Parameterized Domains of Computation System with Support for Signature Functions. Proceedings of DISCO'93. pp. 81-94. LNCS 722, Springer Verlag. 1994.

[San95] P. S. Santas. A Type System for Computer Algebra. Journal of Symbolic Computation. (19). pp 79-109. 1995.

[Tha88] S. Thate. Type Inference with Partial Types. In Automata, Languages and Programming. LNCS, 317. July 1988.

[WBDI+] S.M. Watt, P.A. Broadbery, S.S. Dooley, P. Iglio, S.C. Morrison, J.M. Steinbach, R.S. Sutor. *A♯* User's Guide. NAG. 1995.

Parameterizing Object Specifications

Martin Gogolla

Universität Bremen
Fachbereich Mathematik und Informatik
Arbeitsgruppe Datenbanksysteme
Postfach 330440, D-28334 Bremen, Germany
e-mail: gogolla@informatik.uni-bremen.de

Abstract. We present a proposal for parameterized object specifications allowing especially objects sorts in the parameter. These object specifications permit to describe the part of the world to be modeled as an object community of concurrently existing and communicating objects. Our proposal for parameter passing works well on the syntactical level by means of a pushout construction. On the semantic level we use free constructions and corresponding forgetful functors.

1 Motivation

In the software development process usually two important phases are distinguished, the requirements engineering and the design engineering phase. At the end of the requirements engineering phase, a first formal description of the system in mind should be the "contract" for further development. Therefore, the task of delivering this formal specification is of central concern.
Formal specification calls for suitable specification techniques. It is an open debate whether there exists a general description technique suitable for any system development, or whether special systems call for special specification techniques with made-to-order expressive power.
In order to support the specification of reliable information systems in an adequate way, we developed a simple language for the description of structural and dynamic properties of information system objects, called TROLL *light* [CGH92, GCH93, HCG94]. The main advantage of following the object paradigm is the fact that all relevant information concerning one object can be found within one single unit and is not distributed over a variety of locations.
TROLL *light*, a dialect of OBLOG [SSE87] and TROLL [JHSS91], incorporates ideas from algebraic specification [EM85, EM90, Wir90], semantic data models [HK87, PM88], and process theory [Hoa85, Hen88, Mil89]. Object properties are formalized in the context of so-called templates (or object types) explained here by means of `Author` objects.

```
TEMPLATE Author
  ...
  ATTRIBUTES
    Name:string; DateOfBirth:date;
    Books:LIST(book);
```

```
    ...
    EVENTS
      BIRTH create(Name:string, DateOfBirth:date);
      publishBook(Title:string,Publisher:string);
    ...
  END TEMPLATE;
```

In the database context objects of the same type are usually collected in certain containers called classes[1]. It is a distinguishing feature of TROLL *light* that the basic template concept is also used for the description of those classes, namely in the form of composite objects.

```
  TEMPLATE AuthorContainer
    ...
    SUBOBJECTS
      Authors(No:nat):author;
    ...
    EVENTS
      ...
      addAuthor(No:nat);
      removeAuthor(No:nat);
      ...
  END TEMPLATE;
```

Among other things, the given specification tells us that in the context of an object of type `AuthorContainer` authors are distinguished by natural numbers. Hence `No:nat` serves as a key.

It is clear that in the description of large object communities many containers of the given kind will arise which mainly differ only in the type of the contained objects and the used keys. This directly leads to the idea of parameterized object specifications. The aim of this paper is to find a formal foundation for such specifications.

The structure of the rest of the paper is as follows. In Sect. 2 we briefly explain non-parameterized object specifications examined in detail in [GH95]. In Sect. 3 we consider parameterization, a well-investigated technique in the area of algebraic specification. The main results are on the syntactical side a pushout construction for the application of parameterized specifications to actual ones and a combination of free and forgetful functor constructions for the semantic part. Section 4 puts our approach into the context of other recent research.

In Sect. 5 we give some concluding remarks and mention future work to be done. Due to space limitations we cannot go into details of a comparision of our approach to other recent research [GM87, GR91, Reg91, FM92, SSC92, EDS93, AZ94, DG94, EO94, GD94, PPP94] but must refer for this topic to [GH95].

[1] This notion of class is different from the notion of class in object-oriented programming where object types are called classes.

2 Non-Parameterized Object Specifications

Non-parameterized object specifications were examined in detail in [GH95]. The basic idea consists in viewing object states as attribute algebras and transitions between attribute algebras as so-called event algebras. This section reviews the main ideas.

2.1 Syntax

We first consider the syntax of (TROLL *light*) object specifications. Objects are described on the basis of conventional data type specifications. The basic difference between data values as instances of data types and objects as instances of object types is that values are regarded as stateless items while objects are associated with an inherent notion of state.
Technically speaking object signatures are built of a data signature part (consisting of data sorts and corresponding data operators) and object specific items like object (identifier) sorts, attribute symbols and event symbols.

Definition 1. An **object signature** $O\Sigma = (DS, \Omega, OS, A, E)$ consists of

- a set DS of data sorts,
- a family of sets of operation symbols $\Omega = \langle \Omega_{dw,ds} \rangle_{dw \in DS^*, ds \in DS}$,
- a set OS of object sorts,
- a family of sets of attribute symbols $A = \langle A_{os,w,s} \rangle_{os \in OS, w \in S^*, s \in S}$, and
- a family of sets of event symbols $E = \langle E_{os,w} \rangle_{os \in OS, w \in S^*}$

where S refers to the union of data and object sorts: $S = DS \cup OS$.

Example 2. The template **Author** induces the following object signature.

$$\begin{aligned}
DS &= \{\, string, date \} \\
\Omega &= \{\, concat : string \times string \rightarrow string, \\
&\qquad day : date \rightarrow nat, \ldots \} \\
OS &= \{\, author, book \} \\
A &= \{\, Name : author \rightarrow string, \\
&\qquad DateOfBirth : author \rightarrow date, \ldots \} \\
E &= \{\, create : author \times string \times date, \\
&\qquad publishBook : author \times string \times string, \ldots \}
\end{aligned}$$

In the language TROLL *light* every template induces a corresponding object identifier sort. The relation is demonstrated by the convention that the same names are used for templates and corresponding object identifier sorts with the exception that a template name begins with an upper case letter while the corresponding object identifier sort starts with a lower case letter. With respect to subobject symbols as found in **AuthorContainer** we assume that these are treated just like attribute symbols. The subtle distinction is discussed in [GH95].
The interpretation of data sorts, data operations, object identifier sorts, and attribute symbols characterize the states of an object community. Event symbols are associated with state transitions. Hence we define the notions of data, attribute, and event signatures as follows.

Definition 3. Let an object signature $O\Sigma$ be given. The **data signature** induced by $O\Sigma$ consists of the data sorts and the data operations of the object signature, i.e., $D\Sigma = (DS, \Omega)$.
The **attribute signature** $A\Sigma$ induced by $O\Sigma$ consists of S, i.e., the union of data and object sorts, and the union of operation and attribute symbols, i.e., $A\Sigma = (DS \cup OS, \Omega \cup A)$.
The **event signature** $E\Sigma$ induced by $O\Sigma$ consists of the attribute signature extended by one special sort symbol $\hat{e}$ for events and an appropriate family of event symbols $\hat{E}$, i.e., $E\Sigma = (DS \cup OS \cup \{\hat{e}\}, \Omega \cup A \cup \hat{E})$ with $\hat{E} = \langle \hat{E}_{os,w,\hat{e}} \rangle_{os \in OS, w \in S^*}$ and $\hat{E}_{os,w,\hat{e}} := E_{os,w}$.

The above signatures are signatures in the classical sense. Thus, they induce corresponding classes of algebras. We will use the notions data, attribute, and event algebra to refer to them.
Besides the signature part TROLL *light* templates further include certain axioms restricting the interpretation of object signatures.

Definition 4. Let an object signature $O\Sigma$ be given. The **axioms** for object community specification are divided into constraints, derivation rules, valuation rules, interaction rules, and behavior definitions.

- A **constraint** φ is a formula restricting the class of possible attribute algebras $ALG_{A\Sigma}$.
- A **derivation rule** is of the form $\alpha = t_\alpha$ where α is an attribute and t_α is a term with the same sort as the attribute.
- A **valuation rule** is of the form $\{\varphi\}[t_{\hat{e}}]\alpha = t_\alpha$, where φ is a formula, $t_{\hat{e}}$ is an event term, α is an attribute, and t_α is a term with the same sort as the attribute α.
- An **interaction rule** is of the form $\{\varphi\}\ t_{\hat{e}} \gg t'_{\hat{e}}$ where φ is a formula, and $t_{\hat{e}}$ and $t'_{\hat{e}}$ are event terms.
- A **behavior definition** for a template T consists of a set of process definitions of the following form:
  $\Pi_0 = (\{\varphi_1\}t_{\hat{e}_1} \rightarrow \Pi_1 \mid \{\varphi_2\}t_{\hat{e}_2} \rightarrow \Pi_2 \mid \ldots \mid \{\varphi_n\}t_{\hat{e}_n} \rightarrow \Pi_n)$
  Here the Π_i's are process names, the φ_i's formulas, and the $t_{\hat{e}_i}$'s event terms.

Informally speaking, valuation rules are used to express the effect of events on attributes, interaction rules allow to synchronize events in different objects thereby supporting communication between objects, and with behavior definitions allowed event sequences can be specified.

2.2 Semantics

Object signatures are interpreted by so-called object communities. This name was chosen because in object communities different objects co-exist and communicate with each other.

Definition 5. Let an object signature $O\Sigma$ be given. $MOD_{O\Sigma}$ denotes the **class** of all **$O\Sigma$-models or $O\Sigma$-object communities**:

$$MOD_{O\Sigma} = \{M_{O\Sigma} \mid M_{O\Sigma} \subseteq (ALG_{A\Sigma} \times ALG_{E\Sigma} \times ALG_{A\Sigma})\}$$

We may assume the interpretation of the data signature to be fixed. It is also possible to restrict the interpretation of object identifier sorts in the sense that object identifiers will be chosen from a term algebra induced by the attribute signature. Then, only attributes contribute to states. For details we have to refer to [GH95] again.
Intuitively, triples $\langle A_L, \hat{A}, A_R \rangle$ in $M_{O\Sigma}$ express that there is a state transition from attribute algebra A_L to attribute algebra A_R via the occurence of the events in $\hat{A}$. A step in an object's life can be traced along such triples: For example, an object $obj \in A_{L,os}$ for some $os \in OS$ can have for attribute α the value c_1 in A_L $[\alpha_{A_L}(obj, ...) = c_1]$; this can change in A_R to c_2 $[\alpha_{A_R}(obj, ...) = c_2]$ due to the occurrence of event e in $\hat{A}$ $[e_{\hat{A}}(obj, ...) \neq \perp]$. Of course, triples $\langle A_L, \hat{A}, A_R \rangle$ in $M_{O\Sigma}$ can also be regarded as one algebra whose signature is the disjoint union of $A\Sigma$, $E\Sigma$, and $A\Sigma$.
The possible state transition relation must obey the specified axioms. Therefore one defines the validity of axioms as follows.

Definition 6. Let an object signature $O\Sigma$, an $O\Sigma$-model $M_{O\Sigma}$ and axioms over $O\Sigma$ be given. The **validity** of axioms is defined as follows.

- A constraint φ is **valid** in $M_{O\Sigma}$ iff for all triples $\langle A_L, \hat{A}, A_R \rangle$ in $M_{O\Sigma}$ the formula φ holds in A_L and A_R.
- A derivation rule $\alpha = t_\alpha$ is **valid** in $M_{O\Sigma}$ iff for all triples $\langle A_L, \hat{A}, A_R \rangle$ in $M_{O\Sigma}$ attribute α evaluates in A_L and A_R to the value of t_α.
- A valuation rule $\{\varphi\}[t_{\hat{e}}]\alpha = t_\alpha$ is **valid** in $M_{O\Sigma}$ iff for all triples $\langle A_L, \hat{A}, A_R \rangle$ in $M_{O\Sigma}$ the following is true: If the formula φ holds in A_L and the event $t_{\hat{e}}$ occurs[2] in $\hat{A}$, then the attribute α evaluates in A_R to the value which t_α had in A_L.
- An interaction rule $\{\varphi\}\ t_{\hat{e}} \gg t'_{\hat{e}}$ is **valid** in $M_{O\Sigma}$ iff for all triples $\langle A_L, \hat{A}, A_R \rangle$ in $M_{O\Sigma}$ the following holds: If φ is true in A_L and $t_{\hat{e}}$ occurs in $\hat{A}$, then $t'_{\hat{e}}$ must also occur in $\hat{A}$.
- A behavior definition
  $\Pi_0 = (\{\varphi_1\}t_{\hat{e}_1} \rightarrow \Pi_1 \mid \{\varphi_2\}t_{\hat{e}_2} \rightarrow \Pi_2 \mid ... \mid \{\varphi_n\}t_{\hat{e}_n} \rightarrow \Pi_n)$
  for template T is **valid** in $M_{O\Sigma}$ iff for all triples $\langle A_L, \hat{A}, A_R \rangle$ in $M_{O\Sigma}$ the following holds: If an event e of template T occurs in $\hat{A}$ for an object o of object sort t, then there is a process definition and an index j such that $t_{\hat{e}_j}$ evaluates in $\hat{A}$ to e, φ_j is true in A_L, $process\_state(o) = \Pi_0$ in A_L, and $process\_state(o) = \Pi_j$ in A_R. In order to handle behavior definitions in a correct way, the object signature $O\Sigma$ has to be extended by attributes $process\_state : os \rightarrow process\_state\_sort$ for every object sort $os \in OS$.

Definition 7. Let an object signature $O\Sigma$ and a set AX of axioms over $O\Sigma$ be given. The **class of all $O\Sigma$-models satisfying the axioms** AX is denoted by $MOD_{O\Sigma,AX}$.

In order to explain later in Def. 19 the semantics of parameterized object specifications we need a notion of morphism between object communities.

Definition 8. Let an object signature $O\Sigma$ and object communities $M, M' \in MOD_{O\Sigma}$ be given. A triple $\langle f_L, \hat{f}, A_R \rangle$ of mappings $f_L : M_L \rightarrow M'_L$, $\hat{f} : \hat{M} \rightarrow \hat{M}'$, and

[2] We say an event e occurs in $\hat{A}$ if e evaluates to something different from $\perp$ in $\hat{A}$.

$f_R : M_R \to M'_R$ is called an **object community morphism** iff $\langle A_L, \hat{A}, A_R \rangle \in M$ implies $\langle f_L(A_L), \hat{f}(\hat{A}), f_R(A_R) \rangle \in M'$. Here M_L denotes the restriction of M to the first component, i.e., $M_L := \{A_L \mid \langle A_L, \hat{A}, A_R \rangle \in M\}$. $\hat{M}$ and M_R are defined analogously.

3 Parameterized Object Specifications

One of the main motivation for considering parameterized object specifications is the possibility of reusing specifications. A parameterized object specification consists of a formal parameter part and a body part. A parameterized specification can be applied to different actual parameters yielding different versions of the body specification. Another main advantage of employing parameterized specifications is modularity. This means that one can split large specifications into smaller manageable pieces.

3.1 Syntax

In order to compare different specifications, for instance parameter specification and body specification or parameter specification and actual specification, specifications must become in some sense comparable. The technical tool for this are signature morphisms.

Definition 9. Let object signatures $O\Sigma_1$ and $O\Sigma_2$ be given. A **signature morphism** $f : O\Sigma_1 \to O\Sigma_2$ is given by mappings

- $f_{DS} : DS_1 \to DS_2$,
- $f_\Omega : \Omega_1 \to \Omega_2$ compatible with f_{DS},
- $f_{OS} : OS_1 \to OS_2$,
- $f_A : A_1 \to A_2$ compatible with f_{DS} and f_{OS}, and
- $f_E : E_1 \to E_2$ compatible with f_{DS} and f_{OS}.

Fact 10. The category with object signatures as (category) objects and signature morphism as (category) morphisms has pushouts.

Definition 11. A **parameterized signature** is a pair of signatures $PO\Sigma = (PDS, P\Omega, POS, PA, PE) \subseteq BO\Sigma = (BDS, B\Omega, BOS, BA, BE)$ where the parameter signature $PO\Sigma$ is included in the body signature $BO\Sigma$.

Definition 12. A **parameterized object specification** $(PO\Sigma, PAX) \subseteq (BO\Sigma, BAX)$ is given by a parameterized signature and parameter axioms PAX and body axioms BAX.

Example 13 (Container). The first example for a parameterized object specification introduces a container which holds objects belonging to the following parameter signature.

$$\begin{aligned}
DS &= \{data_sort\} \\
\Omega &= \emptyset \\
OS &= \{entry_sort\} \\
A &= \emptyset \\
E &= \{create_entry : entry_sort, destroy_entry : entry_sort\}
\end{aligned}$$

The body signature gives an object sort for container objects and provides events for insertion and deletion.

$$\begin{aligned}
DS &= \emptyset \\
\Omega &= \emptyset \\
OS &= \{container_sort\} \\
A &= \{entries : container_sort \times data_sort \to entry_sort\} \\
E &= \{ create_container : container_sort, \\
&\qquad destroy_container : container_sort, \\
&\qquad add_entry : container_sort \times data_sort, \\
&\qquad remove_entry : container_sort \times data_sort\}
\end{aligned}$$

It should be clear that the respectives body parts show only the differences from the parameter parts. Thus for instance, in the body we really have $OS = \{entry\_sort, container\_sort\}$. The body axioms synchronize the insertion and deletion events in the container with the given parameter events which are assumed to be birth and death events for entry objects.

$$\begin{aligned}
&add_entry(self, x) \gg create_entry(entries(self, x)) \\
&remove_entry(self, x) \gg destroy_entry(entries(self, x))
\end{aligned}$$

The *data_sort* of the parameter serves in the body as a kind of key attribute (in the sense of semantic data models) because entry objects are identified by its container c and a data value v belonging to the *data_sort*: $entries(c, v)$ gives an entry object.

Example 14 (Synchronization). Quite another example for a parameterized object specification is the following one which mainly serves to synchronize two events. The parameter signature gives these two events.

$$\begin{aligned}
DS &= \emptyset \\
\Omega &= \emptyset \\
OS &= \{obj_sort_1, obj_sort_2\} \\
A &= \emptyset \\
E &= \{event_1 : obj_sort_1, event_2 : obj_sort_2\}
\end{aligned}$$

The body signature adds a new object sort *synchronizer* which is intended to serve as a common super-object.

$$\begin{aligned}
DS &= \emptyset \\
\Omega &= \emptyset \\
OS &= \{synchronizer\} \\
A &= \{ obj_1 : synchronizer \to obj_sort_1, \\
&\qquad obj_2 : synchronizer \to obj_sort_2\} \\
E &= \{create_synchronizer : synchronizer \times obj_sort_1 \times obj_sort_2\}
\end{aligned}$$

The body axioms simply require that $event_1$ occurs in the first object if and only if $event_2$ in the second object occurs.

$$\begin{aligned}
&[create_synchronizer(self, x_1, x_2)]\ obj_1(self) = x_1, obj_2(self) = x_2 \\
&event_1(obj_1(self)) \gg event_2(obj_2(self)) \\
&event_2(obj_2(self)) \gg event_1(obj_1(self))
\end{aligned}$$

Example 15 (Specialization). An example taken from the area of semantic data models is the specialization specification presented hereafter. The parameter signature provides an object sort, an attribute, and an event for this sort.

$$
\begin{aligned}
DS &= \{data\_sort_1, data\_sort_2\} \\
\Omega &= \emptyset \\
OS &= \{original\_sort\} \\
A &= \{original\_attribute : original\_sort \rightarrow data\_sort_1\} \\
E &= \{original\_event : original\_sort\}
\end{aligned}
$$

The body signature introduces an object-valued attribute, which will be pointing to the object to be specialized, the specialized attribute, the specialized event, a new attribute, and a new event.

$$
\begin{aligned}
DS &= \emptyset \\
\Omega &= \emptyset \\
OS &= \{specialized\_sort\} \\
A &= \{\, who\_am\_i : specialized\_sort \rightarrow original\_sort, \\
&\qquad specialized\_attribute : specialized\_sort \rightarrow data\_sort_1, \\
&\qquad new\_attribute : specialized\_sort \rightarrow data\_sort_2\} \\
E &= \{\, create\_specialization : specialized\_sort \times original\_sort, \\
&\qquad specialized\_event : specialized\_sort, \\
&\qquad new\_event : specialized\_sort \times data\_sort_2\}
\end{aligned}
$$

The body axioms serve to inherit the original attribute and the original event into the specialized object. The main technical tool here is a derivation rule for attribute inheritance and an event calling rule for event inheritance.

$$
\begin{aligned}
&[create\_specialization(self, x)]\ who\_am\_i(self) = x \\
&[new\_event(self, y)]\ new\_attribute(self) = y \\
&specialized\_attribute(self) = original\_attribute(who\_am\_i(self)) \\
&specialized\_event(self) \gg original\_event(who\_am\_i(self))
\end{aligned}
$$

Let us emphasize one point here: What the parameterized object specification above achieves is more or less a formal specification of an inheritance mechanism. Thus it may be regarded as as the formal semantics of this inheritance concept. Of course, specifications like the above one are not assumed to be written down in a comfortable specification language but more a means to express semantics of suitable language features.

We now come to the application of parameterized object specifications to actual ones. This is done in the spirit of the classical pushout approach of [EKT+84].

Definition 16. Let a parameterized object specification $(PO\Sigma, PAX) \subseteq (BO\Sigma, BAX)$ and an actual object specification $(AO\Sigma, AAX)$ together with a signature morphism $f : PO\Sigma \rightarrow AO\Sigma$ be given. The result of the **application** of $(PO\Sigma, PAX) \subseteq (BO\Sigma, BAX)$ to $(AO\Sigma, AAX)$ via the parameter passing morphism $f : PO\Sigma \rightarrow AO\Sigma$ is the pushout object specification $(RO\Sigma, RAX)$ of the following diagram.

$$
\begin{array}{ccc}
(PO\Sigma, PAX) & \stackrel{i}{\hookrightarrow} & (BO\Sigma, BAX) \\
f \big\downarrow & po & f' \big\downarrow \\
(AO\Sigma, AAX) & \stackrel{i'}{\rightarrow} & (RO\Sigma, RAX)
\end{array}
$$

Example 17 (Application of a parameterized object specification). Suppose that we have appropriate non-parameterized object specifications for authors and books.

Then we can apply the parameterized object specification container to the author specification by the signature morphism f:

$$\begin{aligned} f: \mathit{data\_sort} &\mapsto \mathit{nat} \\ \mathit{entry\_sort} &\mapsto \mathit{author} \\ \mathit{create\_entry} &\mapsto \mathit{create\_author} \\ \mathit{destroy\_entry} &\mapsto \mathit{destroy\_author} \end{aligned}$$

And we can also apply the container specification to the book specification by the signature morphism f':

$$\begin{aligned} f': \mathit{data\_sort} &\mapsto \mathit{string} \\ \mathit{entry\_sort} &\mapsto \mathit{book} \\ \mathit{create\_entry} &\mapsto \mathit{create\_book} \\ \mathit{destroy\_entry} &\mapsto \mathit{destroy\_book} \end{aligned}$$

Thus in the respective containers, author objects are identified by natural numbers and book objects are identified by strings.

3.2 Semantics

Fact 18. A signature morphism $f: O\Sigma_1 \to O\Sigma_2$ induces a corresponding translation $\mathcal{U}_f : MOD_{O\Sigma_2} \to MOD_{O\Sigma_1}$ on models by defining for $M_{O\Sigma_2} \in MOD_{O\Sigma_2}$

$$\mathcal{U}_f(M_{O\Sigma_2}) := \{ \langle \mathcal{U}_{A\Sigma_2 \to A\Sigma_1}(A_L), \mathcal{U}_{E\Sigma_2 \to E\Sigma_1}(\hat{A}), \mathcal{U}_{A\Sigma_2 \to A\Sigma_1}(A_R) \rangle \mid \langle A_L, \hat{A}, A_R \rangle \in M_{O\Sigma_2} \}$$

Definition 19. The **semantics of a parameterized object specification** is given by a function $\mathcal{F} : MOD_{PO\Sigma,PAX} \to MOD_{BO\Sigma,BAX}$ defined as follows.

- Because the parameter attribute signature is included in body attribute signature $PA\Sigma \hookrightarrow BA\Sigma$ and the parameter event signature is included in body event signature $PE\Sigma \hookrightarrow BE\Sigma$, we have free constructions from parameter attribute algebras to body attribute algebras and from parameter event algebras to body event algebras.

$$\mathcal{F}_{PA\Sigma \to BA\Sigma} : ALG_{PA\Sigma} \to ALG_{BA\Sigma}$$
$$\mathcal{F}_{PE\Sigma \to BE\Sigma} : ALG_{PE\Sigma} \to ALG_{BE\Sigma}$$

- But in general, the carrier sets of object sorts appearing only in the parameter will be empty due to the freeness of the construction. Therefore, we have to add means to generate object identities. We do this by enlarging the body signature by special operations for each object sort os appearing in the body but not in the parameter:

$$\begin{aligned} \mathit{first}_{os} &: \to os \\ \mathit{last}_{os} &: os \to os \end{aligned}$$

  The obtained signature is called $BA\Sigma'$. Now we have a free construction $\mathcal{F}_{PA\Sigma \to BA\Sigma'}$ from parameter attribute algebras to enlarged body algebras. The opposite to the free construction $\mathcal{F}_{BA\Sigma \to BA\Sigma'}$ from body algebras to enlarged body algebras is given by the respective forgetful functor $\mathcal{U}_{BA\Sigma' \to BA\Sigma}$ from enlarged body algebras to body algebras. Analogous constructions work for event algebras. In the total we get the following constructions.

$$\mathcal{F}_{PA\Sigma} := \mathcal{F}_{PA\Sigma \to BA\Sigma'} \circ \mathcal{U}_{BA\Sigma' \to BA\Sigma}$$
$$\mathcal{F}_{PE\Sigma} := \mathcal{F}_{PE\Sigma \to PA\Sigma'} \circ \mathcal{U}_{PA\Sigma' \to PA\Sigma}$$

- For a given $(PO\Sigma, PAX)$-model $M_{PO\Sigma,PAX}$ we can therefore define the following classes of possible $BA\Sigma$- and $BE\Sigma$-algebras.

$$POS_{BA\Sigma} := \{\, \mathcal{F}_{PA\Sigma}(A_L) \mid \langle A_L, \hat{A}, A_R\rangle \in M_{PO\Sigma,PAX}\} \cup \{\, \mathcal{F}_{PA\Sigma}(A_R) \mid \langle A_L, \hat{A}, A_R\rangle \in M_{PO\Sigma,PAX}\}$$

$$POS_{BE\Sigma} := \{\, \mathcal{F}_{PE\Sigma}(\hat{A}) \mid \langle A_L, \hat{A}, A_R\rangle \in M_{PO\Sigma,PAX}\}$$

- Employing these classes we can define a class of possible body models satisfying the axioms as follows.

$$POS_{BO\Sigma,BAX} := \{M \mid M \subseteq (POS_{BA\Sigma} \times POS_{BE\Sigma} \times POS_{BA\Sigma}) \wedge M \models BAX\}$$

- The semantics of the complete construction is the least model in this set (if it exists)

$$\mathcal{F}(M_{PO\Sigma,PAX}) := M_{least}$$

such that there exists an object community morphism $f : M \rightarrow M_{least}$ for all $M \in POS_{BO\Sigma,BAX}$.

Remark 20. In general, the function $\mathcal{F}$ will not be persistent in the sense that a given parameter model $M_{PO\Sigma,PAX}$ will resist in $\mathcal{F}(M_{PO\Sigma,PAX})$. Speaking in more technical terms, $M_{PO\Sigma,PAX}$ and $\mathcal{U}_i(\mathcal{F}(M_{PO\Sigma,PAX}))$ will in general be not isomorphic ($\mathcal{U}_i$ refers to the translation of $BO\Sigma$-models to $PO\Sigma$-models). This is demonstrated by Example 14 for synchronization. There the parameter object society will be restricted in its behavior by the additional synchronization rule, and possible life cycles of the parameter object society are forbidden in the resulting body object society. Another property of $\mathcal{F}$ is partiality. It is easy to write inconsistent object specifications due to the existence of constraints restricting the possible attribute algebras. In the extreme case one could require $x \neq x$ as a constraint resulting in an empty model class for the object specification. If such a constraint would be part of the body axioms one would get for $\mathcal{F}$ the completely undefined function. On the other hand, one can also imagine situations where $\mathcal{F}$ is defined for a model $M_{PO\Sigma,PAX}$ but undefined for $M'_{PO\Sigma,PAX}$ (perhaps due to some constraint which holds in $\mathcal{F}(M_{PO\Sigma,PAX})$ but not in $\mathcal{F}(M'_{PO\Sigma,PAX})$).

Definition 21. Let a parameterized object specification $(PO\Sigma, PAX) \subseteq (BO\Sigma, BAX)$ and an actual object specification $(AO\Sigma, AAX)$ together with a signature morphism $f : PO\Sigma \rightarrow AO\Sigma$ be given. Parameter passing is called **correct** if $\mathcal{U}_f(MOD_{AO\Sigma,AAX}) \subseteq MOD_{PO\Sigma,PAX}$. Thus the parameter passing morphism f is correct, if the formal axioms hold in the actual specification.

Remark 22. A simple sufficient criterion for correctness of parameter passing is $PAX = \emptyset$. This was almost the case in our examples for parameterized object specifications, with the exception of some simple constraints for booleans and natural numbers.

4 Conclusion and Future Work

We have presented a proposal for parameterized object specifications allowing especially objects sorts in the parameter. As far as we know this has not been studied before. Our proposal for parameter passing works well on the syntactical level by

means of a pushout construction. It thereby establishes a link between well-known techniques from algebraic specification [EKT+84] and recent developments in information systems design with object-oriented concepts [GHC+94].
For the semantic level we have considered free constructions and corresponding forgetful functors. This last construction is rather tricky and calls for further study. It is open what kind of further properties this construction has. Another interesting question is under what conditions certain properties of the given actual object specification are preserved by the application of a parameterized object specification. For instance, one could ask whether all possible communications in the actual parameter are also possible in the resulting specification.

[AZ94] E. Astesiano and E. Zucca. D-Oids: A Model for Dynamic Data-Types. *Mathematical Structures in Computer Science*, 1994.

[CGH92] S. Conrad, M. Gogolla, and R. Herzig. TROLL *light*: A Core Language for Specifying Objects. Informatik-Bericht 92–02, TU Braunschweig, 1992.

[DG94] P. Dauchy and M.-C. Gaudel. Algebraic Specifications with Implicit State. Technical Report 887, Université de Paris-Sud, 1994.

[EDS93] H.-D. Ehrich, G. Denker, and A. Sernadas. Constructing Systems as Object Communities. In M.-C. Gaudel and J.-P. Jouannaud, editors, *Proc. Theory and Practice of Software Development (TAPSOFT'93)*, pages 453–467. Springer, Berlin, LNCS 668, 1993.

[EKT+84] H. Ehrig, H.-J. Kreowski, J. Thatcher, E. Wagner, and J. Wright. Parameter Passing in Algebraic Specification Languages. *Theoretical Computer Science*, 28:45–81, 1984.

[EM85] H. Ehrig and B. Mahr. *Fundamentals of Algebraic Specification 1: Equations and Initial Semantics*. Springer, Berlin, 1985.

[EM90] H. Ehrig and B. Mahr. *Fundamentals of Algebraic Specification 2: Modules and Constraints*. Springer, Berlin, 1990.

[EO94] H. Ehrig and F. Orejas. Dynamic Abstract Data Types: An Informal Proposal. *EATCS Bulletin*, 53:162–169, 1994.

[FM92] J. Fiadeiro and T. Maibaum. Temporal Theories as Modularisation Units for Concurrent System Specification. *Formal Aspects of Computing*, 4(3):239–272, 1992.

[GCH93] M. Gogolla, S. Conrad, and R. Herzig. Sketching Concepts and Computational Model of TROLL *light*. In A. Miola, editor, *Proc. 3rd Int. Conf. Design and Implementation of Symbolic Computation Systems (DISCO'93)*, pages 17–32. Springer, Berlin, LNCS 722, 1993.

[GD94] J.A. Goguen and R. Diaconescu. Towards an Algebraic Semantics for the Object Paradigm. In H. Ehrig and F. Orejas, editors, *Proc. 9th Workshop on Abstract Data Types (ADT'92)*, pages 1–29. Springer, Berlin, LNCS 785, 1994.

[GH95] M. Gogolla and R. Herzig. An Algebraic Semantics for the Object Specification Language TROLL *light*. In E. Astesiano, G. Reggio, and A. Tarlecki, editors, *Proc. 10th Workshop on Abstract Data Types (ADT'94)*, pages 288–304. Springer, Berlin, LNCS 906, 1995.

[GHC+94] M. Gogolla, R. Herzig, S. Conrad, G. Denker, and N. Vlachantonis. Integrating the ER Approach in an OO Environment. In R. Elmasri, V. Kouramajian, and B. Thalheim, editors, *Proc. 12th Int. Conf. on the Entity-Relationship Approach (ER'93)*, pages 376–389. Springer, Berlin, LNCS 823, 1994.

[GKS91] G. Gottlob, G. Kappel, and M. Schrefl. Semantics of Object-Oriented Data Models – The Evolving Algebra Approach. In J.W. Schmidt and A.A. Stogny, editors, *Proc. 1st Int. East-West Database Workshop*, pages 144–160. Springer, Berlin, LNCS 504, 1991.

[GM87] J.A. Goguen and J. Meseguer. Unifying Functional, Object-Oriented and Relational Programming with Logical Semantics. In B. Shriver and P. Wegner, editors, *Research Directions in Object-Oriented Programming*, pages 417–477. MIT Press, 1987.

[GR91] M. Grosse-Rhode. Towards Object-Oriented Algebraic Specifications. In H. Ehrig, K.P. Jantke, F. Orejas, and H. Reichel, editors, *Recent Trends in Data Type Specification (WADT'90)*, pages 98–116. Springer, Berlin, LNCS 534, 1991.

[Gur88] Y. Gurevich. Logic and the Challenge of Computer Science. In E. Börger, editor, *Trends in Theoretical Computer Science*, pages 1–57. Computer Science Press, Rockville (MD), 1988.

[HCG94] R. Herzig, S. Conrad, and M. Gogolla. Compositional Description of Object Communities with TROLL *light*. In C. Chrisment, editor, *Proc. Basque Int. Workshop on Information Technology (BIWIT'94)*, pages 183–194. Cépaduès-Éditions, Toulouse, 1994.

[Hen88] M. Hennessy. *Algebraic Theory of Processes*. MIT Press, Cambridge (MA), 1988.

[HK87] R. Hull and R. King. Semantic Database Modelling: Survey, Applications, and Research Issues. *ACM Computing Surveys*, 19(3):201–260, 1987.

[Hoa85] C.A.R. Hoare. *Communicating Sequential Processes*. Prentice-Hall, Englewood Cliffs (NJ), 1985.

[JHSS91] R. Jungclaus, T. Hartmann, G. Saake, and C. Sernadas. Introduction to TROLL – A Language for Object-Oriented Specification of Information Systems. In G. Saake and A. Sernadas, editors, *Information Systems – Correctness and Reusability*, pages 97–128. TU Braunschweig, Informatik Bericht 91-03, 1991.

[Mil89] R. Milner. *Communication and Concurrency*. Prentice-Hall, Englewood Cliffs (NJ), 1989.

[PM88] J. Peckham and F. Maryanski. Semantic Data Models. *ACM Computing Surveys*, 20(3):153–189, 1988.

[PPP94] F. Parisi-Presicce and A. Pierantonio. Structured Inheritance for Algebraic Class Specifications. In H. Ehrig and F. Orejas, editors, *Recent Trends in Data Type Specification (WADT'92)*, pages 295–309. Springer, Berlin, LNCS 785, 1994.

[Reg91] G. Reggio. Entities: An Institution for Dynamic Systems. In H. Ehrig, K.P. Jantke, F. Orejas, and H. Reichel, editors, *Recent Trends in Data Type Specification (WADT'90)*, pages 246–265. Springer, LNCS 534, 1991.

[SSC92] A. Sernadas, C. Sernadas, and J.F. Costa. Object Specification Logic. Internal Report, INESC, University of Lisbon, 1992. To appear in Journal of Logic and Computation.

[SSE87] A. Sernadas, C. Sernadas, and H.-D. Ehrich. Object-Oriented Specification of Databases: An Algebraic Approach. In P.M. Stocker and W. Kent, editors, *Proc. 13th Int. Conf. on Very Large Data Bases (VLDB'87)*, pages 107–116. Morgan-Kaufmann, Palo Alto, 1987.

[Wir90] M. Wirsing. Algebraic Specification. In J. Van Leeuwen, editor, *Handbook of Theoretical Computer Science, Vol. B*, pages 677–788. North-Holland, Amsterdam, 1990.

Analyzing the Dynamics of a Z Specification

Paolo Ciancarini and Cecilia Mascolo

Dipartimento di Scienze dell'Informazione,
Università di Bologna,
e-mail: {cianca,mascolo}@cs.unibo.it

Abstract. We present a method for analyzing the dynamics of a Z document describing a non-sequential system. First a formal operational semantics based on the chemical metaphor is given to Z. Then, some Unity-like temporal logic constructs are defined on such a formal operational semantics in order to allow the specification and analysis of dynamic and temporal properties of concurrent systems, such as safety and liveness properties.

1 Introduction

The introduction of formal methods increased the usefulness of software specification documents by allowing to automatically check them and to formally reason on them. The Z notation [18] is currently widely used as a non executable specification language to formally describe and analyze the requirements and the architectures of software systems. However, Z has been mostly used for the specification of sequential systems. In fact, even if in the recent years it has been used for specifying concurrent, reactive, or even distributed systems, in general non-sequential systems are difficult to be perfectly described and then analyzed using Z.

Even if Z is not executable several researchers have tried to improve the ability of Z to express and support the analysis of dynamic features of non-sequential systems. The integration of Z with operational notations like CSP [1] or Petri Nets [10, 12], or the use of temporal logic for analyzing Z documents [16, 5, 9, 14] are some of the approaches suggested. These approaches all suffer from the same problem: the integration of different notations in a uniform specification method is not formalized because a clear and consistent integration is in general difficult to accomplish.

The approach we introduce here is new, insofar as we formally define in a unified framework both an operational semantics and a logic based on such a semantics to reason on Z documents. The operational semantics we introduce is based on the *chemical metaphor* embedded in the notation of the Chemical Abstract Machine (Cham) [2], The logic includes a number of constructs which allow the definition of dynamic properties of a system specification. We have chosen some Unity-like [6] logic constructs because of their expressiveness (a similar approach can be found in [9]) and because it has already been proved suitable to be the proof system basis for Swarm language [8]: a multiset transformation based language like the Cham.

The semantics of the constructs is defined on an *execution model* based on the operational semantics.

This paper has the following structure: in Sect. 2 a specification style and an interpretation of Z documents as sets of state and operation schemas are given; Sect. 3 describes the operational semantics based on the chemical metaphor we adopted. Sect. 4 contains the definition of the execution model imposed on the operational semantics; Sect. 5 introduces the new logic constructs inherited from the Unity language while the final section contains comparisons and conclusions.

2 A specification style and its semantics

For conciseness, our specification style considers a restricted version of Z; we specify such a fragment using Z itself, thus following the Z tradition [17, 11].

The elementary components of a Z specification are State schemas and Operation schemas. A schema is defined as follows:

SCHEMA_STATE

$name : NAME$
$schema\_imp : \mathbb{P}\, NAME$
$decl : \mathbb{P}\, VAR$
$imported : \mathbb{P}\, VAR$

$name \notin schema\_imp$
$\forall d : VAR \mid d \in imported \bullet \exists s : NAME \mid s \in schema\_imp$
$\quad d \in stateschema(s).decl$

where $[NAME]$ is a basic type specifying names of variables or schemas, *name* identifies the schema, *schema_imp* is a set of imported schemas names [1], *decl* is a set of identifier declarations, *imported* is the set of imported identifiers; the predicate states that every imported identifier should be declared in one of the imported schemas.

Intuitively, we only consider State schemas without predicative part since we will be able to express these predicates as invariant properties using the logic defined in Sect.5.

Semantically, a State schema s can be seen as the set of all its possible instantiations [17]. A schema instance is an instantiated State schema:

SCHEMA_INSTANCE

$sch : NAME$
$values : VAR \rightarrowtail IDENT$

$\forall v : VAR \bullet (v \in sch.decl \lor v \in sch.imported) \Leftrightarrow v \in \operatorname{dom} values$

where *VAR* has a name and a type and *IDENT* is a bound variable. The predicate in *SCHEMA_INSTANCE* ensures that every variable is mapped on an identifier with same name and type.

[1] We only consider two levels of imported schemas.

An Operation schema is:

```
─SCHEMA_OP─────────────────────────────
│ name : NAME_OP
│ delta : NAME
│ environment : ENV
│ inputs : P IDENT
│ precondition : P PRE
│ postcondition : P POST
├──────────────
│ ∃ s : SCHEMA_STATE | s.name = delta •
│     ∀ id : IDENT | id ∈ environment.decl •
│     ∃ v : VAR | v.name_id = id.name_id ∧
│     v.type_id = id.type_id • v ∈ s.decl
└──────────────────────────────────────
```

where ***name*** is the operation name, ***delta*** is the name of the State schema on which the operation acts, ***environment*** is the environment of the variables declaration, ***inputs*** are the inputs of the schema, ***precondition*** is the precondition predicates set and ***postcondition*** is the postcondition predicates set.

The initialization operations are represented by particular operation schemas without preconditions.

3 Operational semantics

The standard Z semantics [17, 4] does not offer formalization for concurrency. Thus, we have defined a new operational semantics based on the concurrency offered by the chemical model.

3.1 The chemical metaphor

In the Chemical Abstract Machine model [2, 3] *Molecules*, *Solutions*, and *Rules* are the fundamental elements. A Chemical Abstract Machine is a triple (G, C, R) where G is a grammar, C is a set of configurations (the language generated by the grammar) or molecules, and R is a set of the rules $condition(C) \times \text{bag } C \times \text{bag } C$. A solution is a multiset of molecules: bag C; $\{\!|\ |\!\}$ symbols usually delimit a solution. Solutions are considered the Abstract Machine states. They can be composed of other subsolutions using $\uplus$: $S = S_1 \uplus S_2$.

There are some general laws valid for any Cham:

- **Reaction Law**: an instance of the right-hand side of a rule can replace the corresponding instance of its left-hand side if conditions on the molecules hold. Given a rule
  $condition(m_1, m_2..m_k) \rightarrow m_1, m_2..m_k \Rightarrow m'_1, m'_2..m'_l$
  if $M_1, M_2..M_k, M'_1, M'_2..M'_l$ are instances of the m_i's and the m'_j's by a common substitution, then
  $condition(M_1, M_2..M_k) \rightarrow \{\!|\ M_1, M_2..M_k\ |\!\} \Rightarrow \{\!|\ M'_1, M'_2..M'_l\ |\!\}$

- **Chemical Law**: reactions can freely happen in a solution

$$\frac{S \Rightarrow S1}{S \uplus S2 \Rightarrow S1 \uplus S2}$$

- **Membrane Law**: a subsolution evolves freely in every context

$$\frac{S \Rightarrow S1}{\{\!| C[S] |\!\} \Rightarrow \{\!| C[S1] |\!\}}$$

where $C[\,]$ indicates a context.

In a Cham two instances of rules can fire concurrently if they do not need the same molecules to react on; so many instances of rules can progress simultaneously on a solution. If two instances of rules conflict, in the sense that they "consume" the same molecules, the choice of which to let react is non deterministic.

We consider a *fair* Cham where repeatedly enabled rules will eventually be fired: in this way it is possible to prove properties defined using Unity logic constructs (Sect. 5).

3.2 An Operational Semantics for Z

We introduce an interpretation of Z specifications which allows us to deal with concurrent dynamics. Intuitively, an instance of a State schema (*inst*) is associated with a solution where, in some way, each variable is a subsolution (in many cases a single molecule). Instead, an operation schema corresponds to a chemical rule where pretuples and posttuples are solutions composed of pre and post conditions of the operation.

A molecule is represented as a tuple including a name, a type, and a value; a solution is a bag of molecules; a rule is composed of a conditional part which defines the applicability of the rule, and two solutions, to indicate molecules to be deleted and added, respectively, to the state solution:

$MOLECULE == NAME \times TYPE \times VALUE$

$SOLUTION == \mathrm{bag}\, MOLECULE$

$RULE == CONDITION \times SOLUTION \times SOLUTION$

We call the first $SOLUTION$ "pretuples" and the second "posttuples" to avoid ambiguities. A rule is applicable to a solution if the solution contains molecules that satisfy the conditional parts ($CONDITION$) of the rule and molecules that match the pretuples of the rule.

The semantic function *FSem* associates a solution to a schema_instance:

$Fsem : SCHEMA\_INSTANCE \rightarrow SOLUTION$

Every identifier in the schema instance is associated with a subsolution (not necessarily a single molecule). We remark that Z sets and bags are decomposed by this function in several molecules to increase potential concurrency.

Fsem_op associates a rule to an operation schema [2]:

[2] A similar function can be defined for the initialization operation, where no preconditions are present [3].

Fsem_op : *SCHEMA_OP* → *RULE*
Fsem_op **associates to pre and postcondition different part of the rule:**

- Every Z schema postcondition that specifies the removal of an element from a set or bag is mapped on a pretuple of the rule (molecule to be deleted).
- Every postcondition that specifies the insertion of an element in a set or bag is mapped on a posttuple of the rule (molecules to be added).
- Every Z precondition that defines a membership (∈, in) is mapped on a pretuple (a removal) and also on a posttuple (reinsertion) if the Z postcondition does not contain an indication of removal of that element: in other words, a check of membership is evaluated as a removal followed by a reinsertion.
- Postconditions containing mathematical operators (+, −,..) on naturals are encoded deleting one molecule and adding the updated molecule.
  Example: $x' = x + 1$ is evaluated as (x, N, v) in pretuples and $(x, \mathsf{N}, v + 1)$ in posttuples of the rule.
- Preconditions containing relational operators (<,>) are encoded as conditions, but the molecule corresponding to the variable is deleted and readded as already described [3].
  Example: $x < 5$ is seen as $v < 5 \rightarrow (x, \mathsf{N}, v) \Rightarrow (x, \mathsf{N}, v)$

Now, thanks to the chemical laws, rules can fire concurrently when they are enabled by conditions and non conflicting on pretuples molecules.

3.3 A simple example

Consider the classic dining philosophers problem. What follows is its formalization in our style.

$FORK ::= fork1 \mid fork2 \mid fork3 \mid fork4 \mid fork5$
$PHILO ::= philo1 \mid philo2 \mid philo3 \mid philo4 \mid philo5$

The following schema illustrates the basic State schema of the system:

System
think : $\mathbb{P}\, PHILO$
eat : $\mathbb{P}\, PHILO$
have_right : $\mathbb{P}\, PHILO$
available : $\mathbb{P}\, FORK$
left, *right* : $\mathbb{P}(PHILO \times FORK)$

where: *eat* denotes the set of eating philosophers; *have_right* is the set of philosophers who got the right fork, and wait for the left one; *think* is the set of thinking philosophers; *available* is the set of available forks; *left* and *right* indicate for every philosopher respectively the left and right fork.

[3] This is done according to the chemical semantics where conditions can only be stated on the local molecules involved in the rule [3].

$$\begin{array}{l}
\textit{Init_system} \\
\hline
\textit{System}' \\
\hline
eat' = \varnothing \\
have_right' = \varnothing \\
think' = \{philo1, philo2, philo3, philo4, philo5\} \\
available' = \{fork1, fork2, fork3, fork4, fork5\} \\
right' = \{(philo1, fork1), (philo2, fork2), (philo3, fork3), \\
\qquad (philo4, fork4), (philo5, fork5)\} \\
left' = \{(philo1, fork5), (philo2, fork1), (philo3, fork2), \\
\qquad (philo4, fork3), (philo5, fork4)\} \\
\hline
\end{array}$$

Initially, all philosophers are thinking and all forks are available.
We now define the operations for philosophers:

$$\begin{array}{l}
\textit{Right_Request} \\
\hline
\Delta System \\
ph? : PHILO \\
f? : FORK \\
\hline
ph? \in think \\
f? \in available \\
(ph?, f?) \in right \\
have_right' = have_right \cup \{ph?\} \\
available' = available \setminus \{f?\} \\
think' = think \setminus \{ph?\} \\
\hline
\end{array}$$

The schema *RightRequest* defines the operation of taking the right fork. Operation *Left_Request* is similar: we do no specify it formally.

When *ph?* has the right fork he can ask for the left one: if the fork is available it is assigned to him.

$$\begin{array}{l}
\textit{Thinking} \\
\hline
\Delta System \\
ph? : PHILO \\
f?, ff? : FORK \\
\hline
ph? \in eat \\
(ph?, f?) \in right \\
(ph?, ff?) \in left \\
think' = think \cup \{ph?\} \\
eat' = eat \setminus \{ph?\} \\
available' = available \cup \{f?, ff?\} \\
\hline
\end{array}$$

If *ph?* quits eating he puts down both forks and begins thinking again.

The initialization operation (*Init_System*) is mapped on a chemical rule having no conditions and no pretuples and as posttuples the following solution:

$(think, \mathbb{P}\, PHILO, philo1), .., (think, \mathbb{P}\, PHILO, philo5),$
$(available, \mathbb{P}\, FORK, fork1), .., (available, \mathbb{P}\, FORK, fork5),$
$(right, \mathbb{P}(PHILO \times FORK), (philo1, fork1)), ..,$
$(right, \mathbb{P}(PHILO \times FORK), (philo5, fork5)),$
$(left, \mathbb{P}(PHILO \times FORK), (philo1, fork5)), ..,$
$(left, \mathbb{P}(PHILO \times FORK), (philo5, fork4))$

The State schema instance obtained after the application of the operation is the same solution presented above. The rule associated with the operation schema *RightRequest* has the following pretuples:

$(think, \mathbb{P}\, PHILO, ph?), (available, \mathbb{P}\, FORK, f?),$
$(right, \mathbb{P}(PHILO \times FORK), (ph?, f?)),$

and posttuples:

$(have\_right, \mathbb{P}\, PHILO, ph?), (right, \mathbb{P}(PHILO \times FORK), (ph?, f?))$

The rule corresponding to the operation *LeftRequest* is similar.
The operation *Thinking* corresponds to the following rule with pretuples:

$(eat, \mathbb{P}\, PHILO, ph?), (right, \mathbb{P}(PHILOF \times ORK), (ph?, f?)),$
$(left, \mathbb{P}(PHILO \times FORK), (ph?, ff?))$

and posttuples:

$(available, \mathbb{P}\, FORK, f?), (available, \mathbb{P}\, FORK, ff?),$
$(think, \mathbb{P}\, PHILO, ph?), (right, \mathbb{P}(PHILO \times FORK), (ph?, f?)),$
$(left, \mathbb{P}(PHILO \times FORK), (ph?, ff?))$

Because of the concurrent interpretation of Z that we are going to give, we make the following assumption: all variables not explicitly mentioned in the postconditions of an operation schema may change (i.e. they have not to be invariant: other operations can concurrently modify them).

This assumption is needed in our interpretation and allows concurrency of the operations. In some papers the assumption introduced is exactly the contrary: "Variables not mentioned in the schemas are considered unchanged" e.g. [16] but this is not standard Z too.

4 The execution model

We make the operational semantics (defined in Sect.3) explicit, to build an *execution model*, namely a way of abstractly executing a Z specification document written according to the style outlined in Sect.2. The execution model is defined on the semantics just described and it represents the unfolding of the application of the semantics rules. From every State schema *s* a tree (execution model) can be constructed in the following way:

- the root node is void;
- the first operation applied is the initialization operation without any preconditions;
- from every node several different applicable operation sets can exist, (chosen among all the enabled operations on that node), thus introducing non determinism in the choice of the operations being in conflict.
- Each branch corresponds to the application of a group of enabled operations which could be applied without conflicts, as dictated by the Cham model.

In order to allow the specification of the Unity logic constructs using Z as meta-language, we introduce a concept of *execution tree*:

$$\begin{aligned} TREE ::= {} & Void\_tree \\ \mid {} & fork\langle\langle PAIR \times \mathrm{seq}\ TREE \rangle\rangle \end{aligned}$$

where

$$PAIR == SCHEMA\_INSTANCE \times \mathrm{seq}\,\mathbb{P}\,SCHEMA\_OP$$

The function *Exec* maps every State schema on an execution tree with particular properties (we omit the Z specification of the function); the chemical interpretation imposes that for every node label (s, seq), where s is an instance and seq is a sequence of operations sets:

- all the operations in the sets belonging to the sequence seq must be enabled on s;
- all the operations in the sets belonging to the sequence seq must act on the State schema of which s is an instance;
- each set, member of the sequence seq, must contain operations that can concur (that is without conflicts);
- for every s', label of one of the children of the node labeled (s, seq), there must hold the postconditions of all the operations in the operations set applied to reach that node (sequence structure help to keep link between nodes and operations set).

5 The logic

Liveness (namely "a good thing will eventually occur") or safety (namely "a bad thing never happens") properties can be expressed. Properties are predicates (as the ones in the operation schemas) built using some logic operators ($\wedge$, $\vee$, $\neg$, $\Leftrightarrow$, $\Rightarrow$) and Unity constructs. Properties have a chemical interpretation as well, so that we can analyze the truth of them on the execution model, based on the chemical metaphor too. In order to be able to reason on dynamic properties, we borrow a few constructs from the Unity logic:

- p **unless** q says that whenever p is true during the execution, surely either q will become true or p continues to hold. In particular, on the tree: if p is true on some nodes then on their children q is true or p still holds.

- **Stable** is an alias for p **unless** false, that is when p becomes true it will hold forever. On the tree: if p is true on a node it will remain true for all the subtree of that node.
- **Invariant** p says that p is true forever. That is, for every node of the execution tree p is valid.
- The meaning of p **ensures** q is that when p becomes true then eventually q will hold and before that moment p is still valid. That is, if p is true on a node N, then each branch through N there is a node M below N where q holds and on nodes between nodes N and M in the path, p holds.
- p **leads_to** q has quite the same meaning as **ensures** except that it does not ensure that p is valid until q becomes true. On the tree: if p is true on a node q will eventually hold on a node in all its sub-branches.

The following axiomatic schema shows how we formalize the meaning of the logic constructs on the execution model. We report only the **ensures** definitions:

$$ensures : PROPERTY \times PROPERTY \rightarrow \mathbb{B}$$

$$\begin{array}{l}
\forall p, q : PROPERTY \bullet ensures(p, q) = \text{true} \Leftrightarrow \\
\quad ((unless(p, q) = \text{true}) \wedge \\
\quad (\forall e : TREE;\ e1 : \text{seq}\ TREE; \\
\quad schema : SCHEMA_STATE;\ set : \text{seq}\,\mathbb{P}\,SCHEMA_OP; \\
\quad inst : SCHEMA_INSTANCE \mid \\
\quad subtree(Exec(schema), e) = \text{true} \wedge fork(inst, set, e1) = e \\
\quad \wedge\ valid(inst, and(p, not(q))) = \text{true} \bullet \\
\quad (\exists e3 : \text{seq}\ TREE;\ set' : \text{seq}\,\mathbb{P}\,SCHEMA_OP; \\
\quad inst' : SCHEMA_INSTANCE;\ e2 : TREE \mid \\
\quad subtree(e2, e) \wedge fork(inst', set', e3) = e2 \\
\quad \bullet\ valid(inst', q)) = \text{true}))
\end{array}$$

where function *valid* indicates when a property holds on an instance state. Intuitively this is done considering every property as a solution and analyzing the matching with the state solution like what has been done for rules.

This formalization of *ensures* derives from *unless*; in fact, p **ensures** q if p **unless** q and there exists a branch of the tree that from a state where p is valid (and not q) leads to a state where q holds.

Example We state some properties about the dining philosophers system:

- **Theorem 1** $philo1 \in think$ **ensures** $philo1 \in have\_right$
- **Theorem 2 stable** $(philo1 \in have\_right \wedge philo2 \in have\_right \wedge philo3 \in have\_right \wedge philo4 \in have\_right \wedge philo5 \in have\_right)$

The first property states that if *philo*1 is thinking, he will eventually get the right fork; this property can be stated for all other philosophers as well. The second property defines a deadlock: when every philosopher has got the right fork then the system cannot proceed.

Proof of Theorem 1: If p **ensures** q (where p is $philo1 \in think$ and q is $philo1 \in have\_right$) must be valid, first p **unless** q has to hold (see **ensures** formalization). This means that for every enabled operations set on the solution containing the molecule $(think, \mathbb{P}\, PHILO, philo1)$, the application leads to a state where either molecule $(have\_right, \mathbb{P}\, PHILO, philo1)$ is present or the previous molecule is still in the solution (this is the **unless** formalization of our execution model).

Considering our Z specification, we notice that for all the enabled operations sets that we could choose, each of them modifies the instance solution leaving the molecule (in case we choose only operations acting on other philosophers) or $(have\_right, \mathbb{P}\, PHILO, philo1)$ is inserted (in case the operations set contains *RightRequest* that is the only operation enabled for $philo1$ on the state solution considered). Hence, p **unless** q holds.

To prove p **ensures** q is now necessary (following the *ensures* formalization) to ensure that, given a state where $p \land \neg\, q$ holds exists an enabled operations set applicable, that leads to a state on which q holds. In our specification we have to prove that on a state where p holds exists an enabled (also not continuously) operation set that leads to q. This set contains the instance of the operation *RightRequest* on $philo1$ and other operations acting on other philosophers. Then remembering that our Cham is fair (Sect. 3.1) we can state that the set will eventually be applied. This completes the proof.

6 Comparisons and conclusions

We have defined a chemical semantics for a fragment of Z, and showed that it offers a good basis for the formalization of logic constructs which allow the expression and the analysis of concurrent properties.

We are studying the possibility to map Z schema inclusion using membranes of Cham and airlock. Some other dynamic aspects could be treated such as execution order, synchronization and communication; we are also studying the possibility to introduce real time in our model.

Formalizing dynamics of concurrent and distributed systems is one of the topical challenges to formal languages. Z has been used in this sense several times; the simplest solution consists of considering an intuitively concurrent semantics for schemas: operations are considered atomic and non determinism guides the choice of the operation to apply. Such a model produces a specification whose analysis can hardly expose concurrent properties. In [13] a specification of the distributed IBM Customer Information Control System (CICS) is presented: although no formalization of Z dynamics can be found in the paper, this is considered one of the most successful applications of Z in this sense, because of the reduction of production costs that the Z specification involved. The paper [15] contains the formal specification of a reactive dialog system: Z schemas are used in order to state invariant properties and a formal interpretation of the behavior of the system is given. However, the approach described in the paper is weak in term of formalization of concurrency and semantics.

More formal approaches integrate Z with other notations; for instance, Benjamin [1] integrates Z schemas with CSP notation. CSP is used to specify an abstract system while Z defines more detailed aspects of the design. The integration is minimal and not formally specified. In [12] Petri Nets are used to formalize control flows, causal relations, and dynamic behavior of systems statically defined using Z; nevertheless the formalization of the interaction between the two notations is also minimal. [10] studies a more formal model of integration of Z with Petri Nets: Petri Nets are mapped on Z specifications so that graphical representation given by Petri Nets can be used to animate Z documents, yet we think a visualization cannot replace formal semantics.

In [16] a formalism based on temporal logic is used to integrate Z schemas. The use of temporal logic has offered good starting points to the study of the dynamics of Z specifications however integration is not supported by semantics. Something more formal has been done for Object-Z [5]: a sequential execution model is introduced, defining a notion of abstract trace as a sequence of pairs (states and operations), and using some temporal logic operators ($\diamond$, $\square$, $\bigcirc$) to reason on such a model. TLA has been proposed to be integrated with Z as well, however in this case Z is only used to define *actions* specification [14]. In [9] a Unity like logic is used to formalize properties on the behavior of systems; an interleaving model with atomic operation interpretation is given but not formalized. The simplicity of Unity logic constructs fit quite well with the purpose of effectively specifying systems dynamics.

We remark that the use of Unity logic on a model based on multiset transformation is not new, in fact it has been applied to the Swarm language in order to provide a proof system to a parallel language [8]. The Swarm experience, in which the idea consists of mapping Unity-like constructs on a coordination language similar to Linda, inspired us to make some experiments in concurrent animation of Z. In fact, our semantic model based on multiset transformation offers a good basis for the parallel animator of specifications described in [7]. The animator can compile the Z language into programs written in a coordination language so as to allow a truly concurrent animation.

Acknowledgments. Partial support for this work was provided by the Commission of European Union under ESPRIT Programme Basic Research Project 9102 (COORDINATION), and by the Italian MURST 40%- "Progetto Ingegneria del Software".

References

1. M. Benjamin. A Message Passing System. An example of combining Z and CSP. In J. Nicholls, editor, *Proc. 4th Z Users Workshop (ZUM89)*, Workshops in Computing, pages 221–228, Oxford, 1989. Springer-Verlag, Berlin.
2. G. Berry and G. Boudol. The Chemical Abstract Machine. *Theoretical Computer Science*, 96:217–248, 1992.
3. G. Boudol. Some Chemical Abstract Machines. In J. deBakker, W. deRoever, and G. Rozenberg, editors, *A Decade of Concurrency*, volume 803 of *Lecture Notes in Computer Science*, pages 92–123. Springer-Verlag, Berlin, 1993.

4. P. Breuer and J. Bowen. Towards Correct Executable Semantics for Z. In J. Bowen and J. Hall, editors, *Proc. 8th Z Users Workshop (ZUM94)*, Workshops in Computing, pages 185–212, Cambridge, 1994. Springer-Verlag, Berlin.
5. D. Carrington, D. Duke, R. Duke, P. King, G. Rose, and G. Smith. Object-Z: an Object-Oriented Extension to Z. In *Formal Description Techniques (FORTE 89)*, pages 281–296. North-Holland, 1989.
6. K. M. Chandy and J. Misra. *Parallel Programming Design*. Addison-Wesley, 1988.
7. P. Ciancarini, S. Cimato, and C. Mascolo. Engineering Formal Requirements: Analysis and Testing. In *Proc. 8th Int. Conf. on Sw. Eng. and Knowledge Eng. (SEKE)*, Lake Tahoe, Ca, June 1996.
8. H. Cunningham and G. Roman. A Unity-Style Programming Logic for Shared Dataspace Programs. *IEEE Transactions on Parallel and Distributed Systems*, 1(3):365–376, July 1990.
9. A. Evans. Specifying and Verifying Concurrent Systems Using Z. In M. Bertran, T. Denvir, and M. Naftalin, editors, *Proc. FME'94 Industrial Benefits of Formal Methods*, volume 873 of *Lecture Notes in Computer Science*, pages 366–380. Springer-Verlag, Berlin, 1994.
10. A. Evans. Visualizing Concurrent Z Specifications. In J. Bowen and J. Hall, editors, *Proc. 8th Z Users Workshop (ZUM94)*, Workshops in Computing, pages 269–281, Cambridge, 1994. Springer-Verlag, Berlin.
11. P. Gardiner, P. Lupton, and J. Woodcock. A simpler semantics for Z. In J. Nicholls, editor, *Proc. 5th Z Users Workshop*, Workshops in Computing, pages 3–11. Springer-Verlag, Berlin, 1990.
12. X. He. PZ Nets: A Formal Method Integrating Petri Nets with Z. In *Proc. 7th Int. Conf. on Software Engineering and Knowledge Engineering*, pages 173–180, Rockville, Maryland, 1995. Knowledge Systems Institute.
13. I. Houston and M. Josephs. Specifying distributed CICS in Z; accessing local and remote resources. *Formal Aspects of Computing*, 6(6):569–579, 1994.
14. L. Lamport. TLZ. In J. Bowen and J. Hall, editors, *Proc. 8th Z Users Workshop (ZUM94)*, Workshops in Computing, pages 267–268, Cambridge, 1994. Springer-Verlag, Berlin.
15. K. Narayama and S. Dharap. Invariant Properties in a Dialog System. In M. Moriconi, editor, *Proc. ACM SIGSOFT Int. Workshop on Formal Methods in Software Development*, volume 15:4 of *ACM SIGSOFT Software Engineering Notes*, pages 67–79, 1990.
16. D. Richardson, S. Aha, and T. O'Malley. Specification-based Test Oracles for Reactive Systems. In *Proc. 14th IEEE Int. Conf. on Software Engineering*, pages 105–118, Melbourne, Australia, 1992.
17. J. Spivey. *Understanding Z*. Cambridge Tracts in Theoretical Computer Science. Cambridge University Press, 1988.
18. J. Spivey. *The Z Notation. A Reference Manual.* Prentice-Hall, 2 edition, 1992.

Walking Faster

Beatrice Amrhein Oliver Gloor Wolfgang Küchlin

Wilhelm Schickard Institute for Computer Science
University of Tübingen, Germany

{amrhein,gloor,kuechlin}@informatik.uni-tuebingen.de
http://www-sr.informatik.uni-tuebingen.de

Abstract. The Gröbner Walk is an algorithm that converts a given Gröbner basis of a polynomial ideal I of arbitrary dimension to a Gröbner basis of I with respect to another term order. The conversion is done in several steps (the walk) following a path in the Gröbner fan of I. We report on our experiences with an implementation of the walk. We discuss several algorithmic variations as well as important implementation techniques whose combined effect is to elevate the walk to a new level of performance.

1 Introduction

The form and size of a Gröbner basis of a polynomial ideal [3] depends heavily on the underlying term order. The same is true for the complexity of its computation by the Buchberger algorithm. Unfortunately, the lexicographic term orders as well as similar ones that enable the elimination of variables (and hence can be used for polynomial system solving) are particularly bad in this respect. Therefore, it is often advantageous to use the Buchberger algorithm only to compute a total degree order Gröbner basis, and then to convert to a lexicographic basis by a special basis conversion algorithm. This combined process can be considerably less expensive than the direct computation of a lexicographic Gröbner basis.

There are several conversion methods, for example the well-known FGLM search method [6], which in its original form is limited to zero-dimensional ideals. (To compute ideals of higher dimensions, a degree bound for the search has to be determined.) Collart, Kalkbrener, and Mall [5] recently developed a different approach, called the *Gröbner Walk*, which is independent of the dimension of the ideal.

The Gröbner walk algorithm takes as input two term orders $<, \ll$ and the Gröbner basis G of I with respect to $<$. It constructs a finite number of term orders $< = \prec_0, \dots, \prec_m = \ll$ and bases $G = G_1, \dots, G_m$, such that G_k is a Gröbner basis of I with respect to $\prec_k$. As G_{k+1} lies in the *neighborhood* of G_k, that is the corresponding cones of the Gröbner Fan [8] of I are adjacent, G_{k+1} is computed from G_k with relative ease.

Our implementation of the Gröbner walk is in C, using SACLIB [2] and some functions of the GRÖBNER package [4]. This allows us a migration path to a future parallel implementation in PARSAC-2 [7].

Our first implementation of the walk [1] confirmed the results obtained by the experimental implementation [5] on top of *Mathematica*. While for small ($< 20s$) examples the walk may present some (seconds) overhead, for larger examples it is much faster to compute a total degree Gröbner basis followed by a walk than to directly compute a lexicographic Gröbner basis. We have achieved speed-ups of 10–100 (a drop from minutes to seconds) for medium examples, and large examples which GRÖBNER[1] failed to do in hours directly could be done in minutes by walking. Note that in all the bigger cases which we investigated in [1] (*trinks1, canny, canny2, jhd5, s6*), the time for the first basis computation (with respect to a total degree ordering) accounted for only about 1% of the time spent walking home from the first base.

The purpose of this paper is to introduce and discuss important improvements, both in the algorithm and in its implementation, such as *path perturbation*, *interreduction*, *integral weight vectors*, and *special initial Gröbner basis computation*. Second, we present the first empirical comparison between the walk and the FGLM method.

The main results of this paper are, first, that the *combination* of the above mentioned improvements yields additional speed-ups on the order of 10–100 for the walk. By the above remark, these speed-ups carry over to the computation of lex Gröbner bases in many examples. Second, in our installation and on our sample of results the walk is (apart from very simple cases) at least as fast as the FGLM conversion. In fact, as the dimension of $K[x_1, \ldots, x_n] / I$ grows, the walk clearly has an advantage.

The remainder of the paper is organized as follows. Section 2 recalls from [5] some definitions and facts about term orders and weight vectors and the crucial observation which enable us to pass from the Gröbner basis over a given cone to the Gröbner basis over an adjacent cone. Section 3 describes the algorithm in some detail. Sections 4 and 5 discuss our present algorithmic variations and implementation techniques, respectively. Section 6 presents empirical results. We conclude with remarks on future work in Section 7. Readers familiar with Gröbner walk theory may skip ahead to Section 4; Sections 4–6 contain the main contributions of this paper.

For further references and related work, the reader is referred to [5].

2 Theoretical Background

Let I be an ideal of the polynomial ring $R := K[x_1, \ldots, x_n]$, where K is an arbitrary field. Let f be a polynomial in R, G a subset of R, and $\langle G \rangle$ the ideal generated by G. For a term order $<$, the initial monomial (or head) of f is denoted by $in_<(f)$ and $G_<$ is the set $\{in_<(g) \mid g \in G\}$. If G is a Gröbner basis of I, then $G_<$ is a Gröbner basis of the monomial ideal $\langle I_< \rangle$.

A weight vector ω is an element of

$$\Omega^n := \{(\psi_1, \ldots, \psi_n) \in \mathbb{Q}^n \mid \psi_i \geq 0 \text{ for every } 1 \leq i \leq n\}.$$

[1] We improved GRÖBNER somewhat, e.g. by adding the *sugar* strategy and by modifying the reduction algorithm. It can probably be speeded up further by a faster coefficient arithmetic.

ω defines for each monomial $t = ax_1^{e_1} \cdots x_n^{e_n}$ in R an ω-degree by

$$deg_\omega(t) := \sum_{i=1}^{n} e_i \omega_i$$

and so induces a transitive relation on the terms. The ω-degree of a non-zero polynomial f is the maximum of the ω-degrees of the monomials in f.

The initial form of f with respect to ω, abbreviated $in_\omega(f)$, consists of all monomials in f with maximum ω-degree. For a subset G of R, we denote the set $\{in_\omega(g) \mid g \in G\}$ by G_ω. For the term order $\prec$ and the weight vector ω, we define the term order $\omega^\prec$ by

$$t_1 \, \omega^\prec \, t_2 \quad :\Longleftrightarrow \quad deg_\omega(t_1) < deg_\omega(t_2) \ \vee \ (deg_\omega(t_1) = deg_\omega(t_2) \wedge t_1 \prec t_2).$$

We denote by $C_\prec(I)$ the usual topological closure of $\{\omega \in \Omega^n \mid \langle I_\prec \rangle = \langle I_\omega \rangle\}$. This is a convex cone in $\mathbb{Q}^n$ with nonempty interior. The *Gröbner Fan* of I [8] is the set $F(I) = \{C_\prec(I) \mid \prec \text{ a term order}\}$.

Let $<$ and $\ll$ be two term orders which cones contain a common weight vector ω (i.e., $\omega \in C_<(I) \cap C_\ll(I)$) and $G = \{g_1, \ldots, g_r\}$ a reduced Gröbner basis of I with respect to $<$. The following lemma shows how the Gröbner basis of I with respect to $\ll$ (i.e., the *neighboring Gröbner basis*) can be obtained from G.

Lemma 1 [5]. *Let $\{m_1, \ldots, m_s\}$ be the reduced Gröbner basis of $\langle I_\omega \rangle$ with respect to $\ll$. Then, for all $i = 1, \ldots s$, there is a representation*

$$m_i = \sum_{j=1}^{r} h_{ij} \, in_\omega(g_j)$$

with ω-homogeneous polynomials $h_{i1}, \ldots, h_{ir}$. Furthermore, the set $\{f_1, \ldots, f_s\}$ with $f_i := \sum_{j=1}^{r} h_{ij} g_j$ is a Gröbner basis of I with respect to $\ll$.

This immediately leads to one step in the Walk. Namely, starting with a Gröbner basis G with respect to $<$, we determine G_ω and compute the Gröbner basis $\{m_1, \ldots, m_s\}$ of $\langle G_\omega \rangle$ with respect to $\ll$. Applying Lemma 1, we lift $\{m_1, \ldots, m_s\}$ to a Gröbner basis of I with respect to $\ll$.

3 The Algorithm

We assume that we are given a Gröbner basis G of I with respect to a given term order $<$. Our aim is to compute a Gröbner basis of I with respect to another term order $\ll$.

First, we determine weight vectors $\sigma \in C_<(I)$ and $\tau \in C_\ll(I)$. As the weight vectors can be seen as projective coordinates in the fan, for the geometric view it may be best to choose σ and τ such $||\sigma|| = ||\tau|| = 1$. Then, $\sigma = (1/n, \ldots, 1/n)$

and $\tau = (1, 0, \ldots, 0)$ may be the input for a walk from a total degree ordering Gröbner basis to a lexicographic ordering Gröbner basis, for example.

We *walk* on the Gröbner Fan of I, passing through a succession of cones (cf. Figure 2). In terms of weight vectors ω, we walk along the line segment $\overline{\sigma\tau}$, where intermediate weight vectors are denoted by

$$\sigma =: \omega_1, \ldots, \omega_m := \tau.$$

To obtain a uniform designation, we introduce the notation

$$\prec_0 := < = \sigma^<, \ \prec_1 := \omega_1^{\ll} = \sigma^{\ll}, \ \prec_k := \omega_k^{\ll}, \ \prec_m := \omega_m^{\ll} = \tau^{\ll} = \ll .$$

By $G(J, \prec)$, we denote a Gröbner basis of the ideal J with respect to $\prec$. Then, the Gröbner basis of I over the cone $C_k = C_{\prec_k}(I)$ is denoted by $G(I, \prec_k)$.

Given the Gröbner basis $G(I, \prec_{k-1})$ and the weight vector ω_k, we walk from cone C_{k-1} to C_k. We first determine $(G(I, \prec_{k-1}))_{\omega_k} = \{in_{\omega_k}(g) : g \in G^{k-1}\}$ (cf. ❷ in Table 1), which is a Gröbner basis of $\langle I_{\omega_k}\rangle$ with respect to $\prec_{k-1}$. A reduced Gröbner basis $G(\langle I_{\omega_k}\rangle, \prec_k) = \{m_1, \ldots, m_s\}$ is computed by applying Buchberger's algorithm ❹ with the term order $\prec_k$. This is usually a very short task, as most of the initials in $\{in_{\omega_k}(g) \mid g \in G_{k-1}\}$ consist of only one monomial.

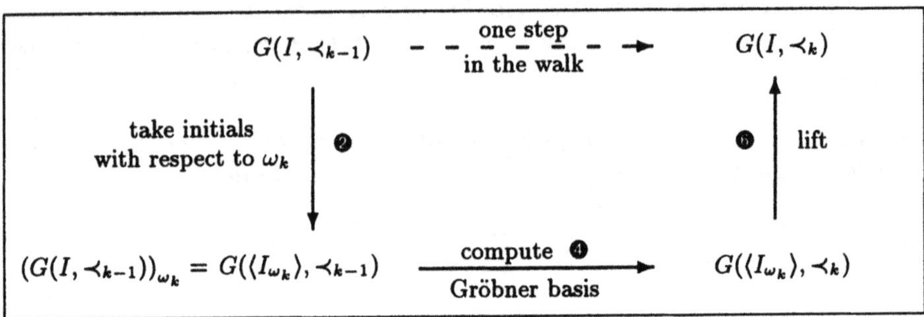

Fig. 1. Step k of the Gröbner Walk

Afterwards, we construct $G(I, \prec_k)$ according to Lemma 1 (❺). For that, we reduce each polynomial $m_i \in \{m_1, \ldots, m_s\} = G(\langle I_{\omega_k}\rangle, \prec_k)$ modulo the Gröbner basis $G(\langle I_{\omega_k}\rangle, \prec_{k-1})$ and subsequently obtain a representation

$$m_i = \sum_{j=1}^{r} h_{ij} in_{\omega_k}(g_j)$$

with $g_j \in G(I, \prec_{k-1})$. Since $\omega_k \in \left(C_{\prec_{k-1}}(I) \cap C_{\prec_k}(I)\right)$, the set $F := \{f_1, \ldots, f_s\}$ with

$$f_i := \sum_{j=1}^{r} h_{ij} g_j$$

is a Gröbner basis of I with respect to $\prec_k$.

It remains to determine the next weight vector ❾, that is, the point on our walk where we leave the cone C_k. For that, we apply the following

Lemma 2 [5]. *Let $<_1$ and $<_2$ be two term orders and $G = \{g_1, \ldots, g_r\}$ the reduced Gröbner basis of I with respect to $<_1$. Then, $C_{<_1}(I) = C_{<_2}(I)$ if and only if $in_{<_1}(g_i) = in_{<_2}(g_i)$ for every $i \in \{1, \ldots, r\}$.*

To detect a change in initial forms, we determine the first weight vector $\omega(t) := \omega_k + t(\tau - \omega_k)$, $0 < t \leq 1$, on the directed path $\overline{\omega_k \tau}$ with the following property.

$$t = \min\left(\{s : deg_{\omega(s)}(p_1) = deg_{\omega(s)}(p_i), g = p_1 + \ldots + p_n, g \in G(I, \prec_k)\} \cap (0,1]\right)$$

This can be done by the calculation of one scalar product in $\mathbb{Q}^n$ and one rational quotient per monomial. Thus, if t is defined, it is a positive rational number.

- If $t = 1$ is the minimum, the next conversion is the last step of the walk.
- If there is no t with $0 < t \leq 1$, then $G_{\prec_k} = G_{\tau\ll}$ and we already determined the final Gröbner basis.

In this way, we successively compute Gröbner bases and weight vectors

$$G(I, \prec_1),\ \omega_2,\ G(I, \prec_2),\ \ldots,\ \omega_m,\ G(I, \prec_m).$$

The termination of the algorithm follows from

Lemma 3 [8]. *The Gröbner fan of a polynomial ideal has finite cardinality.*

```
Groebner_Walk
Input:   <, ≪: orders
         G: reduced Gröbner Basis with respect to <
         σ ∈ C<(⟨G⟩): starting weight vector
         τ ∈ C≪(⟨G⟩): target weight vector
Output:  G: Gröbner Basis with respect to ≪

≺, ≺+: current and next order
ω: current weight vector

❶ ω = σ; ≺ = ω<; ≺+ = ω≪;
❷ Gω = initials(G, ω);                // Take initials of G
❸ Gω+ = sort(Gω, ≺+);                 // Sort initials according new order
❹ Gω+ = init_gb(Gω+, ≺+);             // Compute Groebner Basis of initials
❺ Gω+ = interreduce(Gω+, ≺+);         // Interreduce Gω+
❻ G = lift(Gω+, ≺+, Gω, G, ≺);        // Lift Gω+ to a full Groebner Basis wrt. ≺+
❼ G = interreduce(G, ≺+);             // Interreduce G
❽ if( ω = τ ) return(G);              // Stop if target order reached
❾ t = determine_border(ω, τ, G);      // Determine the next parameter
  if( undefined(t) ) return(G);       // No further conversion needed
  ω = (1-t)*ω + t*τ;                  // Determine the next border
❿ ≺ = ≺+; ≺+ = ω≪; Goto ❷;
```

Table 1. The Gröbner Walk Algorithm

4 Algorithmic Variations

4.1 Path Perturbation

Whenever the path leaves a cone of the Gröbner Fan, some of the initial monomials $in_\omega(g)$ of the Gröbner basis with respect to the weight vector ω become initial forms (true *poly*nomials). Adjacent cones meet in (hyper-)planes (planes or edges in three dimensions). At points of such intersections of several cones, either several monomials in a polynomial have the same maximal weight and become the initial form, or several polynomials have initial forms containing more than one monomial. Hence, the initial forms become larger. Especially in a complicated fan, meeting-points where several cones adjoin are frequent. Moreover, if the walk moves along the intersection of two or even more cones (along a surface in three dimensions), there are monomials which keep the same maximal weight, and therefore remain in the initial form of a polynomial on this line. Anyway, both cases cause the initial forms to be unnecessarily heavy during the walk.

We can avoid walking through meeting-points or along (hyper-)surfaces of cones if we slightly perturb the starting point (the starting weight vector σ) and the end-point (the target weight vector τ) of the walk, making sure we stay within their cones. Then, the path passes through a sequence of maximal adjacent cones. The initial forms are shorter and the individual tasks of converting their Gröbner bases from $\prec_{k-1}$ to $\prec_k$ likewise become much smaller. However, we may have to compute more Gröbner bases since we may have to walk through more cones on the perturbed path.

Figure 2 shows a slice of a sample fan of an ideal in three variables (x, y, z) as intersection with the plane $x + y + z = 1$. Path segment ① goes through a common edge of three cones (a point in the slice), path segment ② runs along a surface of two cones (a line in the slice).

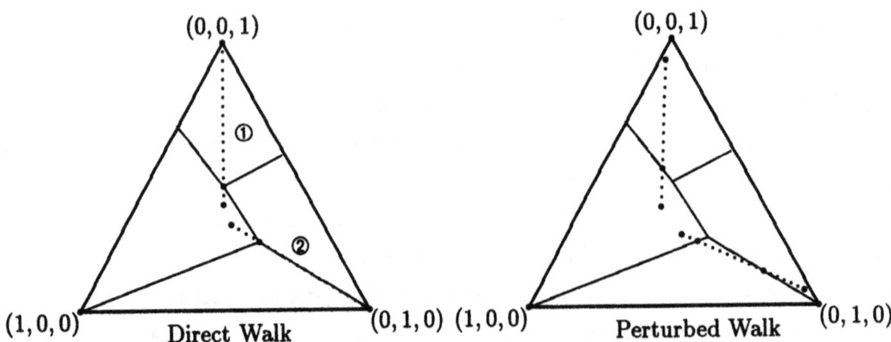

Fig. 2. A Slice of a Gröbner Fan

In the first versions of our implementation [1], it was rarely profitable to perform path perturbation. However, relying now on a very fast Gröbner basis algorithm for the special case of short polynomials, path perturbation became a very important and powerful algorithmic variation (see the results section).

4.2 Interreduction of Intermediate Gröbner Bases

As mentioned in [5], Lemma 1 remains valid if we replace "reduced" by "minimal." Furthermore, if the Gröbner basis $\{m_1, \ldots, m_s\}$ of $\langle I_\omega \rangle$ is minimal, then the lifted basis $\{f_1, \ldots, f_s\}$ is also minimal. Therefore, the lifting process does not rely on the interreductions ❼.

Usually, we compute the next border ❾ using Lemma 2. If we do not have a reduced Gröbner basis, Lemma 2 does not hold. Still, whenever we change the cone, the initial forms of the Gröbner basis change. Therefore, we can use the same technique to determine the next weight vector, which, however, may still belong to the same cone.

As a consequence of this weakening, we may erroneously conclude that we are leaving a cone (because an initial form of a non-reduced polynomial changed). All the same, we have to adapt the Gröbner basis at this point:

Example *For all orders, the reduced Gröbner basis of* $I = \langle\{x,y\}\rangle \subseteq \mathbb{Q}[x,y]$ *is* $G = \{\ x,\ y\ \}$ *and hence the fan consists of only one cone.*
$G_1 = \{\ x+y,\ y\ \}$ *is a minimal (non-reduced) Gröbner basis of* I *for* $x > y$ *(with initials* $\{x,y\}$*). Walking from* $\sigma = (1,0)$ *towards* $\tau = (0,1)$*, we find at* $\omega = (1/2, 1/2)$ *the new initial forms* $\{x+y, y\}$*. We are neither at the border of nor are we leaving the cone, but nevertheless we have to transform the basis since* $G_2 = \{\ y+x,\ y\ \}$ *is no Gröbner basis of* I *for* $y > x$*.*

This transformation can be done by a usual step of the walk. Such an additional step results in the desired Gröbner basis and is performed very quickly.

Omitting interreduction, we may compute extra Gröbner bases. Note that with minimal bases, extra basis computation can happen only finitely many times within each cone (and hence termination of the walk is still guaranteed).

In the first version of our implementation [1], the extra cost of basis computation was more than offset by the savings in reductions. However, using a better representation of the weight vectors lowers the cost for the interreduction as well as for the sorting of polynomials. Now, it seems that we gain more by the elimination of the extra Gröbner basis computations than we loose by the interreduction. For examples see the result section.

5 Implementation Techniques: Our Gait

5.1 Integral Weight Vectors

Given the order $\omega^<$, every comparison of two monomials involves costly rational arithmetic with the rational weight vector ω. Particularly in the perturbed walk, the representation of the rational numbers in the weight vectors may become very long. Neither the algorithm nor the implementation requires the walk to stay on the (hyper-)plane $\sum x_i = 1$. As only the direction of the weight vectors is needed, we can scale them until they are integral vectors. Then, all comparisons become much cheaper, mainly because they don't allocate heap space any more.

The first step for this modification of the implementation is to choose an integral starting vector and an integral target vector. Moreover, all intermediate vectors can be chosen integral as well. From the geometric point of view, the walk is then on a zigzag course in the plane E through starting point, target point, and the zero of the fan (see Figure 3). Its projection onto the (hyper-)plane $\sum x_i = 1$ gives the original walk, which also lies on plane E.

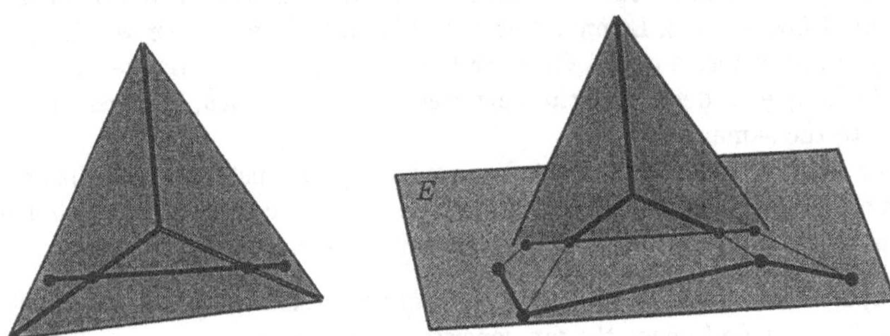

Fig. 3. The Zigzag Walk

5.2 The Specialized Buchberger Algorithm for Initial Forms

A further analysis of the initial forms that are used for the Gröbner basis computation in step ❹ reveals the need for some adaptation for this special case.

First of all, in the perturbed walk most of the initial forms are monomials. In our examples, typically only one or two initials out of 60 or more polynomials are not monomials. As the critical pair of two monomials is unnecessary (its S-polynomial is always zero), we have to take care that we don't even consider forming such pairs.

In addition, our empirical data showed that Gröbner bases of two maximal adjacent cones (as we deal with in the perturbed walk) are very similar. That is, the computation of the initial Gröbner basis usually adds only one or two new polynomials to the basis. As most of the S-polynomials reduce to zero in one step, there is no use spending time on sophisticated selection strategies.

6 Empirical Results

For the present work, we used a SPARCstation 10 (90MHz HyperSPARC) with 64MB RAM, of which we used 16MB for the heap (SAC-cells).

6.1 Algorithmic Variations

We can summarize the results as follows.

- Perturbation alone does not speed up the walk, as long as the Buchberger algorithm for the initial forms ❹ is not specialized.
- Interreduction ❼ is sometimes helpful, but interreduction alone does not achieve the big breakthrough.
- Using the specialized Buchberger algorithm, we obtain always better timings. However, the gain is often only moderate.
- The real breakthrough is achieved only by the combined methods. The specialized Buchberger algorithm, for example, becomes important only if the walk is perturbed. In general, we obtain best timings if we combine all three methods: we take a perturbed path with integral weight vectors, interreduce after every lifting step, and choose a specialized Buchberger algorithm for initial forms.

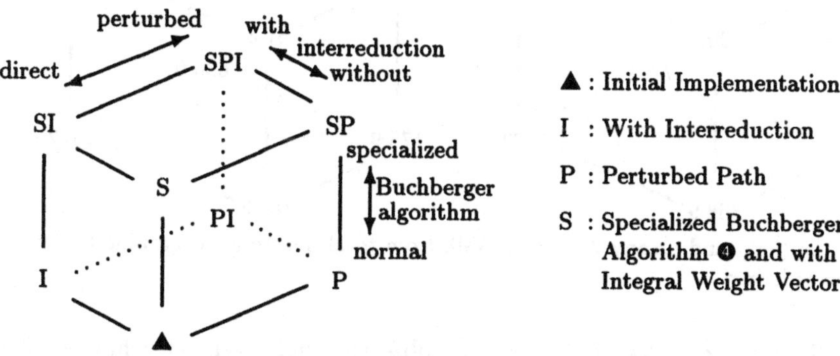

Fig. 4. The Legend for the Cube of Timings

As the various variations of the algorithm are independent, we represent the different runtimes in a cube (see Figure 4). Each vertex stands for a setup of the algorithm. In this way, we can show the impact of the different methods. In earlier work [1], we evaluated only the setups ▲ and, to some extent, P.

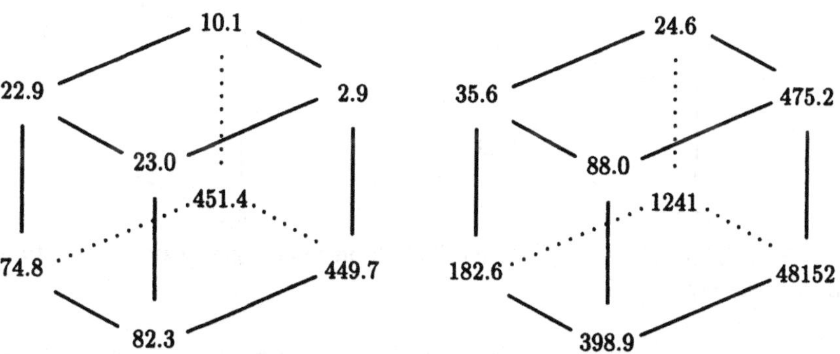

Fig. 5. The Canny example. Walk from total degree lexicographic ordering (left) and graded reverse lexicographic ordering (right) to lexicographic ordering.

In Figure 5 left, the interreduction does not much affect the runtimes. In the perturbed path, the walk takes 6 steps in the non-interreduced case and 4 steps in the interreduced case. The interreduction needs more time than we gain by saving extra steps.

Perturbation of the path ↙↗ often slows down the methods with non-specialized Gröbner basis computations and sometimes even the specialized Gröbner basis algorithm; see for example Figure 5, right. Here, the enormous differences of the timings originate from the large span of the number of steps: the interreduction lowers the steps from 10 to 2 in the direct path and from 618 to 23 in the perturbed path.

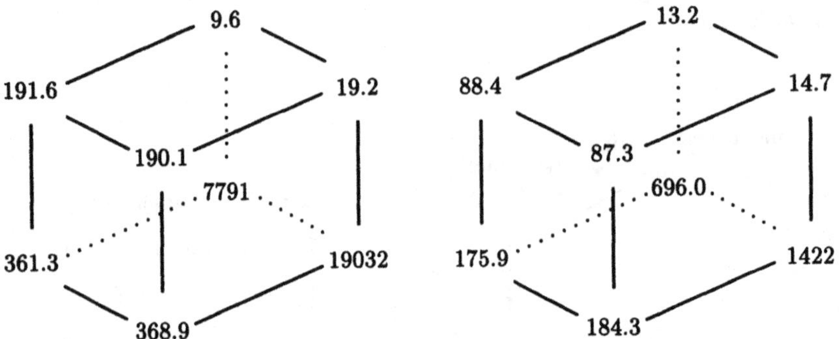

Fig. 6. s6 (left) and Canny-2 (right). Walk from total degree lexicographic to lexicographic ordering.

In the two examples of Figure 6, we observe a quite typical behavior. The perturbation severely slows down the walk in the non-specialized implementation. However, using the specialized Buchberger algorithm and integral weight vectors, we achieve remarkable speed-ups.

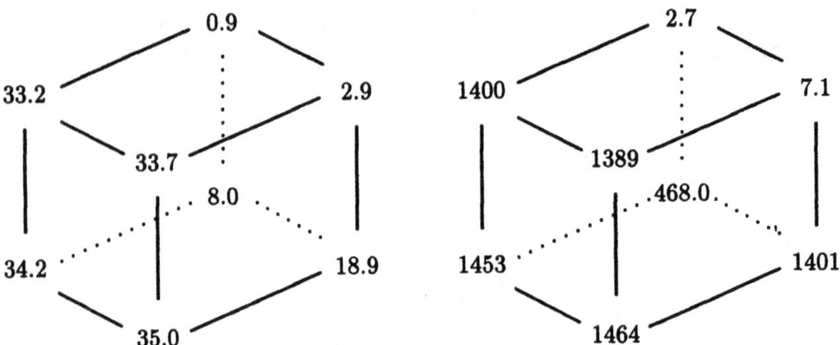

Fig. 7. Trinks 1 (left) and jhd 5 (right). Walk from total degree lexicographic to lexicographic ordering.

In the examples of Figure 7, using the specialized Buchberger algorithm ↕ in the direct path does not show much profit. The reason for this behavior is that the direct path takes only very few steps (2-6), which are therefore relatively hard. Thus, they are not significantly accelerated by the specialized Buchberger algorithm.

Using the specialized Buchberger algorithm $\updownarrow$ is usually most profitable if the path is perturbed and interreduction is performed.

6.2 Comparison with FGLM

We compared our timings with the ones we obtained by our implementation of the FGLM basis conversion algorithm. In our installation and on our sample of problems the walk is (apart from very simple cases) at least as fast as the FGLM conversion. In fact, with growing dimension of the vector space $K[x_1, \dots, x_n] / I$ the walk is much faster. The following tables show some typical examples.

	Walk	FGLM
Ex1 grev → lex	15.3 s	113.4 s
Ex2 grev → lex	119.8 s	523.4 s
Ex3 grev → lex	71.6 s	677.9 s
s6 grev → lex	15.8 s	55.1 s
s7 grev → lex	164.9 s	739.4 s

	Walk	FGLM
Ex1 tot → lex	15.1 s	143.0 s
Ex2 tot → lex	119.6 s	481.4 s
Ex3 tot → lex	71.4 s	912.8 s
s6 tot → lex	5.7 s	55.0 s
s7 tot → lex	64.1 s	769.1 s

7 Conclusions

For larger examples, it is much faster to compute a total degree Gröbner basis followed by a walk than to directly compute a lexicographic Gröbner basis. With our initial implemention, we already achieved speed-ups of up to a factor 10 over the direct lexicographic Gröbner basis computation. The *combination* of the further improvements yields additional speed-ups on the order of 10–100 for the walk.

	grev → lex		tot → lex	
	▲	SPI	▲	SPI
Ex1	1688 s	15.3 s	1690 s	15.1 s
Ex2	> 120 min	119.8 s	> 120 min	119.6 s
Ex3	> 75 min	71.6 s	> 75 min	71.4 s

Furthermore, we have observed that, apart from very simple cases, the walk is always faster than the FGLM conversion.

Acknowledgements

We are indebted to Stéphane Collart, Michael Kalkbrener, and Daniel Mall for many helpful discussions and suggestions. In particular, the perturbed walk was also suggested by Kalkbrener. This research is supported by grant Ku 966/2-1 from Deutsche Forschungsgemeinschaft.

Appendix

For our tables, we chose the following examples:

Ex1	$x < y < z$	$xy^3 + y^4 + yz^2 - z^3 - 2xz^3$
	$2x^2y + x^3y + 2xy^2z$	$2 - 3x^2y + 2x^3y + yz^3$
Ex2	$x < y < z$	$x + 3xy^3 + y^4 + yz^2$
	$-x^2z + 2y^3z + z^2 + 2yz^2 + 3xyz^2$	$3x^3 + xy^2 + yz^2 - 2xz^3$
Ex3	$x < y < z$	$x^2 + y^4 + x^3z + yz - 2xz^3$
	$x^2y^2 + y^3z + z^3 + 3yz^3$	$y^4 - x^2z + 2y^2z - 2xyz^2$
s6	$x_6 < x_5 < x_4 < x_3 < x_2 < x_1$	$2x_6x_2 + 2x_5x_3 + x_4^2 + x_1^2 + x_1$
		$2x_6x_3 + 2x_5x_4 + 2x_2x_1 + x_2$
	$2x_6x_4 + x_5^2 + 2x_3x_1 + x_3 + x_2^2$	$2x_6x_5 + 2x_4x_1 + x_4 + 2x_3x_2$
	$x_6^2 + 2x_5x_1 + x_5 + 2x_4x_2 + x_3^2$	$2x_6x_1 + x_6 + 2x_5x_2 + 2x_4x_3$
s7	$x_7 < x_6 < x_5 < x_4 < x_3 < x_2 < x_1$	$2x_7x_2 + 2x_6x_3 + 2x_5x_4 + x_1^2 + x_1$
	$2x_7x_3 + 2x_6x_4 + x_5^2 + 2x_2x_1 + x_2$	$2x_7x_4 + 2x_6x_5 + 2x_3x_1 + x_3 + x_2^2$
	$2x_7x_5 + x_6^2 + 2x_4x_1 + x_4 + 2x_3x_2$	$2x_7x_6 + 2x_5x_1 + x_5 + 2x_4x_2 + x_3^2$
	$x_7^2 + 2x_6x_1 + x_6 + 2x_5x_2 + 2x_4x_3$	$2x_7x_1 + x_7 + 2x_6x_2 + 2x_5x_3 + x_4^2$

The other examples originate from the PoSSo [9] examples list.

References

1. B. Amrhein, O. Gloor, and W. Küchlin. How Fast Does the Walk Run? 5th Rhine Workshop for Computer Algebra, St. Louis, France, 1996.
2. Buchberger, Collins, Encarnación, Hong, Johnson, Krandick, Loos, Mandache, Neubacher, and Vielhaber. SACLIB User's Guide, 1993. On-line documentation.
3. B. Buchberger. Gröbner bases: An algorithmic method in polynomial ideal theory. In N. K. Bose, editor, *Recent Trends in Multidimensional Systems Theory*, chapter 6. Reidel, 1985. (Also Report CAMP-83.29, U. Linz, 1983).
4. B. Buchberger and W. Windsteiger. GRÖBNER: A library for computing Gröbner bases based on SACLIB, 1993. Manual for Version 2.0.
5. S. Collart, M. Kalkbrener, and D. Mall. Converting bases with the Gröbner Walk. *JSC*, 1996. In print.
6. J. Faugère, P. Gianni, D. Lazard, and T. Mora. Efficient computation of zero-dimensional Gröbner Bases by change of ordering. *JSC*, 16:329–344, 1993.
7. W. W. Küchlin. PARSAC-2: Parallel computer algebra on the desk-top. In J. Fleischer, J. Grabmeier, F. Hehl, and W. Küchlin, editors, *Computer Algebra in Science and Engineering*, pages 24–43, Singapore, 1995. World Scientific.
8. T. Mora and L. Robbiano. The Gröbner Fan of an ideal. *JSC*, 6:183–208, 1988.
9. PoSSo. Polynomial systems library. `ftp: posso.dm.unipi.it`.

Integer and Rational Arithmetic on MasPar

Tudor Jebelean*

Research Institute for Symbolic Computation
A-4232 Hagenberg, Austria (Europe)
tudor@risc.uni-linz.ac.at

Abstract. The speed of integer and rational arithmetic increases significantly by systolic implementation on a SIMD architecture. For multiplication of integers one obtains linear speed-up (up to 29 times), using a serial–parallel scheme. A two-dimensional algorithm for multiplication of polynomials gives half-linear speed-up (up to 383 times). We also implement multiprecision rational arithmetic using known systolic algorithms for addition and multiplication, as well as recent algorithms for exact division and GCD computation. All algorithms work in "least-significant digits first" pipelined manner, hence they can be well aggregated together. The practical experiments show that the timings depend linearly on the input length, demonstrating the effectiveness of the systolic paradigm for multiple precision arithmetic.

1 Introduction

Systolic parallelization of multiprecision arithmetic in the "most-significant digits first" (MSF) pipelined manner was considered by [13] and other authors (see [12], chapter 3), using the signed-digit redundant representation.

Our approach is different: we use "least-significant digits first" (LSF) algorithms, because this allows pipelined aggregation of the various operations. Also, these algorithms use standard representation of multiprecision integers in an arbitrary radix (typically a power of 2), which makes them suitable for implementation on multiprocessor architectures.

SIMD parallelization of computer algebra algorithms did not receive much attention in the literature. [15] reports a 45 times speed–up of Gröbner Basis computations by parallelizing multiprecision algorithms on the SIMD like Convex vector processor, but most of the speed-up is due to some improvements in list processing operations and to the use of 64-bit arithmetic. Univariate polynomial multiplication by parallelizing both the level of coefficient operations and binary-digit operations was considered by [1] on the ICL DAP computer (SIMD architecture).

The MasPar computer – shortly presented in Section 2 – is particularly suitable for the implementation of systolic algorithms. Of paramount importance for

* Supported by the Austrian *Fonds zur Förderung der wissenschaftlichen Forschung*, project P10002 MAT

our application are: the fast communication between adjacent processors, and the high efficiency of global broadcasting.

In Sections 3 and 4 we describe the implementation of long integer and of univariate integral polynomial multiplication. We use multiprecision variants of serial / parallel multipliers which can be easily derived from the "school method". Apparently these algorithms were the first to be considered for hardware multiplication – see [2]. In both algorithms, one of the operands is present in the array at the beginning of computation, while the elements of the other one are broadcasted to the parallel processors, one at each step. The first algorithm pipelines the result out via the first processor, while the second algorithm leaves the result in the array. The second algorithm is suitable for embedding into polynomial multiplication scheme, yielding an algorithm with two–level systolic parallelism, which maps naturally onto the two–dimensional architecture of MasPar.

For multiplication of multiprecision integers we obtain almost linear speed-up over the classical sequential algorithm (29 times for 30 digit integers, efficiency 95%). The two-dimensional algorithm for multiplication of univariate integral polynomials gives almost half-linear speed-up (383 times for polynomials of degree 29 with multiprecision coefficients of 15 digits, efficiency 43%).

In Sections 5 – 9 we present the systolic implementation of a rational operation which is widely used in typical algebraic computations – e. g. in Gröbner Bases [4]. Besides multiplication, one also uses addition, division, and GCD computation. Theoretically, **addition** in standard representation cannot be improved by systolic parallelization, but practically it runs in constant time, because the carry chain seldom exceeds two digits. Since **division** is with null remainder, one can use the exact division algorithm recently introduced in [7] - some systolic parallelizations of exact division are described in [8] - we choose the one which is most suitable in the present context. The most complicated operation is the **computation of the GCD**, which is implemented using the systolic parallelization [9] of the recently developed algorithm from [6, 14].

The most important conclusion of the practical experiments is that one obtains linear timings - that is, running time depends linearly on the length of the input numbers. This demonstrates the effectiveness of the systolic paradigm for the implementation of long integer and long rational arithmetic.

2 The MasPar Computer

We present here only those features which are relevant for our approach.

MasPar is a SIMD distributed memory machine, with 1024 Processing Elements (PE's), arranged in a 32 by 32 mesh (torus). The PE's are driven by a sequential Array Control Unit (ACU). The device can be programmed in the language C, with some special extensions for handling MasPar parallelism.

Data: The ACU and each PE have their own internal memory for data. In C language, one has to use `plural` to declare the variables which are allocated on the parallel PE's. A `plural` variable will have one instance on each PE, possibly containing different values. The variables which are not `plural` are called *singular* and are allocated on ACU.

Program flow: The operations involving only *singular* variables are executed sequentially on the ACU. The operations involving `plural` variables are executed in parallel on the PE's. All PE's execute the same instructions synchronously. However, at certain moments some of the PE's may be "masked" (by conditional instructions), and then they execute nothing.

Data communication between ACU and PE's: The ACU accesses all the PE's in parallel. Hence, data can be broadcasted to all the PE's in one step. For instance, if `a` is *singular* and `b` is `plural`, then the assignment "`b = a;`" will send the value of `a` to all active PE's. The reverse operation is not possible, but one can use "`a = globalor(b);`", which performs a bitwise logical OR on all `b`'s in the active PE's. Also, a plural variable (say `b`) on a particular PE can be accessed using `proc[i].b` (linear addressing, $0 \leq \mathtt{i} \leq 1023$) or `proc[y][x].b` (2D addressing, $0 \leq \mathtt{x}, \mathtt{y} \leq 31$). The `proc` construct may be used in either left or right hand side of assignments, thus yielding the means to store/load values to/from particular PE's.

Data communication between PE's: Data may be moved between adjacent PE's by using the `xnet` construct. For instance, if `b` and `c` are plural variables, then "`xnetW[1].b = c;`" means "store the value of `c` into `b` of the left neighbor". This is also executed in parallel on all active PE's.

3 Long Integer Multiplication

The inputs A, B and the output $C = A*B$ are multiprecision integers represented as lists of positive digits in radix β:

$$A = a_0 + a_1 * \beta + \ldots + a_{n-1} * \beta^{n-1}, \quad B = b_0 + b_1 * \beta + \ldots + b_{m-1} * \beta^{m-1}.$$

$$C = c_0 + c_1 * \beta + \ldots + c_{n+m-1} * \beta^{n+m-1}.$$

The classical algorithm for multiplication consists of a double loop whose innermost instruction is:

$$(carry, c_{i+j}) \leftarrow c_{i+j} + b_j * a_i + carry.$$

This algorithm is inherently sequential, because each step uses the *carry* produced by the previous step. We change this by using a list $Y = (y_0, y_1, \ldots, y_{n+m-1})$ to hold the carries produced at each step. This list has as many elements as C has. The computation then proceeds in two stages:

- Stage 1: The additions and the multiplications are performed and the carries are produced. Since the outermost loop (over A) is performed sequentially, the carries produced in one step may be used in the following step.
- Stage 2: The carries are absorbed into C, by adding each y_k to c_{k+1}. These additions may produce new carries, which are again stored in Y list, and the absorption stage is repeated until all the carries become zero.

This scheme allows the parallelization of the inner loop (over B), leading to the systolic algorithm shown in Fig. 1. The local variables on each processor are denoted by $\overline{B} = (\overline{b}_0, \overline{b}_1, \ldots, \overline{b}_m)$, $\overline{C} = (\overline{c}_0, \overline{c}_1, \ldots, \overline{c}_m)$, $\overline{Y} = (\overline{y}_0, \overline{y}_1, \ldots, \overline{y}_m)$. The vector A is not stored in the processors. Rather, at each iteration of the main loop, one element of A is send to all the processors for the computation in line { 6}. During Stage 1, the m processors act as a window which moves along the vector C, one element at a time. In other words, C is piped through the string of m processors. During Stage 2, the window is fixed on the last m elements of C.

In practice we use an $m + 1^{th}$ processor whose $\overline{b}_m, \overline{c}_m$, and $\overline{y}_m$ are zero all the time. This boundary processor does not participate in the computation, but its presence avoids boundary tests.

{ 0} $C \leftarrow$ **IntSysMul.1**(A, B) [Systolic integer multiplication, version 1]
{ 1} $\quad \overline{B}, \overline{C}, \overline{Y} \leftarrow (0, \ldots, 0)$ [$m+1$ positions]
{ 2} $\quad$ **for** $j = 0, 1, \ldots, m-1$ **do** [load B sequentially]
{ 3} $\quad\quad \overline{b}_j \leftarrow b_j$
{ 4} $\quad$ **for** $i = 0, 1, \ldots, n-1$ **do** {Stage 1: add and multiply]
{ 5} $\quad\quad$ **for** $j = 0, 1, \ldots, m-1$ **in parallel do**
{ 6} $\quad\quad\quad (\overline{y}_j, \overline{c}_j) \leftarrow \overline{c}_j + \overline{b}_j * a_i + \overline{y}_j$ [compute]
{ 7} $\quad\quad\quad c_i \leftarrow \overline{c}_0$ [extract next digit of C]
{ 8} $\quad\quad$ **for** $j = 0, 1, \ldots, m-1$ **in parallel do**
{ 9} $\quad\quad\quad \overline{c}_j \leftarrow \overline{c}_{j+1}$ [shift $\overline{C}$ left]
{10} $\quad$ **while globalor**$(\overline{Y})$ **do** [Stage 2: absorb carries]
{11} $\quad\quad$ **for** $j = 0, 1, \ldots, m-1$ **in parallel do**
{12] $\quad\quad\quad (\overline{y}_{j+1}, \overline{c}_j) \leftarrow \overline{c}_j + \overline{y}_j$
{13} $\quad\quad \overline{y}_0 \leftarrow 0$
{14} $\quad$ **for** $j = 0, 1, \ldots, m-1$ **do** [extract rest of C sequentially]
{15} $\quad\quad c_{n+j} \leftarrow \overline{c}_j$

Fig. 1. Systolic multiprecision multiplication, version 1.

The parallel loops {5}, {8}, {11} and the initialization {1} require constant time. The other loops are {2}: n steps, {4}: n steps, {10}: at most m steps, and {14}: m steps. Hence $T_{\text{systolic}} = O(n+m)$. For balanced–length operands: $T_{\text{systolic}} = O(n)$.

We implemented the algorithm **IntSysMul.1** on MasPar MP1, using only the first row of 32 processors, and the classic algorithm, using only one processor. Fig. 2 shows the timings in milliseconds for the two algorithms (systolic timings are scaled by 10). The speed–up is linear w.r.t. input length and ranges between 4 (at 5 digits) and 29 (at 30 digits). The efficiency ranges between 81% and 96%.

Second version: If C is to be used in subsequent computations (as it is the case in polynomial multiplication), then it is useful to leave it in the array,

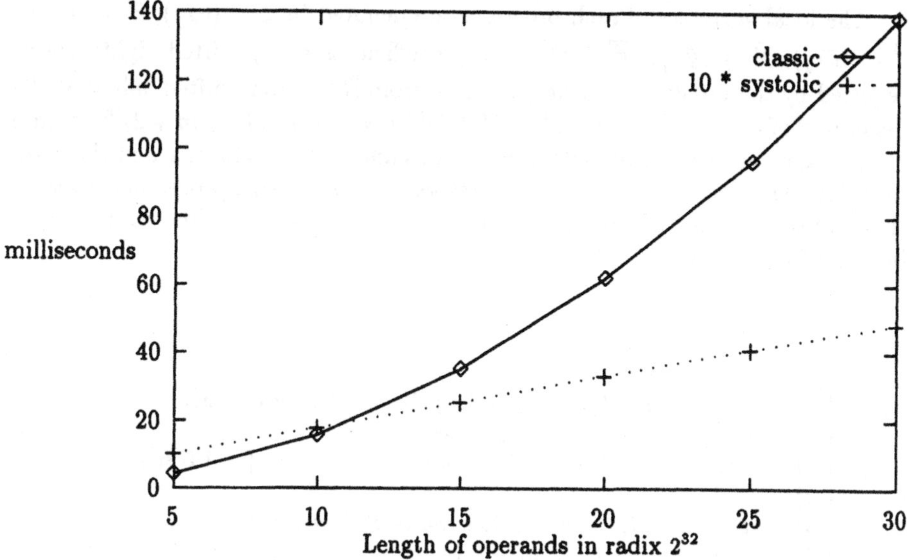

Fig. 2. Comparative timings for multiprecision multiplication.

instead of pipelining/extracting it. Using $n+m+1$ processors, C can be stored in $\overline{C}$, and then $\overline{B}$ and $\overline{Y}$ must be shifted rightward one position at each step. However, most of the time only m of the $n+m+1$ processors do useful work. This results in a lower efficiency of parallelism (1/2 for balanced-length operands).

4 Polynomial Multiplication

The inputs $\mathcal{A}, \mathcal{B}$ and the output $\mathcal{C} = \mathcal{A} * \mathcal{B}$ are integral univariate polynomials represented as lists of multiprecision integers:

$$\mathcal{A} = A_0 + A_1 * x + \ldots + A_{N-1} * x^{N-1}, \quad \mathcal{B} = B_0 + B_1 * x + \ldots + B_{M-1} * x^{M-1}.$$

$$\mathcal{C} = C_0 + C_1 * x + \ldots + C_{N+M-2} * x^{N+M-2}.$$

The innermost loop of the classical algorithm performs the operation $C+A*B$ over long integers. In this case there is no problem in parallelizing the inner loop of the algorithm, using M computing units. As in the case of `IntSysMul.1` (fig. 1), $\mathcal{C}$ is piped through the string of these M computing units. However, each of these computing units must be able to compute $C + B * A$ on long integers. This is exactly what is done by the second version of the integer systolic algorithm, if the initialization of C is removed. Therefore, one can use a row of processors for each of the above M computing units. Overall, one needs a matrix of $(M+1)*(n+m+2)$ processors, where n, m are the maximum lengths of the coefficients of $\mathcal{A}, \mathcal{B}$.

The local variables of each processor are denoted by $\overline{\mathcal{B}} = (\bar{b}_{J,j})$, $\overline{\mathcal{B}}' = (\overline{b'}_{J,j})$, $\overline{\mathcal{C}} = (\bar{c}_{J,j})$, $\overline{\mathcal{Y}} = (\bar{y}_{J,j})$. $\overline{\mathcal{B}}'$ contains the coefficients of $\mathcal{B}$, shifted rightward as required by the integer algorithm, and gets from $\overline{\mathcal{B}}$ the non–shifted values at the beginning of each main cycle. The $(M+1)^{th}$ row and the $(n+m+2)^{th}$ column of processors ensure the boundary conditions and do not participate in the computation. The coefficients of $\mathcal{A}$ are not stored in the parallel processors. Rather, they are send to all the processors, one digit at each step, for the computation in line { 9} (see Fig.3).

{ 0} $\mathcal{C} \leftarrow$ **PolySysMul**$(\mathcal{A}, \mathcal{B})$ [Systolic polynomial multiplication]
{ 1} $\overline{\mathcal{C}}, \overline{\mathcal{B}}, \overline{\mathcal{Y}} \leftarrow (0, \ldots, 0)$ [in parallel]
{ 2} **for** $(J, j) = (0,0), (0,1), \ldots, (M-1, m-1)$ **do** [load $\mathcal{B}$ sequentially]
{ 3} $\bar{b}_{J,j} \leftarrow b_{J,j}$
{ 4} **for** $I = 0, 1, \ldots, N-1$ **do** [scan coefficients of $\mathcal{A}$]
{ 5} **for** $J = 0, 1, \ldots, M-1$ **in parallel do** $[C_{I+J} \leftarrow C_{I+J} + B_J * A_I]$
{ 6} $\bar{b}'_{J,j} \leftarrow \bar{b}_{J,j}$ [restore $\overline{\mathcal{B}}$]
{ 7} **for** $i = 0, 1, \ldots, n-1$ **do** [Stage 1: add and multiply]
{ 8} **for** $j = 0, 1, \ldots, n+m$ **in parallel do**
{ 9} $(\bar{y}_{J,j+1}, \bar{c}_{J,j}) \leftarrow \bar{c}_{J,j} + \overline{b'}_{J,j} * a_{I,i} + \bar{y}_{J,j}$ [compute, shift $\overline{\mathcal{Y}}$]
{10} $\overline{b'}_{J,j+1} \leftarrow \overline{b'}_{J,j}$ [shift $\overline{\mathcal{B}}$]
{11} **while globalor**$(\overline{\mathcal{Y}})$ **do** [Stage 2: absorb carries]
{12} **for** $j = 0, 1, \ldots, n+m$ **in parallel do**
{13} $(\bar{y}_{J,j+1}, \bar{c}_{J,j}) \leftarrow \bar{c}_{J,j} + \bar{y}_{J,j}$
{14} **for** $j = 0, 1, \ldots, n+m$ **do** [extract C_I sequentially]
{15} $c_{I,j} \leftarrow \bar{c}_{0,j}$
{16} **for** $(J, j) = (0,0), (0,1), \ldots, (M-1, m-1)$ **in parallel do**
{17} $\bar{c}_{J,j} \leftarrow \bar{c}_{J+1,j}$ [shift $\overline{\mathcal{C}}$ upwards]
{16} **for** $J = 0, 1, \ldots, M-1$ **do** [extract rest of $\mathcal{C}$ sequentially]
{17} **for** $j = 0, 1, \ldots, n+m$ **do** [extract C_{N+J} sequentially]
{18} $c_{N+J,j} \leftarrow \bar{c}_{J,j}$

Fig. 3. Systolic polynomial multiplication.

Time complexity: We do not count the parallel loops {5}, {8}, {12} and {16}. The loop {2} is performed $M * m$ times, the loop {4} (N times) has several inner loops: {7} n times, {11} at most m times, and {14} $n+m$ times, hence $N * (n+m)$ is dominating. Finally, the loop {16} is repeated $M * (n+m)$ times. All in all: $T_{\text{systolic}} = O((N+M) * (n+m))$. For balanced–length operands: $T_{\text{systolic}} = O(N * n)$.

We implemented the algorithm **PolySysMul** on MasPar MP1, using the matrix of 32 by 32 processors, and the classical algorithm, using only one processor. The experiments used polynomials with 5 to 30 coefficients, whose size ranges between 5 and 15 words. The timings of the classical algorithm show the characteristic parabola, and grow up to 32 seconds. Figure 4 shows the speed-up – the maximum is 383. The efficiency ranges between 38% and 43%, approaching the 50% theoretical limit.

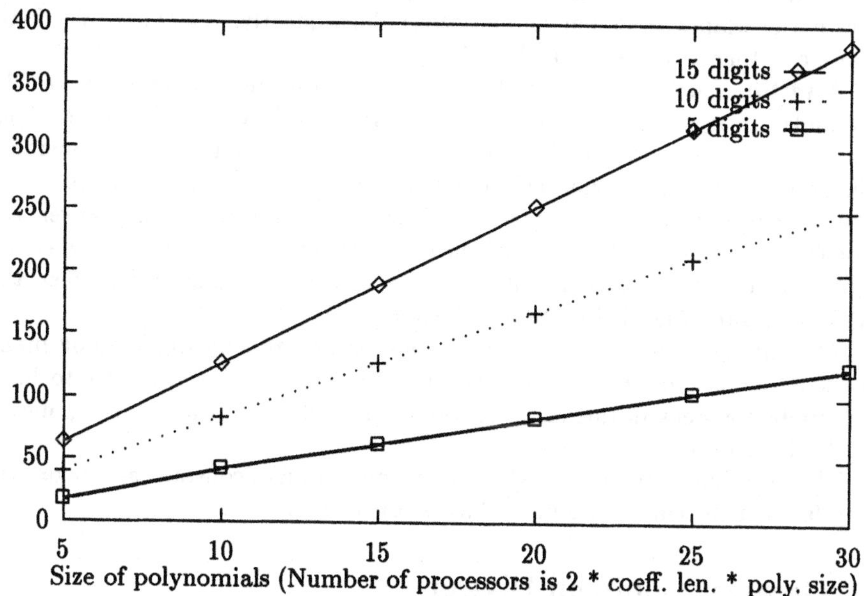

Fig. 4. Speed-up of systolic polynomial multiplication.

5 Rational Arithmetic

The operation which we implement is *rational reduction*, that is, given rational numbers $\frac{A}{B}$, $\frac{C}{D}$, $\frac{X}{Y}$, find $\frac{E}{F} = \frac{A}{B} - \frac{X}{Y} * \frac{C}{D}$, where E/F is normalized. This operation is heavily used, for instance, in Gröbner bases computation [4], for inter-reduction of polynomials. All the basic operations with long integers are involved in this reduction: $E' = A * Y * D - B * X * C$, $F' = B * Y * D$, $G = GCD(E', F')$, $E = E'/G, F = F'/G$, where the last two divisions are *exact divisions*. Note that we do not use here Henrici's approach [5], which gives poor results in the context of systolic parallelization.

The input/output is intermingled with the computations as follows: the operands are loaded into the array during multiplication, exact division outputs the result during computation.

6 Multiplication and Addition

Each **multiplication** is performed according to the second variant of the systolic algorithm presented in Section 3. This scheme requires one operand to be present in the array, while the other is loaded during multiplication. Therefore, first C and D are sequentially loaded into the array (this is the only I/O operation which does not overlap with actual computation). Subsequently, $Y * D$ and $X * C$ are computed, then $A * Y * D$, $B * Y * D$ and $B * X * C$.

Addition (subtraction) is performed using complement representation. The operands are represented by filling-up the array with additional words (*sign-words*) which equal zero for a positive operand and $2^{32} - 1$ for a negative operand. The actual addition is performed in digit-parallel fashion, using the ripple-carry scheme. This scheme has a linear-time worst-case complexity, but, however, the probability of the worst-case situation is extremely low when using high-radix digits. In fact, in our experiments we never encountered a situation when the carry propagation needed more than 2 steps.

The only situation when the carry could systematically ripple along many words is the case of subtraction when the result is positive. In order to limit the number of steps in this case, we use a *mask*, which indicates the significant words of the result.

The experimentally measured running time is in fact constant and takes less than 0.1% of the entire rational reduction operation.

7 Greatest Common Divisor

GCD computation is the most complicated and also the most time-consuming operation. Also, parallelization of the classical Euclidean algorithm – or Lehmer improved scheme [10] – is difficult, because of the carry propagation. An algorithm in which the decisions are taken using the *least-significant* digits of the operands is the *binary* algorithm of [11], which was adapted for systolic computations by [3] – the so called PlusMinus algorithm. These algorithms, however, work at *binary* level, hence they are less suitable for implementation on multiprocessor machines working at *word* level.

Therefore, we parallelize here the *generalized binary* algorithm from [6], which works least-significant digits first, and also at word level. This algorithm needs some further adaptations in order to be suitable for systolic parallelization. Namely, the problem is that the generalized binary algorithm finds an *approximation* G' of the true $G = GCD(A, B)$, which in the sequential version is corrected by computing $G = GCD(A, B, G') = GCD(A \bmod G', B \bmod G', G')$. These computations are difficult to parallelize systolically, hence we want to avoid them. One way would be to replace the division with remainder by exact division, whose result is also suitable for finding the true GCD, and then continue the computation using the systolic PlusMinus algorithm or an improved version for high-radix computation. We do not use this approach here, but an exact version of the generalized binary algorithm as described in [9]. The only cause for this is the simplicity of the implementation.

```
G ← IntSysGCD(A, B) [C ← GCD(A, B)]
    (A, B) ← ShiftTwo(A, B) [shift common zeroes]
    while B ≠ 0 [main loop]
        A ← ShiftOne(A) [shift A]
        B ← ShiftOne(B) [shift B]
        (x, y, x', y', s) ← Cofactors(a0, b0) [compute cofactors]
        A' ← LinComb(x, A, s, y, B) [first linear combination]
        B' ← LinComb(x', A, 1 − s, y', B) [second linear combination]
        if A' ≠ 0 [replace]
            then (A, B) ← (A', B')
            else (A, B) ← (B', A')
    [end of main loop: B is 0, A is the GCD]
    G ← ComplIfNeg(C) [complement G if negative]
```

Fig. 5. Systolic multiprecision GCD computation.

The outline of the algorithm is presented in Fig. 5. The routine `ShiftOne`(X) shifts the least-significant bits out of the nonzero X. The routine `ShiftTwo`(X, Y) shifts out the *common* least-significant bits from its arguments (of which at least one must be nonzero). Both routines operate in (almost) constant time, because the probability that many least-significant *words* are null is very small.

The routine `LinComb`(x, X, s, y, Y) computes the linear combination $x * X \pm y * Y$ (parameter s indicates $+$ or $-$). The routine works for negative operands also, using complement representation. A mask is used in order to indicate the range of correct values.

In order to avoid rippling the carries at each step, X, Y and the result of the linear combination are represented by *two* arrays of values, one array containing the actual digits, and one containing the carries which are not propagated yet. During each linear combination, the carries are propagated only 1 step, after which each carry becomes at most 1 (this decreases the cost of the next multiplication). Note that the least-significant digit of the result (needed for the next reduction step) is always correct.

After the main loop, G is complemented if negative. In fact, G should be also shifted with the same number of binary positions which were shifted out from the inputs at the beginning. However, in the actual implementation we perform `ShiftTwo(...)` *before* calling the GCD routine, thus the normalization is still correctly done. Indeed, the GCD computation is needed for the *normalization* of the rational fraction E'/F'. We shift out of E', F' the common trailing binary zeroes, obtaining $E", F"$. Then the GCD algorithm is used to find $G" = GCD(E", F")$, and then $E = E"/G"$ and $F = F"/G"$ are found by exact division. Note that $G"$ is always odd, which suits well the needs of the exact division algorithm.

The main reduction scheme works only if the operands are multiprecision. If the GCD is single precision, then at some moment both operands A, B might also become single precision. From this moment the [single precision] Euclidean algorithm is used for finding the GCD.

8 Exact Division

The final stage of computation consists in performing the exact divisions by the GCD. As explained in the previous section, the divisor is already odd, hence the exact division algorithm introduced in [7] can be applied without any preprocessing. In [8] several systolic variants of this algorithm are described. We choose for implementation the version which suits well the particular characteristics of this application – see Fig. 6. Namely, the algorithm is simpler because global communication can be used and also the digits of the result are pushed out *during* the computation.

```
B ← IntSysEDIV(C, A) [B ← C/A]
   a' ← ModInv(a) [find a0^-1 mod 2^32]
   while C ≠ 0 [main loop]
      b ← (c0 * a') mod 2^32 [find next digit of the quotient ...]
      B_next ← b [...and push it out]
      C ← LinComb(1, C, 1, a', A) [C ← C - a' * A]
      for i = 0, 1, ... in parallel do [shift C left]
         c_i ← c_i+1
   [end of main loop]
```

Fig. 6. Systolic multiprecision exact division.

The vector B in this algorithm is external to the processor array – it represents the output of the algorithm. The function **ModInv**(a) is based on the recursion developed in [7], hence we avoid the (expensive) extended Euclidean algorithm. A simplified version of the function **LinComb** from the GCD algorithm is used for performing the operation $C - a' * A$. Again the carries are not propagated at each step, because only the correct value of the least-significant digit of C is needed for continuing the computation.

9 Experimental Results

The algorithms were implemented on a computer MasPar MP-1 having an array of 32 by 32 processors. The programs handle this two-dimensional array as an one-dimensional array of 1024 processors, virtually connecting the rows at their edges.

A straightforward calculation of the time complexity of all the algorithms will be similar to the one for multiplication (see Sect. 3) and will reveal linear complexity. For practical purposes, however, direct timing of the algorithms is even more relevant. We timed the execution for random inputs having length up to 100 32-bit words. That means GCD is computed for operands having (roughly) 300 words, while its output is usually small (single precision). The timings are presented in fig. 7. The times consumed for addition and `ShiftTwo` before GCD computation are 0.70 and 0.35 milliseconds, respectively, and are not shown in the figure.

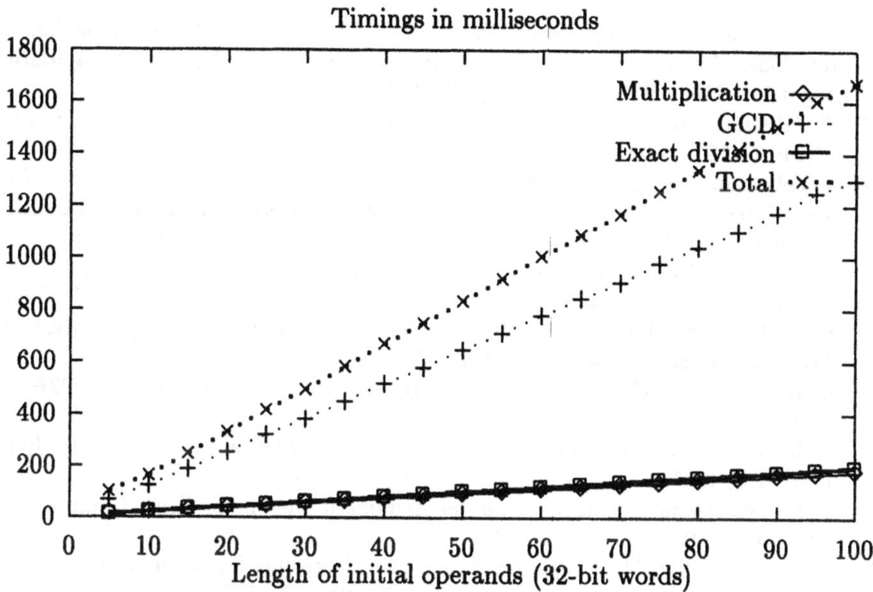

Fig. 7. Timings of the rational reduction and its components.

The most important characteristic of the timing is the *linear dependence* of the lengths of the input. This shows that the systolic model can be effectively used on MasPar architecture for implementing long integer arithmetic.

Further work includes improving the efficiency of the implementation, – especially that of the GCD computation, which takes most of the time – and embedding the rational reduction algorithm in higher-level algebraic computations.

References

1. R. Beardsworth. On the application of array processors to symbol manipulation. In *SYMSAC'81*, 1981.
2. A. D. Booth. A signed binary multiplication technique. *Q. J. Mech. Appl. Math.*, 4:236–240, 1951.
3. R. P. Brent and H. T. Kung. A systolic algorithm for integer GCD computation. In K. Hwang, editor, *Procs. of the 7th Symp. on Computer Arithmetic*, pages 118–125. IEEE Computer Society, June 1985.
4. B. Buchberger. Gröbner Bases: An Algorithmic Method in Polynomial Ideal Theory. In Bose and Reidel, editors, *Recent trends in Multidimensional Systems*, pages 184–232, Dordrecht-Boston-Lancaster, 1985. D. Reidel Publishing Company.
5. P. Henrici. A subroutine for computations with rational numbers. *Journal of the ACM*, 3:6–9, 1956.
6. T. Jebelean. A generalization of the binary GCD algorithm. In M. Bronstein, editor, *ISSAC'93: International Symposium on Symbolic and Algebraic Computation*, pages 111–116, Kiev, Ukraine, July 1993. ACM Press.
7. T. Jebelean. An algorithm for exact division. *Journal of Symbolic Computation*, 15(2):169–180, February 1993.
8. T. Jebelean. Systolic Algorithms for Exact Division. In *Workshop on Fine Grain and Massive Parallelism*, pages 40–50, Dresden, Germany, April 1993. Published in *Mitteilngen–Gesellschaft für Informatik e. V. Parallel Algorithmen und Rechnerstrukturen*, Nr. 12, July 1993.
9. T. Jebelean. Systolic algorithms for long integer GCD computation. In J. Volkert B. Buchberger, editor, *CONPAR 94 - VAPP VI, Linz, Austria, September*, pages 241–252. Springer Verlag LNCS 854, 1994.
10. D. H. Lehmer. Euclid's algorithm for large numbers. *Am. Math. Mon.*, 45:227–233, 1938.
11. J. Stein. Computational problems associated with Racah algebra. *J. Comp. Phys.*, 1:397–405, 1967.
12. E. E. Swartzlander, editor. *Computer Arithmetic*, volume 2. IEEE Computer Society Press, 1990.
13. K. S. Trivedi and M. D. Ercegovac. On-line algorithms for division and multiplication. *IEEE Trans. on Computers*, C-26(7):681–687, 1977.
14. K. Weber. The accelerated integer GCD algorithm. *ACM Trans. on Math. Software*, 21(1):111–122, March 1995.
15. D. Weeks. Adaptation of SAC-1 algorithms for an SIMD machine. In J. Della Dora and J. Fitch, editors, *Computer Algebra and Parallelism*, pages 167–177. Academic Press, 1989.

Parallel 3-Primes FFT Algorithm

Giovanni Cesari[1,2] and Roman Maeder[2]

[1] Universerstità degli Studi di Trieste
DEEI - 34100 Trieste, Italy
[2] Institute of Theoretical Computer Science
ETH Zurich, CH-8092 Zurich, Switzerland

Abstract. Multiple precision arithmetic forms the core of symbolic computation systems. A speedup of the basic algorithms can therefore improve all higher-level algorithms. Several packages have been developed for sequential processors. Although they were implemented carefully, they are unsatisfactory for large inputs. Performance can be improved by chosing algorithms with better asymptotic behavior or by using other computing models such as parallel machines.

In this paper we present a parallel implementation of the three primes fast Fourier transform algorithm and compare its efficiency with the parallel Karatsuba algorithm. The target machine is the distributed memory Paragon.

1 Introduction

Multiple precision arithmetic forms the core of symbolic computation systems. The performance of all higher-level algorithms which use the basic arithmetic depends on its efficiency. We work on the implementation of a parallel computer algebra library. Several multiple precision packages have been developed (e.g. [2, 7, 12]). However, they have been designed for sequential processors, and therefore they cannot be integrated in our parallel library without loss of efficiency.

In this paper we study a parallel implementation of an algorithm of fundamental importance for the underlying arithmetic: the three primes fast Fourier transform for integer multiplication. We limit our analysis to distributed memory machines and report on an implementation on a Paragon. This algorithm has been integrated in our parallel computer algebra library, which is now used as computational engine by AlgBench [15]. This is a symbolic computation system developed in our group and designed originally for sequential architectures.

In the follows we use the standard representation of arbitrary size integer (big-integer). An unsigned big-integer is stored as an array of digits in a positional number system with base B. The array $a_0, a_1, \ldots, a_{n-1}$ of length n represents the number $a = \sum_0^{n-1} a_i B^i$, where the digits are in the range $0 \leq a_i < B$ (the sign is stored separately).

2 Multiple Precision Arithmetic

2.1 Optimization of the Sequential Code

Three algorithms are generally used to implement the multiplication [10]: the classical naive multiplication algorithm [9], the Karatsuba algorithm [8, 14], and FFT-based methods [16, 13]. An efficient implementation of the multiplication is crucial for many algorithms over the integers, including the Newton method for computing reciprocals on which division can be based [9].

The optimization of the sequential code is machine dependent. On 32-bit processors with hardware multiplication and division instructions a good choice for the base B is 2^{32}. Because C compilers do not use these instructions, it is necessary to write some assembler code. Most multi-precision packages (e.g. the GNU-MP V.1.3.2.) use such assembly routines for each processor type. On a Sparc 5 (V8 architecture) we have observed a gain of a factor two in the basic operations like multiplication and division over a pure C version. Further performance enhancement (about a factor 1.5) can be obtained by writing in assembler also the inner loops which operate on arrays of words. This is done for example in the new Beta version 2.0 of the GNU-MP package.

The threshold for switching from one method to the other should be chosen experimentally. For many processors, including Sparc with V8 architecture and Intel 860 (which is the processor used by the Paragon), a value of 40 bytes is optimal (the minimum is rather flat) for switching from the trivial multiplication method to the Karatsuba algorithm. The threshold between Karatsuba and FFT for our implementation can be derived from Table 1.

	GNU-MP	bigint		
size	v.1.3.2	school book	karatsuba	FFT
1024	0.015	0.042	0.016	0.068
2048	0.05	0.16	0.052	0.14
4096	0.15	0.72	0.158	0.33
8192	0.45	3.20	0.485	0.78
16384	1.35	12.8	1.51	1.53
32768	3.92	51.3	4.44	3.30
65536	12.3	209	13.6	7.01
131072	35.9	815	45.7	14.6
262144	104	3471	130	32.8
524288	313	13217	365	67.1
1048576	1100	-	1097	146

Table 1. Timings in seconds of the multiplication algorithm. The measurements are performed on a Sparc 5 (V8 architecture). In the first column are the results of the GNU-MP package which uses the Karatsuba algorithm. In *bigint* is our implementation running sequentially. The three methods use assembler routines to perform the basic 32-bit multiplication. FFT uses the 3-primes algorithm. The input size is in bytes.

2.2 Parallel Implementation of the Karatsuba Algorithm

A parallel implementation of the multiplication algorithm needs both Karatsuba and FFT methods to be efficient for all input size.

The Karatsuba algorithm divides the input numbers a and b of size n into two halves. The product can be formed by only three recursive multiplications of size $n/2$ and some additions and subtractions. The sequential time $T_s(n)$ satisfies the following recurrence: $T_s(n) = 3T(n/2) + O(n)$. This gives a complexity of $T_s(n) = O(n^{\log_2 3})$ digit multiplications. Let $m = \lceil n/2 \rceil$. We can write $a = a_l + a_h B^m$, $b = b_l + b_h B^m$, where $a_l = \sum_0^{m-1} a_i B^i$, $a_h = \sum_0^{m-1} a_{i+m} B^i$. The product $r = ab$ can be expressed as:

$$r = a_l b_l + ((a_l + a_h)(b_l + b_h) - a_l b_l - a_h b_h) B^m + a_h b_h B^{2m} \tag{1}$$

We have parallelized the Karatsuba algorithm following a fork-join approach [4, 3]: one processor starts at the root and recursively assigns separate nodes in the tree to different processors. Each node assigns three subnodes for the three recursive multiplications and combines the results according to equation (1) after the three subnodes have completed their computation. (One of the threes subtasks is performed by the parent process, which otherwise waits idly for the completion of the three suprocesses). When a certain cutoff criterion is reached, no further splitting occurs: the work remaining in a given task is done sequentially.

Let the number of processors be $p = 3^l$. We assume a message passing model of computation. Let $T_c(n) = c_1 + c_2\, n$ the time necessary to communicate between two processors an integer of length n and let $T_a(n) = c_3\, n$ the time to add two integers of length n. The sequential Karatsuba time is given by $T_{sk} = k_1\, n^{k_2}$. The total parallel time can be approximated well in the following way:

$$\begin{aligned} T_k(n) &= T_{sk}\left(\frac{n}{2^l}\right) + 4\sum_{i=0}^{l-1} T_c\left(\frac{n}{2^i}\right) + 5\sum_{i=0}^{l-1} T_a\left(\frac{n}{2^i}\right) \\ &\approx \frac{T_{sk}(n)}{p} + cn + 4c_1 l \end{aligned} \tag{2}$$

The three terms represent, respectively, the theoretically optimal time, the sequential part due to communications and additions, and the total communication latency. The last term can be ignored for $p \ll n$. We put $c = 2(4c_2 + 5c_3)$.

Both, speedup and efficiency, approach the theoretical maximum for $n \to \infty$.

$$S_k(n) = \frac{T_{sk}(n)}{T_k(n)} \approx p - \frac{cp^2}{k_1 n^{0.585}} \tag{3}$$

$$E_k(n) = \frac{T_{sk}(n)}{pT_k(n)} \approx 1 - \frac{cp}{k_1 n^{0.585}} \tag{4}$$

3 Parallel Implementation of the 3-Primes FFT

The three primes algorithm is a fast integer multiplication algorithm based on the discrete fast Fourier transform (FFT) and on the computation with multiple homomorphic images. It has a complexity of $O(n \log(n))$ (see [13] for all algebraic methods).

The algorithm treats big-integers as arrays of coefficients of a polynomial. Let $2n$ be the length of the result, and $N = 2^m$, where $m = \lceil \log(2n) \rceil$. For three carefully chosen primes p, the polynomials are evaluated modulo p at N points using the FFT. The corresponding values are then multiplied pointwise to get three new sets of points, which are interpolated with the fast Fourier inverse (FFI) to get three new polynomials.

After applying the Chinese remainder algorithm to the polynomials in Z_{p_i}, we have one polynomial in Z that is the unique solution as long as each original coefficient was less than the minimum of the three primes. The polynomial is then evaluated at B and converted back into a big-integer. The primes are of the form $p_i = 2^{e_i}\, l + 1$. The three primes algorithm is a subasymptotic algorithm: the size n of the problem which can be handled is bounded by $N \leq B$ and $N \leq 2^E$ where E is $\min\{e_i\}$ (the latter bound follows from the requirement that primitive roots of unity of order N exist). Moreover, to guarantee the correctness of the Chinese remainder algorithm, it is necessary that $NB^2 \geq p_1 p_2 p_3$. The primes we used were $3 \times 2^{30} + 1$, $13 \times 2^{28} + 1$, and $29 \times 2^{27} + 1$. They are the largest primes fitting in a word of length $B = 2^{32} - 1$ and fullfilling the requirements described above. They allow us to multiply two integers of up to 2^{26} digits.

The algorithm can be shortly described (in the C notation) as follows. We use the unspecified data type `digit`; `int prime[3]` contains the three prime numbers. `a` and `b` are the two input arrays of the same size `n` (for simplicity). The result is stored in `c`.

```
void three_primes_fft(digit *a, digit *b, digit *c, int n)
{
  digit ap[N], bp[N], cp[N], *A, *B, *cp, *u[3];
  for (int i = 0; i < 2; i++) {
    for(int j = 0; j < n; j++) {            // step 1
      ap[j] = a[j] % prime[i];
      bp[j] = b[j] % prime[i]
    }
    A = fft(ap, prime[i], N);               // step 2
    B = fft(bp, prime[i], N);
    // pointwise multiplication             // step 3
    cp = mul_mod(A, B, prime[i])
    // interpolation                        // step 4
    u[i] = ffi(cp, prime[i], N);
   }
   // chinese remainder problem
   c = crp(u[0], u[1], u[2], prime[0], prime[1], prime[2]);
}
```

The code above can be parallelized as follows:

- Step 1 (and 3) are parallelized dividing the arrays between p processors and performing the operations locally. This leads to:

$$T_m = m\ \frac{n}{p} \tag{5}$$

where m has the two values m_1 and m_3 for step 1 and 3 respectively.
- We have implemented both FFT and FFI using the binary exchange iterative algorithm [11]. This algorithm performs well when, like in the Paragon, the network bandwidth (c_2) is much larger than the computing time (k_f) (see below). Each processor performs the $\log(N)$ steps of the FFT on a subarray of length N/p with communication occurring only during the first $\log(p)$ steps. After the FFTs have been calculated each processor has the data ready for step 3. The two FFTs on the inputs a and b respectively could be performed in parallel on $p/2$ processors. However, this does not reduce the total communication: the less communication in step 2 is compensated exactly by the data exchange necessary to start step 3.

Let the sequential time to compute an FFT of length n be $k_f n \log(n)$. The parallel time is given then by $T_f\ =\ k_f\ (n/p)\ \log(n) + T_{comm}$, where $T_{comm} = \log(p)\ T_c(n/p)$ is the communication time during the FFT algorithm. T_f can be approximated as follows:

$$T_f \approx \frac{n}{p}\ (k_f\ \log(n) + c_2\ \log(p)) \tag{6}$$

A similar formula holds for the FFI.
- The FFI computes the interpolated polynomials in Z_{p_i}. We can lift the result in Z using the Chinese remainder algorithm: $c = v_0 + v_1 p_0 + v_2 p_0 p_1$. The coefficients v_i are calculated using Garner algorithm [6]. The three terms v_0, $v_1 p_0$, and $v_2 p_0 p_1$ are arrays of size N whose components can be computed independently from the u_is on different processors. We have performed the addition of the three arrays sequentially on one processor. This step could also be parallelized using carry-lookahead techniques.
The time to compute step 4 is then given by

$$T_{cra} \approx \frac{k_{c1} n}{p} + (k_{c2} + c_2)\ n \tag{7}$$

c_2 takes into account the time necessary to send all the data to one processor; k_{c2} characterizes the time necessary to compute c sequentially, while k_{c1} is the constant of Garner's algorithm.

- To avoid the communication necessary to start the Chinese remainder algorithm it is more convenient to compute sequentially the outer loop of the three prime numbers. It is also convenient to pre-compute when the system is initialized all powers of the radix unit necessary for the FFT. In this way we obtain the following total time necessary to compute 3-primes FFT algorithm:

$$\begin{aligned} T_{fft}(n) &= 3\ \ (2T_{m1}(n,p) + T_{m2}(n,p)) + \\ &+\ 3\ (2\ T_f(n,p) + \ T_{fi}(n,p)) + \ T_{cra}(n,p) \\ &\approx \frac{n}{p}\,(\alpha + \beta\log(n) + \gamma\log(p)) + \delta\ n \end{aligned} \tag{8}$$

$\alpha = 6m_1 + 3m_3 + k_{c1}$ represents the part of the algorithm which can be parallelized without communication, $\beta = 6k_f + 3k_{fi}$ is the FFT and FFI parallel part, $\gamma = c_2$ is the communication during FFT and FFI computation, and $\delta = 9c_2 + k_{c2}$ is the sequential part of the algorithm.

Let $T_s = K_f n\log(n)$ be the time to compute sequentially the 3-prime FFT. The expressions for speedup $S_{fft}(n)$ and efficiency $E_{fft}(n)$ are given by the following equations:

$$S_{fft}(n) = \frac{T_s(n)}{T_{fft}(n,p)} \approx p - \frac{\delta p^2}{\beta\ \log(n)} \tag{9}$$

$$E_{fft}(n) = \frac{T_s(n)}{pT_{fft}(n,p)} \approx 1 - \frac{\delta p}{\beta\ \log(n)} \tag{10}$$

Other approaches to the multiprecision multiplication algorithm using complex discrete Fourier transform can be found in [9, 16, 1].

4 Experimental Results

The Karatsuba and the 3-primes FFT algorithms have been implemented on an Intel Paragon at ETH Zurich with 150 computing nodes [5]. This machine is a 2-D mesh with 50 MHz Intel 860 processors. The values (in seconds) of the constants which characterize our analysis are summarized below:

start-up time c_1	$30\ 10^{-6}$
transmission time per short c_2	$0.007\ 10^{-6}$
addition constant c_3	$5\ 10^{-7}$
Karatsuba constant k_1	$6.35\ 10^{-7}$
Karatsuba constant k_2	$1.595\ 10^{-6}$
3 primes constant K_f	$2.5\ 10^{-5}$
fft constant k_f	$2.16\ 10^{-6}$
ffi constant k_i	$2.35\ 10^{-6}$
cra constant k_{c1}	$6\ 10^{-5}$
cra constant k_{c2}	$8\ 10^{-7}$
mod constant k_{m1}	$9\ 10^{-7}$
mod constant k_{m3}	$5\ 10^{-6}$

Table 2 shows 3-primes FFT measurements. In Figure 1 we compare the theoretical results with time and efficiency measurements. We used equations (2) and (8) to compute Karatsuba and FFT times and equations (4) and (10) to compute efficiencies. We can see that there is an input range where the Karatsuba algorithm achieves better performance. If we use the whole Paragon, Karatsuba on 81 processors is faster than 3-primes FFT on 128 processors for input length less than 32768 bytes. After this point the 3-primes FFT algorithm is the method of choice. It is also interesting to note that our parallelization of the Karatsuba algorithm is more efficient than that of the 3-primes FFT. The efficiency of the Karatsuba method is in fact $E_k \approx 1 - 2.5p/n^{0.585}$ (see equation (4)), while the efficiency of the 3-primes FFT is $E_f \approx 1 - 0.5p/\log(n)$ (equation (10)). If the last step of the Chinese remainder problem is parallelized, the 3-primes FFT efficiency becomes superior: $E \approx 1 \ - \ 3.5\ 10^{-4}\log(p)/\log(n)$ which is approximately the efficiency of the Fourier transform.

References

1. D.H. Bailey. A portable high performance multiprecision package. Technical Report RNR-90-022, NAS Applied Reserach Branch, Moffett Field, CA 94035, 1993.
2. C. Batut, D. Bernardi, H. Cohen, and M. Olvier. *User's guide to PARI-GP*, 1995.
3. G. Cesari and R. Maeder. Performance analysis of the parallel karatsuba multiplication algorithm for distributed memory architectures. *Journal of Symbolic Computation, Special Issue on Parallel Symbolic Computation*, to appear.
4. B. Char, J. Johnson, D. Saunders, and A. P. Wack. Some experiments with parallel bignum arithmetic. In Hoon Hong, editor, *Proceedings of the 1st International Symposium on Parallel Symbolic Computation*, pages 94–103, 1994.
5. K. E. Gates and W. P. Petersen. A technical description of some parallel computers. Technical Report IPS 93–15, IPS, ETH–Zentrum, CH–8092 Zurich, 1990.
6. K. O. Geddes, S. R. Czapor, and G. Labahn. *Algorithms for Computer Algebra*. Kluwer Academic Publishers, 1992.
7. T. Granlund. *GNU MP. The GNU Multiple Precision Arithmetic Library*, 1993.

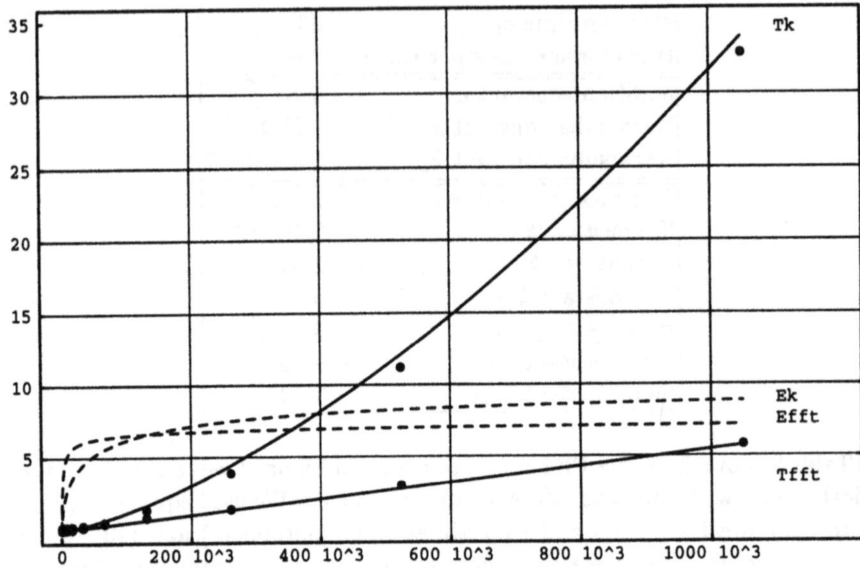

Fig. 1. Parallel time in seconds and efficiency (multiplied by a factor 10) of Karatsuba and 3-primes FFT multiplication algorithms for $p = 81$ and $p = 128$ processors, respectively. T_k and E_k are the Karatsuba time and efficiency respectively (equations (2), (4)); T_{fft} and E_{fft} are the 3-primes FFT time and efficiency (equations (8), (10)).

size	sequential	2 proc	4 proc	8 proc	16 proc	32 proc	64 proc	128proc
1024	0.269	0.142	0.078	0.047	0.040	0.049	0.078	0.167
2048	0.576	0.302	0.161	0.088	0.061	0.056	0.084	0.169
4096	1.23	0.628	0.326	0.173	0.106	0.081	0.093	0.183
8192	2.63	1.33	0.681	0.356	0.201	0.132	0.121	0.191
16384	5.67	2.83	1.43	0.746	0.402	0.241	0.180	0.216
32768	12.0	6.06	3.05	1.57	0.826	0.467	0.306	0.284
65536	25.4	12.80	6.45	3.28	1.70	0.935	0.568	0.453
131072	54.4	27.08	13.6	6.91	3.56	1.90	1.10	0.823
262144	113	56.7	29.6	14.6	7.50	3.93	2.19	1.44
524288	244	120	60.7	30.9	15.8	8.40	4.62	3.02
1048576	517	253	147	83.7	33.2	17.5	9.62	5.84

Table 2. Measurements in seconds of the parallel 3-primes FFT algorithm. The input length is in bytes.

8. A. Karatsuba and Y. Ofman. Multiplication of multidigit numbers on automata. *Soviet Physics doklady*, 7(7):595–596, 1963.
9. D. E. Knuth. *Seminumerical Algorithms*, volume 2 of *The Art of Computer Programming*. Addison-Wesley, second edition, 1981.
10. Wolfgang W. Kuechlin, David Lutz, and Nicholas Nevin. Integer multiplication in PARSAC-2 on stock microprocessors. In *Proc. AAECC-9, New Orleans*, volume 539 of *SLNCS*. Springer Verlag, 1991.
11. V. Kumar, A. Grama, A. Gupta, and G. Karypis. *Introduction to Parallel Computing*. The Benjamin/Cummings Publishing Company, Inc, 1994.
12. A. Lenstra. *Lip*, 1995.
13. J. Lipson. *Elements of Algebra and Algebraic Computing*. Benjamin Cumming, 1984.
14. R. E. Maeder. Storage allocation for the Karatsuba integer multiplication algorithm. In Alfonso Miola, editor, *Design and Implementation of Symbolic Computation Systems (Proceedings of DISCO '93)*, SLNCS, pages 59–65. Springer Verlag, 1993.
15. Roman E. Maeder. Algbench: An object-oriented symbolic core system. In J. P. Fitch, editor, *Design and Implementation of Symbolic Computation Systems (Proceedings of DISCO '92)*, SLNCS. Springer Verlag, 1992.
16. A. Schoenhage, A. Grotefeld, and E. Vetter. *A multitape Turing machine implementation*. BI-Wissenschaftsverlag, 1994.

A Master-Slave Approach to Parallel Term Rewriting on a Hierarchical Multiprocessor

Reinhard Bündgen Manfred Göbel Wolfgang Küchlin

Wilhelm-Schickard-Institut, Universität Tübingen,
Sand 13, D-72076 Tübingen, Germany
E-mail: ⟨{buendgen,goebel,kuechlin}@informatik.uni-tuebingen.de⟩

Abstract. We report on a parallel implementation of an unfailing term completion procedure on a network of multiprocessor workstations. Our parallelization concept integrates distributed search parallelism on the network, based on a master-slave approach, with the parallel execution of each search on a multiprocessor. Both levels of parallelism are realized by a uniform fork / join paradigm using multi-threading. In many of our examples we are able to combine the benefits of distributed and shared-memory approaches for superior overall speed-ups.

1 Introduction

Modern high performance computer architectures are typically parallel architectures. They can be subdivided into shared memory multiprocessors (SMP) and distributed memory multiprocessors (DMP). Their most popular and commonly available representatives are probably parallel workstations, and networks of workstations, respectively. Combining both kinds of architectures in form of a network of SMPs results in a hierarchical multiprocessor (HMP). In this paper, we present both the concept of a parallel term completion procedure geared towards an HMP, and empirical results with its actual realization on a small net of state-of-the-art multi-processor SPARCstations.

There are two major approaches to parallelizing software for symbolic computation. The first approach uses *search parallelism* (*OR-parallelism*). It exploits the fact that commonly there are several strategies for finding the solution, the best of which is not known a priori. In this case, executing several different strategies in parallel, until the first solution is found by any of them, often speeds up the search. This kind of parallelization may even yield super-linear speed-ups, if, on k processors, the best strategy in the pool executes less than k times the work of the sequential default strategy. Due to the coarse granularity of search parallelism, it is an ideal candidate for distribution on a network of workstations.

However, its effect vanishes as the known sequential strategies approach an optimal strategy. In particular, all strategy dependent speed-ups are limited by the performance of an optimal strategy. Therefore it is interesting to speed up each individual strategy by performing its work in parallel (but preserving the strategy).

This second approach to parallelizing symbolic computation software relies on *work parallelism* (*AND-parallelism*), which spreads an amount of work over a number of processors. If only work parallelism is used, then the parallel completion algorithm executes exactly the same strategy no matter how many processors it runs on, and therefore we also speak of a *strategy compliant parallelization*. Due to the high synchronization requirement of a single completion loop with a fixed strategy, and the fine grain of its parallel components, strategy compliant parallelization is typically only profitable on SMPs.

In practice, both approaches to parallel completion have their limits. The speed-ups of search parallelism are limited by the number and relative quality of different selection strategies; work parallelism is limited by the amount of parallel work within each strategy and by the architectural constraints of today's SMPs that allow at most a few dozen processors.

The goal of this work is to show that the *modular* combination of search parallelism and work parallelism on an HMP can yield aggregate speed-ups exceeding those of each isolated scheme. For an empirical validation of our concept we implemented an equational theorem prover, based on an unfailing completion procedure, on a network of multiprocessor SPARCstations. The prover combines a top-level master-slave scheme distributing search parallelism over the net with the strategy compliant parallel completion scheme of PaReDuX [12] on each multi-processor node.

Several implementations of parallel software for symbolic computation have been reported; see [7, 35, 27] for surveys. The scheme for parallel term-rewriting proposed in [22] distributes the subterms of a term across a DMP and performs loosely synchronized rewriting of subterms on each node. We experimented with SMP parallel rewriting of individual terms in the AC case only [11], but found it hard to obtain speed-ups on our inputs. In the present work we never performed parallel reductions within a term.

Most of the completion systems employ search parallelism implicitly by maintaining several loosely synchronized completion loops on different subsets of critical pairs. All seem to be geared for, and implemented on, either SMPs or DMPs. Strategy compliant parallelization of term completion on an SMP has been investigated by the authors [10, 11, 13]. A similar design was applied to distributed Gröbner base completion in [2]. Shared memory search parallelism is exploited e. g., by the ROO prover [29]. On the side of distributed search parallelism we can observe a wide range of solutions with simple competitive parallelism (e. g., [32]) at one end of the scale and systems that support elaborate communication schemes between agents performing different strategies (e. g., Aquarius [5] and DISCOUNT [3]). DISCOUNT even controls strategies by evaluating their success on a meta level using the *Teamwork* method.

Our top-level scheme of search parallelism is inspired by the Teamwork concept and borrows strategies pioneered in that approach. However our master-slave approach lies in between the two extremes and can be parameterized to allow for more or less communication between different strategies. Due to its conceptual simplicity, the master-slave approach need not set up separate communication channels. It can be implemented with the same fork / join paradigm of multi-threaded programming employed in PaReDuX, but now supported over the network by our *Distributed Thread System* (DTS) [8]. This supports a seamless integration of both DMP and SMP parallelism, which is desirable on an HMP computer.

According to our knowledge this paper presents the first parallelization of software for symbolic computation on an HMP. We are also the first to combine in a modular way strategy compliant and search parallel schemes for term-rewriting completion.

2 Parallel Symbolic Computation on an HMP

Hierarchical Multiprocessors are extremely important from a practical point of view. They keep the difficult network part of the communication relatively small and manageable while allowing standard multi-threading techniques on each node. In our example, it is much easier to parallelize and synchronize work between two 4-processor machines than between 8 uniprocessors.

In theory, one might just regard an HMP as a DMP with somewhat faster communication between nodes on the same cluster (the SMP components). Taking this view of our example, we might parallelize for 8 processors using PVM [33] and exploit a fast PVM implementation on the SMPs. This is *not* our approach because we wish to exploit the full power of an SMP with a multi-threaded operating system. Indeed, our results show (cf. Table 1) that with few exceptions our system is faster on one 4-processor machine than on two double-processor machines; this clearly points out the usefulness of an SMP component.

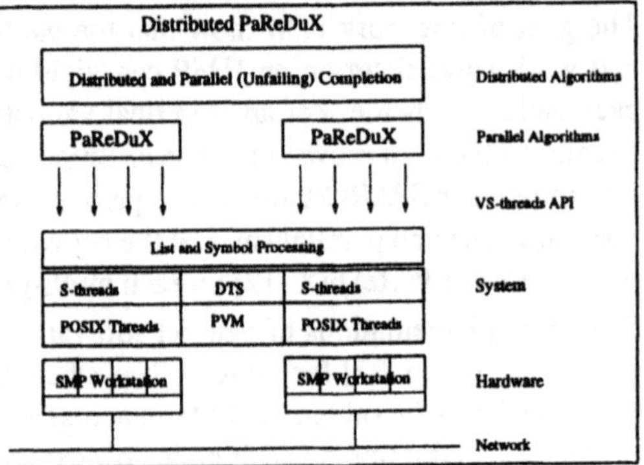

Fig. 1.: The distributed PaReDuX system

Fig. 1 illustrates the overall architecture of *Distributed PaReDuX*.

Our *S-threads* environment [26] enhances POSIX operating system threads to permit parallel list processing and garbage collection. The *VS-threads* extension [28] provides *virtual threads* which map a virtually unlimited amount of logical parallelism onto a small number of concurrent kernel threads. By multi-threading on each SMP we achieve fine-grained parallelization and synchronization, fast load balancing, and shared memory accesses. We only use PVM on the network as a communication substrate for our DTS software [8] which realizes a much higher level of programming abstraction.

On an SMP, a *thread of control* is an execution context for a procedure. A *fork* will cause the procedure to run concurrently, and a *join* will wait for its termination and retrieve its result. All threads live within the same process with shared memory. Standard threads can therefore not migrate across the nodes of a DMP. However, if a thread does not exploit shared memory during its execution, then it is functionally equivalent to an *asynchronous remote procedure call*.

DTS implements the asynchronous remote procedure call abstraction together with some dynamic load balancing and failure recovery on the network. DTS uses PVM to place a copy of a program on each (SMP) node of the network, together with a local DTS representative, the *node manager*. The node manager intercepts thread forks which have been marked *networkable* by the user, and communicates them together with their parameters to other nodes on the net with spare capacity. If a node manager accepts such an remote procedure call, it immediately forks the computation as a local thread. This thread can in turn fork more threads, both local and networkable. If the thread terminates, the server node manager sends its result back to the client. The client node manager accepts the result and delivers it to a joining thread.

The combined S-threads and DTS system forms the "middle ware" between operating system and algorithms. The high-level theorem proving algorithms are all written for its application programming interface which provides functionality, abstraction, and portability. The outcome is a heterogeneous multi-threaded rewrite laboratory running on a non-uniform memory access machine (cf. Fig. 1).
The PaReDuX system component is a collection of multi-threaded programs to experiment with term completion procedures on an SMP [13]. PaReDuX features several variants of Knuth-Bendix completion [10, 11, 31]. These programs are strategy compliant parallel versions of the corresponding ReDuX [9] programs.
The *virtual S-threads* system and DTS were both originally developed for the parallel Computer Algebra library PARSAC-2 [25, 27]. PARSAC-2 contains many parallel algebraic algorithms, including a parallel installation of Buchberger's algorithm for Gröbner basis completion. Thus the scope of our system environment extends well beyond theorem proving.

3 Unfailing Completion

We assume that the reader is familiar with term rewriting systems (TRS) and completion procedures. Our notation follows [16]. Completion without failure is an enhancement of the Knuth-Bendix method [23]. Whereas general completion wishes to construct a complete set of reductions from a set of equational axioms, unfailing completion attempts to find a ground Church-Rosser system, in which every ground term has a unique normal form [20, 4, 6]. Unfailing completion is usually used to prove an equation (*goal*) $a \leftrightarrow b$, and it stops with success as soon as a and b have a common $\mathcal{R} \cup \mathcal{E}$-normal form.
A *term ordering* $\succ$ is a terminating ordering on terms such that if $l \succ r$ for all $l \to r \in \mathcal{R}$ then the reduction relation is terminating. Term orderings can also be used to define a terminating reduction relation by a set of equations $\mathcal{E}$: Let $\mathcal{E}_\succ = \{s\sigma \to t\sigma \mid s \leftrightarrow t \in \mathcal{E} \text{ or } t \leftrightarrow s \in \mathcal{E} \text{ and } s\sigma \succ t\sigma\}$ be the set of all orientable instances of $\mathcal{E}$ where $s\sigma \leftrightarrow t\sigma$ is an instance of $s \leftrightarrow t$. Then $\to_{\mathcal{E}_\succ}$ is the *ordered rewrite relation* defined by $\mathcal{E}_\succ$. Note for $s \leftrightarrow t \in \mathcal{E}$, neither $s \to t$ nor $t \to s$ need to be in $\mathcal{E}_\succ$. Such equations are called *unorientable*. Since it is much easier to rewrite w.r.t. a rule set than w.r.t. a set of equations, orientable equations will in general be turned to a set of rules $\mathcal{R}$. Thus we will in general reduce w.r.t. $\mathcal{R} \cup \mathcal{E}_\succ$.

Rule	Inference	Condition
Delete	$\dfrac{(\mathcal{P}\cup\{s\leftrightarrow s\};\ \mathcal{E};\ \mathcal{R})}{(\mathcal{P};\ \mathcal{E};\ \mathcal{R})}$	
Simplify	$\dfrac{(\mathcal{P}\cup\{s\leftrightarrow t\};\ \mathcal{E};\ \mathcal{R})}{(\mathcal{P}\cup\{s\leftrightarrow u\};\ \mathcal{E};\ \mathcal{R})}$	if $t \to_{\mathcal{R}\cup\mathcal{E}_\succ} u$
Orient	$\dfrac{(\mathcal{P}\cup\{s\leftrightarrow t\};\ \mathcal{E};\ \mathcal{R})}{(\mathcal{P};\ \mathcal{E};\ \mathcal{R}\cup\{s\to t\})}$	if $s \succ t$
Unfail	$\dfrac{(\mathcal{P}\cup\{s\leftrightarrow t\};\ \mathcal{E};\ \mathcal{R})}{(\mathcal{P};\ \mathcal{E}\cup\{s\leftrightarrow t\};\ \mathcal{R})}$	if $s \not\succ t$ and $t \not\succ s$
Compose	$\dfrac{(\mathcal{P};\ \mathcal{E};\ \mathcal{R}\cup\{s\to t\})}{(\mathcal{P};\ \mathcal{E};\ \mathcal{R}\cup\{s\to u\})}$	if $t \to_{\mathcal{R}\cup\mathcal{E}_\succ} u$
Collapse$_1$	$\dfrac{(\mathcal{P};\ \mathcal{E};\ \mathcal{R}\cup\{s\to t\})}{(\mathcal{P}\cup\{u\leftrightarrow t\};\ \mathcal{E};\ \mathcal{R})}$	if $s \to_{\{l\to r\}} u$ where $l \to r \in \mathcal{R}\cup\mathcal{E}_\succ$ and l is not an instance of s.
Collapse$_2$	$\dfrac{(\mathcal{P};\ \mathcal{E}\cup\{s\leftrightarrow t\};\ \mathcal{R})}{(\mathcal{P}\cup\{u\leftrightarrow t\};\ \mathcal{E};\ \mathcal{R})}$	if $s \to_{\{l\to r\}} u$ where $l \to r \in \mathcal{R}\cup\mathcal{E}_\succ$ and l is not an instance of s.
Deduce	$\dfrac{(\mathcal{P};\ \mathcal{E};\ \mathcal{R})}{(\mathcal{P}\cup\{s\leftrightarrow t\};\ \mathcal{E};\ \mathcal{R})}$	if $(s,t) \in E\dot{P}_\succ(\mathcal{R}\cup\mathcal{E})$

Fig. 2.: Inference rules for unfailing completion

Given a set of equations $\mathcal{P}$, an unfailing completion procedure attempts to compute a terminating and (ground) confluent reduction relation $\rightarrow_{\mathcal{R}\cup\mathcal{E}_{\succ}}$ defined by a set of unorientable equations $\mathcal{E}$ and a set of rewrite rules $\mathcal{R}$. This is done by repeatedly transforming a triple $(\mathcal{P}_i; \mathcal{E}_i; \mathcal{R}_i)$ of two equation sets $\mathcal{P}_i$ and $\mathcal{E}_i$ and one rule set $\mathcal{R}_i$ into a triple $(\mathcal{P}_{i+1}; \mathcal{E}_{i+1}; \mathcal{R}_{i+1})$ using the inference rules in Fig. 2. In the *Deduce*-step the set of *extended critical pairs* $EP_{\succ}(\mathcal{R} \cup \mathcal{E})$ is computed. See [4] for a precise definition of $EP_{\succ}(\mathcal{R} \cup \mathcal{E})$. Our implementation gives the highest priority to reductions (Compose, Collapse$_i$, Simplify) and deletions. Thus all equations and rules are always kept completely normalized[1]. Several reductions on a single term are bundled in a normalization procedure that exploits marking techniques to avoid redundant reduction trials. The next priority is given to the computation of extended critical pairs (Deduce) and least priority is granted to the orientation steps (Orient, Unfail). The work performed between two orientation steps is called a *completion cycle*. To guarantee *fairness* of the completion procedure we must ensure that no equation stays in $\mathcal{P}$ forever.

4 Completion Strategies

The exact order in which equations from $\mathcal{P}$ are turned into rules in $\mathcal{R}$, or into equations in $\mathcal{E}$, respectively, is called the *completion strategy*. The completion strategy is determined by a selection function choosing the *best* equation in $\mathcal{P}$. To yield a fair completion the best equation must be minimal in $\mathcal{P}$ w. r. t. some terminating ordering on equations. Small changes in this strategy can have huge effects (positive or negative) on the duration and work profile of a completion or proof task. Fig. 3 shows the completion profile obtained when completing the same input with two different strategies. The x-axis counts the completion cycles and the y-axis shows $|\mathcal{P}|$ measured before the i-th orientation operation.

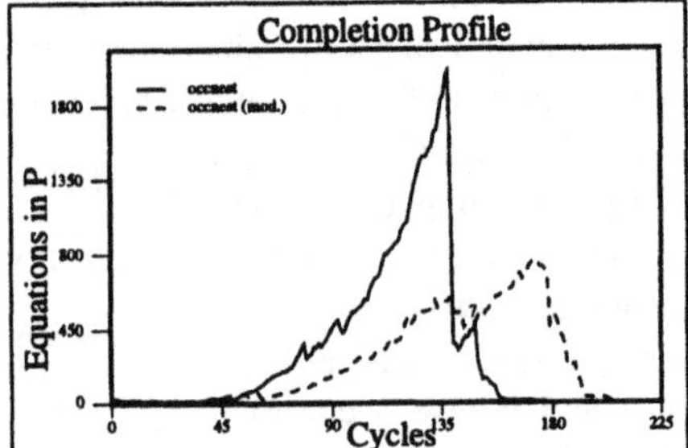

Fig. 3.: Completion profiles for different selection functions

Let $\mathcal{F}$ be the set of operators and let $\mathcal{V}ar$ be the set of variables. The selection function is typically based on a weight function $\phi : \mathcal{F} \cup \mathcal{V}ar \rightarrow \mathbb{N}$ defined by $\phi(x) = c_v$ for each variable $x \in \mathcal{V}ar$ and $\phi(f) = c_f$ for each function symbol $f \in \mathcal{F}$. The weight function can be homomorphically extended to terms $t \in T(\mathcal{F}, \mathcal{V}ar)$, e. g.:

$$\phi(t) = \begin{cases} c_v & \text{if } t \equiv x \in \mathcal{V}ar \\ c_f + \sum_{i=1}^{n} \phi(t_i) & \text{if } t \equiv f(t_1, \ldots, t_n), f \in \mathcal{F} \end{cases}$$

Then for an equation $c \leftrightarrow d$, $\phi(c) + \phi(d)$ (add-weight) or $max\{\phi(c), \phi(d)\}$ (max-weight) define typical selection functions, which have proven to perform well [24].

[1] Experiments with a version that does not completely normalize $\mathcal{P}$ did not perform uniformly better than the version described here. Cf. [31] for these experiments.

Besides weight based selection functions that predominate in standard completion procedures, goal oriented selection functions have been proven to be very successful in unfailing completion procedures [15]. Goal orientation is achieved by comparing measures acquired from a critical pair $c \leftrightarrow d$ with the corresponding measures obtained from the goal.
In our experiments we used the selection function ON_G. Let $occ(f,t)$ and $nest(f,t)$ be the occurrence and nesting of the symbol f in a term t (cf. [18]), let $O^G_{f,t_1,t_2} = G(occ(f,t_1), occ(f,t_2))$ and let $N^G_{f,t_1,t_2} = G(nest(f,t_1), nest(f,t_2))$ for some operation G. Then ON_G is defined as

$$ON_G(c \leftrightarrow d) = (\phi(c) + \phi(d)) \prod_{f \in \mathcal{F}} ((O^G_{f,c,d} \dot{-} O^G_{f,a,b}) + 1)((N^G_{f,c,d} \dot{-} N^G_{f,a,b}) + 1)$$

where ϕ is a weight function, and $a \leftrightarrow b$ is the goal. Then $occnest = ON_{\max}$ is the instance of ON_G defined in [1, 18]. We also use the instance $occnest(mod.) = ON_\times$ (i. e., G is the product) in our experiments (cf. Sec. 7). *Occnest (mod.)* takes into account operator specific information of both terms of an equation and in contrast to *occnest* it gives strong penalties if none of the two terms of a pair is close to the goal.

5 Strategy Compliant Parallel Unfailing Completion

The multi-threaded unfailing completion procedure (PUC) included in PaReDuX is designed to exploit medium grained parallelism on an SMP. It has been derived from the sequential unfailing completion procedure (UC) of ReDuX by multi-threading the normalization of the equations in $\mathcal{P}$ and the computation of critical pairs. The parallelization is based on a divide & conquer paradigm that maps independent work (such as normalizing two sublists of $\mathcal{P}$) to different VS-threads.
Note that in ReDuX normalizing $\mathcal{P}$ and computing critical pairs requires write access to $\mathcal{R}$ and $\mathcal{E}$ because ReDuX stores substitutions in the term-structure as side-effects. In order to avoid rule locking, we made these data-structures for PaReDuX *multi-thread safe* by adding a level of indirection (a switch table) which keeps the writable part thread-local. Many threads may simultaneously use the same global rule or equation for normalization and there is no need to make copies or to synchronize in $\mathcal{R}$ or $\mathcal{E}$ [10]. Finally, the sequential unfailing completion code had to be modified by reorganizing iterative loops (such as normalizing $\mathcal{P}$) as divide & conquer style procedures, and by inserting VS-thread-fork and -join instructions to compute in parallel. The overall parallel architecture of PUC is depicted in Fig. 4.
Besides computing independent tasks (like normalization and critical pairing) in parallel, each of the loops depicted in doubly framed boxes is parallelized, again using divide & conquer. To keep the tasks as coarse-grained as possible, the parallel normalization and deletion procedures are called from within the threads for reducibility tests and critical pair computations, respectively. Thus we can have more threads for normalization and deletion than for reduction tests and critical pair computations, but not fewer.
This parallelization scheme contains a purely sequential part comprising the test in the while loop, the orientation, composition and collapse steps as well as joining the results of the parallel part, and trying to prove the goal. Yet our experience

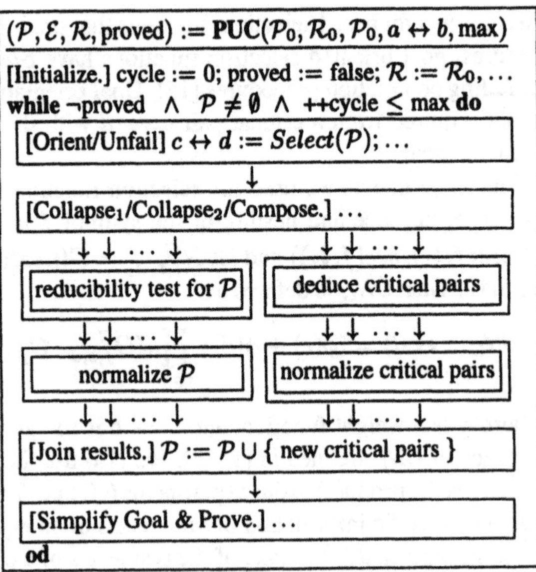

Fig. 4.: Strategy compliant unfailing completion

shows that this sequential part typically accounts only for less than 2% of the whole sequential completion time. Using an appropriate selection function in the Orient/Unfail-step we can guarantee that the parallelized completion is *strategy compliant*, performing exactly the same selection sequence (and work) as its sequential model.
Clearly a strategy compliant parallel completion is fair if its sequential strategy is. The constraint on the selection function is that it must modulo variable renaming unambiguously pick an equation from a (multi)set of equations. This theoretically simple requirement may be non-trivial to realize in practice in a parallel environment because most selection functions give rise to quasi-orderings only. Our solution to this problem is to compose the sub-results (i. e., critical pair lists) of different parallel threads as in the sequential computation and *not* in the order they are available. Note that in the VS-thread system a thread waiting for a join does not idly wait. Therefore due to the rather fine granularity of our parallelization and the logical parallelism available in the problem our strategy to collect the results does not prevent the processors from working.

Our strategy compliant parallelization proved to be scalable and very robust w. r. t. modifications of parallelization parameters. We could find speed-ups of up to 3.5 on 4 processors. For a more detailed description and analysis of the strategy compliant parallel unfailing completion procedure see [10, 31].

6 The Master-Slave Approach to Unfailing Completion

A master-slave unfailing completion procedure consists of one master performing an unfailing completion to find a proof for a goal, and a set of slaves employed by the master to search for rules, equations, and critical pairs, which can be fed into the master's proof process. Thus the slaves serve as 'oracles' providing valid lemmas to the master proof procedure.
Each slave is composed of an unfailing completion procedure that is stopped after a maximum number of cycles, and a procedure that subsequently evaluates the results of the completion and passes the best results back to the master. Each slave uses a specific (not necessarily fair) strategy different from the master strategy. When forked it obtains the following arguments from the master: the maximum number of cycles the slave completion shall run, and the current critical pairs, equa-

tions, rules and the goal of the master. If a slave succeeds in proving the goal this is signaled to the master immediately. Otherwise the results obtained after the slave completion procedure stopped are evaluated. Therefore, during the proof process, every slave counts the number of matches as well as successful *collapse* and *compose* operations for its system of rules and equations. This statistical information is then used to compute a heuristic selecting the five best rules and equations which are sent back to the master.

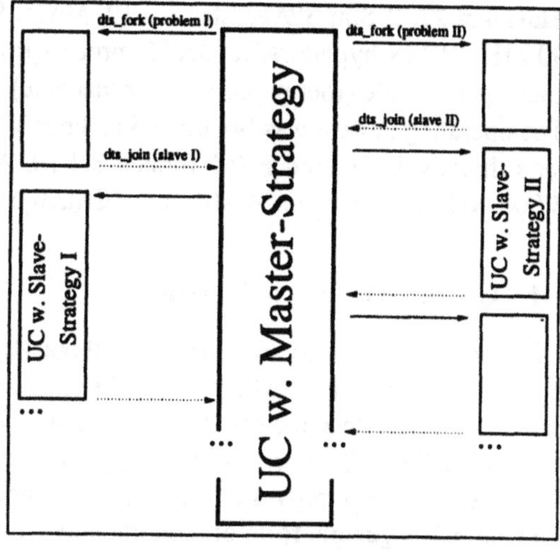

Fig. 5.: The master-slave approach

The master process is an unfailing completion procedure that additionally can fork and join slaves over the network. Forking and joining slaves is done in the sequential part of the completion (e. g., at the beginning of the while loop of Fig. 4). During the initialization the master completion process forks a fixed number of slaves depending on the number of machines in the DTS environment. In the sequel the master checks once in every cycle whether any of its slaves can be joined (i. e., whether a slave has completed its work). If no, the master continues with its own completion task. Otherwise it either terminates with success if a slave could prove the goal, or it inserts the results of the slave into the master's queue of critical pairs with high priority. Subsequently a new slave is forked (with the same strategy as the slave just joined) and the master continues with its own completion work. Fig. 5 sketches the principle of our master-slave approach.

Note that the completion is fair if the master uses a fair strategy, and between joining and forking two slaves the master considers at least one critical pair it has deduced itself. Therefore the work of the slaves (number of cycles to run) should be scaled accordingly. Assuming the same time complexity for each completion cycle a slave should run more than $n \cdot m$ cycles where n is the number of slaves and m is the number of results returned by a slave.

Both the master and the slave can equally well run sequential or strategy compliant parallel completion procedures because the communication and synchronization between the master and the slaves is done in the sequential part of the completion procedure.

7 Empirical Results

We have implemented and tested our parallel unfailing completion procedure on a network of Sun SPARCstations 10. The configuration was as follows:

Platform #1 2 Sun SPARCstations 10, Model 402, 32 Mbyte of main memory, two 40 MHz SuperSPARC processors, no secondary level cache (two processors per machine)

Platform #2: 2 Sun SPARCstations 10, Model 914, 80 Mbyte of main memory, two 90 MHz ROSS hyperSPARC double processor modules[2], 512 Kbyte secondary level cache per module (four processors per machine)
The operating system was Solaris 2.5 and our VS-threads system was installed on Solaris threads. DTS used PVM version 3.3.10. All experiments were performed with 3,000,000 list cells (i. e., 24 MByte) of dynamic memory per workstation.

7.1 Characterization of Experiments

Our problem specifications consist for all but Z_{22} of a set of equations, a reduction ordering, and an equation to be proved. The task for Z_{22} is to find a complete TRS. $glb1a$, $transa$ and $p3b$ describe proof tasks for lattice-ordered groups. These examples were taken from the TPTP library [34] (Version 1.2.0; Problems: GRP138-1.p, GRP162-1.p, GRP170-2.p) $glb1a$ proves the greatest lower bound axiom, $transa$ the transitivity axiom and $p3b$ a general form of monotonicity from the given equational axiomatization. The $Lusk$ examples are taken from [30]. $Lusk_3$ proves that a ring with $x^2 = x$ is commutative. $Lusk_4$ shows that the commutator $h(h(x,y),y) = e$ holds in a group with $x^3 = e$, and $Lusk_5$ proves that $f(x,g(x),y) = y$ holds in a ternary algebra where the third axiom is omitted. The $Luka$ examples are taken from [36]; cf. also [29]. The input equations are an equational axiomatization for propositional calculus by Frege. Lukasiewicz gave another set of axioms, which are the goals in our examples $Luka_2$ and $Luka_3$. Finally, Z_{22} is a problem in combinatorial group theory taken from [29]. The task is to prove that the given finitely presented group with is the cyclic group of order 22.

7.2 Strategy Compliant vs. Master-Slave Unfailing Completion

The strategy compliant experiments were run with *occnest* as selection function. All reported master-slave experiments run with one slave; *occnest* was the strategy of the master and *occnest (mod.)* the strategy of the slave.
Table 1 shows the timings for all our experiments in seconds measured on platforms (PF) #1 and #2. The third column shows the number of completion cycles (Cyc.), the next two columns the sequential times (Time UC), and the speed-ups w.r.t. the multi-threaded strategy compliant parallelization (UC/PUC).
The master-slave experiments were run for sequential (MS(s)) and multi-threaded strategy compliant workers (MS(p)). For both cases, the results consist of the number of completion cycles (Cyc.) of the master, the number (#) of slave forks times the number of completion cycles of a slave per fork (cpf), and finally the speed-ups w.r.t. the different implementations (UC/MS(s), MS(s)/MS(p), PUC/MS(p), UC/MS(p)).
$Lusk_3$, $Lusk_4$, Z_{22} and $transa$ reveal very good speed-ups by strategy effects that can be further improved—in most of the cases—by work parallelism. $Lusk_4$ could not

TRS	PF	Strategy Compliant Cyc.	Time UC	UC/ PUC	Master-Slave (seq.) Master Cyc.	Slaves # * cpf	UC/ MS(s)	Master-Slave (parallel) Master Cyc.	Slaves # * cpf	MS(s)/ MS(p)	PUC/ MS(p)	UC/ MS(p)
glb1a	#1	132	99.0	1.7	138	2 * 40	0.8	142	2 * 40	1.3	0.6	1.1
	#2	132	58.0	2.9	142	2 * 40	0.9	142	2 * 40	2.4	0.7	2.1
transa	#1	56	134.5	1.4	34	2 * 40	4.7	40	2 * 40	0.5	1.6	2.2
	#2	56	81.5	2.6	34	2 * 40	5.0	35	2 * 40	2.1	4.0	10.2
p3b	#1	116	226.3	1.5	126	3 * 40	0.9	114	2 * 40	1.7	1.0	1.6
	#2	116	134.8	2.3	125	3 * 40	0.9	125	3 * 40	3.0	1.2	2.7
$Lusk_3$	#1	57	1098.1	1.7	77	3 * 40	1.6	63	2 * 40	1.8	1.7	2.9
	#2	57	655.9	3.3	75	3 * 40	2.1	61	2 * 40	7.1	4.5	14.5
$Lusk_4$	#1	–	>1200	–	72	2 * 40	72.4†	74	2 * 40	1.3	–	–
	#2	–	>1200	–	71	2 * 40	39.1†	77	2 * 40	2.0	–	–
$Luka_2$	#1	79	55.8	1.7	84	1 * 40	0.9	84	1 * 40	1.4	0.7	1.3
	#2	79	32.2	2.9	84	1 * 40	0.9	84	1 * 40	2.6	0.8	2.5
$Luka_3$	#1	119	182.7	1.8	124	1 * 80	1.0	124	1 * 80	1.4	0.7	1.3
	#2	119	109.6	3.4	124	1 * 80	1.0	124	1 * 80	2.0	0.6	2.0
Z_{22}	#1	176	249.7	1.2	162	4 * 40	2.1	110	2 * 40	2.9	5.0	5.9
	#2	176	152.6	3.2	156	4 * 40	2.1	110	3 * 40	6.5	4.3	13.6

Table1. General statistics for parallel and distributed unfailing completion (†: runtime [s])

superlinear. With the exception of *transa* and $Lusk_4$, all problems are solved faster on one 4-processor machine than on two double-processors (after scaling the processor performance) which underscores the importance of the strategy compliant component.

8 Discussion

The master-slave approach leads in a set of test cases to superlinear speed-ups w.r.t. the sequential implementation. Our system reaps strategy effects—which where first discovered and explored with the DISCOUNT system [3, 14]—by a much simpler communication technique between the different proof processes. Moreover, it allows a simulation of different parallel completion techniques: E.g., we obtain a multi-threaded and strategy compliant theorem prover, if no slave is activated; we obtain a simple competitive theorem prover, if we join the slaves after an indefinite (very large) number of completion cycles; and finally, we obtain a restricted teamwork based theorem prover, where the supervisor can not be changed.

There are some examples where the slave contributions are not as good as expected. Of course, the setup overhead for slaves is not completely negligible, and moreover, slaves can slow down the master by misleading results. To avoid such effects, the setup of slaves and the evaluation of results have to be done very carefully and deserve further investigation. The communication between master and slave could be extended in such a way that good intermediate slave results can reach the master whenever they occur and not only after the termination of a slave. It should also be possible to distribute powerful rules and equations of the master system to all slaves. More elaborate master-slave

communication techniques will reduce the number of slave setups and allow longer slave lifetimes, but go beyond the simple fork / join paradigm.
We have shown that our parallel master-slave unfailing completion procedure in its current form is a conceptionally simple but powerful approach towards the application of distributed and shared-memory based thread-systems for theorem proving. And it is a step forward towards the integration of shared memory multi-threaded components in existing distributed theorem provers.

Acknowledgements

This work was supported by grant Ku 966/1-2 within the DFG focus program "Deduktion". Special thanks to Patrick Maier for implementing the parallel unfailing completion procedure and to Tilmann Bubeck for his support during the integration of DTS and VS-threads. We thank the referees for insightful comments.

References

1. S. Anantharaman and N. Andrianarievelo. Heuristical criteria in refutational theorem proving. In A. Miola, editor, *Design and Implementation of Symbolic Computation Systems*, 1994. Springer-Verlag.
2. G. Attardi and C. Traverso. A strategy-accurate parallel Buchberger algorithm. In Hong [19].
3. J. Avenhaus and J. Denzinger. Distributing equational theorem proving. In Kirchner [21].
4. L. Bachmair, N. Dershowitz, and D. Plaisted. Completion without failure. In H. Aït-Kaci and M. Nivat, editors, *Resolution of Equations in Algebraic Structures*, volume 2 of *Rewriting Techniques*, chapter 1. Academic Press, 1989.
5. M. Bonacina and J. Hsiang. Distributed Deduction by Clause-Diffusion: Distributed Contraction and the Acquarius Prover. *J. Symbolic Computation*, 19(1/2/3), 1995.
6. M. Bonacina and J. Hsiang. Completion procedures as semidecision procedures. In S. Kaplan and M. Okada, editors, *Conditional Term Rewriting Systems*, 1990. Springer-Verlag.
7. M. Bonacina and J. Hsiang. Parallelization of deduction strategies: An analytical study. *J. Automated Reasoning*, 13:1–33, 1994.
8. T. Bubeck, M. Hiller, W. Küchlin, and W. Rosenstiel. Distributed symbolic computation with DTS. In A. Ferreira and J. Rolim, editors, *Parallel Algorithms for Irregularly Structured Problems, 2nd Intl. Workshop, IRREGULAR'95*, 1995.
9. R. Bündgen. Reduce the redex → ReDuX. In Kirchner [21].
10. R. Bündgen, M. Göbel, and W. Küchlin. A fine-grained parallel completion procedure. In J. von zur Gathen and M. Giesbrecht, editors, *Proc. 1994 International Symposium on Symbolic and Algebraic Computation: ISSAC'94*, 1994. ACM Press.
11. R. Bündgen, M. Göbel, and W. Küchlin. Multi-threaded AC term rewriting. In Hong [19].
12. R. Bündgen, M. Göbel, and W. Küchlin. Parallel ReDuX → PaReDuX. In J. Hsiang, editor, *Rewriting Techniques and Applications, 6th Intl. Conf., RTA-95*, 1995. Springer-Verlag.
13. R. Bündgen, M. Göbel, and W. Küchlin. Strategy compliant multi-threaded term completion. *J. Symbolic Computation*, 1996. To appear.
14. J. Denzinger. *Teamwork: Eine Methode zum Entwurf verteilter, wissensbasierter Theorembeweiser*. PhD thesis, Universität Kaiserslautern, 1993.

15. J. Denzinger and D. Fuchs. Goal oriented equational theorem proving using teamwork. In *Proc.* 18th *KI-94*, 1994. Springer-Verlag.
16. N. Dershowitz and J.-P. Jouannaud. Rewrite systems. In *Formal Models and Semantics*, volume 2 of *Handbook of Theoretical Computer Science*, chapter 6. Elsevier, 1990.
17. B. Fronhöfer and G. Wrightson, editors. *Parallelization in Inference Systems*, 1990. Springer-Verlag.
18. M. Fuchs. Learning proof heuristics by adapting parameters. Technical report, Fachbereich Informatik, Universiät Kaiserslautern, 1995.
19. H. Hong, editor. *First Intl. Symp. Parallel Symbolic Computation PASCO'94*, 1994. World Scientific.
20. J. Hsiang and M. Rusinowitch. On word problems in equational theories. In T. Ottmann, editor, *Automata, Languages and Programming*, 1987. Springer-Verlag.
21. C. Kirchner, editor. *Rewriting Techniques and Applications*, volume 690 of *LNCS*, Montreal, Canada, 1993. Springer-Verlag.
22. C. Kirchner and P. Viry. Implementing parallel rewriting. In Fronhöfer and Wrightson [17].
23. D. Knuth and P. Bendix. Simple word problems in universal algebra. In J. Leech, editor, *Computational Problems in Abstract Algebra*. Pergamon Press, 1970.
24. W. Küchlin. A theorem-proving approach to the Knuth-Bendix completion algorithm. In J. Calmet, editor, *Computer Algebra*, 1982. Springer-Verlag.
25. W. Küchlin. PARSAC-2: A parallel SAC-2 based on threads. In S. Sakata, editor, *Applied Algebra, Algebraic Algorithms, and Error-Correcting Codes: 8th International Conference, AAECC-8*, 1990. Springer-Verlag.
26. W. Küchlin. The S-threads environment for parallel symbolic computation. In R. Zippel, editor, *Computer Algebra and Parallelism*, 1992. Springer-Verlag.
27. W. Küchlin. PARSAC-2: Parallel computer algebra on the desk-top. In J. Fleischer, J. Grabmeier, F. Hehl, and W. Küchlin, editors, *Computer Algebra in Science and Engineering*, 1995. World Scientific.
28. W. Küchlin and J. Ward. Experiments with virtual C Threads. In *Proc. Fourth IEEE Symp. on Parallel and Distributed Processing*, 1992. IEEE Press.
29. E. Lusk and W. McCune. Experiments with ROO, a parallel automated deduction system. In Fronhöfer and Wrightson [17].
30. E. Lusk and R. Overbeek. Reasoning about equality. *J. Automated Reasoning*, 1:209–228, 1985.
31. P. Maier, M. Göbel, and R. Bündgen. Multi-threaded unfailing completion. Technical Report WSI 95–06, Universität Tübingen, D-72076 Tübingen, 1995.
32. J. Schumann. SiCoTHEO — simple competive parallel theorem provers based on SETHEO. 1995. Proc. PPAI'95, Montreal, Canada, 1995.
33. V. Sunderam. PVM: A framework for parallel distributed computing. *Concurrency: Practice and Experience*, 2(4):315–339, 1990.
34. G. Sutcliffe, C. Suttner, and T. Yemenis. The TPTP problem library. In A. Bundy, editor, *Automated Deduction—CADE-12*, 1994. Springer-Verlag.
35. C. Suttner and J. Schumann. Parallel automated theorem proving. In L. Kanal, V. Kumar, H. Kitano, and C. Suttner, editors, *Parallel Processing for Artificial Intellegence*. Elsevier, 1994.
36. A. Tarski. *Logic, Semantics, Metamathematics*. Oxford University Press, 1956.

Multi-Agent Cooperation – Concepts and Applications

Hans Haugeneder and Donald Steiner

Siemens AG
Corporate Research and Development
Otto-Hahn-Ring 6, 81730 München, Germany
{Hans.Haugeneder,Donald.Steiner}@zfe.siemens.de

Abstract

In the current discussion the notion of an agent and its descendents like agent-based systems, agent software etc. are used in manifold ways exhibiting both similarities and divergencies. Our approach to this area stems from Distributed Artificial Intelligence (DAI). The central problem underlying the design of agent based systems from a DAI point of view is characterized by the following question:

> How can individually motivated, independently designed computational artefacts, agents, act together to achieve some (at least partial) common goal in a genuinely distributed problem space?

Under this perspective collections of agents and their interaction is of crucial importance rather than single agents. The field of *Multi*-Agent Systems has established itself as one of the two major streams within DAI – the other one being Cooperative Problem Solving – to design methods and architectures dedicated specifically to this problem area.

The paper presented has its focus on three issues,[1] which explicate our general approach to constructing multi-agent systems and its practical application. We will present

1. the conceptual design and specification of an agent architecture and a multi-agent language for implementation of single agents and cooperative interaction among them,
2. a multi-agent environment for constructing cooperative applications based on this model, and
3. two application studies in the area of group scheduling and individualized traffic management.

[1] The work presented here was performed within the ESPRIT Project No. 5362 IMAGINE (Integrated Multi-AGent INteractive Environment) project, which was supported by the Commission of the European Communities as part of the ESPRIT II framework.

Agent software, although a comparatively young area in software design with many different facets and technical instantiations, has a high potential to provide improved solutions for a wide range of applications. We also witness the first steps towards market entry of first generation agent software, both as generic technology and dedicated solutions. Despite of this the field exhibits a drawback originating in the diversity of approaches used to create agent-based solutions. The field still shows some lack of standarisation. There are, however, a number of dimensions where the initiation of a process of convergence seems to be necessary and possible, such as:

1. Agent communication and execution models
2. A general agent service architecture
3. A format for information exchange between agents
4. Agent interaction protocols
5. A generic agent application interface

All steps towards standardisation with respect to these five areas will create another stimulating force for the technical progress of the field of agent software as a whole as well as for its practical impact.

References

[GS94] P. Gaudin and D. Steiner. The MECCA Platform. In H. Haugeneder, editor, *IMAGINE Final Project Report*, pages 125–151. 1994.

[HAM90] H. Hämmäinen, J. Alasuvanto, and R. Mäntylä. Experiences on semi-autonomous user agents. In J. Müller Y. Demazeau, editor, *Decentralized AI*, pages 235–249. Elsevier Science Publishers, 1990.

[Hau94] H. Haugeneder. *IMAGINE Final Project Report*. IMAGINE Technical Report Series, 1994.

[HS94a] H. Haugeneder and D. Steiner. A Multi-Agent Scenario for Cooperation in Urban Traffic. In S. M. Deen, editor, *1993 Proceedings of the SIG on Cooperating Knowledge Based Systems*, pages 83–99. DAKE Centre, University of Keele, 1994.

[HS94b] Hans Haugeneder and Donald Steiner. Ein Mehragentenansatz zur Unterstützung kooperativer Arbeit. In U. Hasenkamp, S. Kirn, and M. Syring, editors, *CSCW - Computer Supported Cooperative Work*, pages 209–229. Addison Wesley, 1994.

[HSM94] Hans Haugeneder, Donald Steiner, and Frank. G. McCabe. IMAGINE: A framework for building multi-agent systems. In S. M. Deen, editor, *Proceedings of the Second International Working Conference on Cooperating Knowledge Based Systems*, pages 31–66. DAKE Centre, University of Keele, 1994.

[LK92] A. Lux and M. Kolb. Linking Humans and Intelligent Systems or: What are User Agents Good for? In *Proceedings of the 16th German AI Conference*, pages 372–385, 1992.

[LS95] A. Lux and D. Steiner. Understanding cooperation: an agent's perspective. In *Proceedings ICMAS-95*, 1995.

[Lux94] A. Lux. MAM - MECCA Appointment Manager. In H. Haugeneder, editor, *IMAGINE Final Project Report*, pages 183–194. 1994.

[SBKL93] D. Steiner, A. Burt, M. Kolb, and Ch. Lerin. The Conceptual Framework of MAI2L. In *Proc. Fifth European Workshop on Modelling Autonomous Agents in a Multi-Agent World*, 1993.

[Sea69] J. R. Searle. *Speech Acts*. Cambridge University Press, 1969.

[Sho93] Y. Shoham. Agent-oriented programming. *Artificial Intelligence*, 60:51–92, 1993.

[SMH90] D. Steiner, D. Mahling, and H. Haugeneder. Human Computer Cooperative Work. In M. Huhns, editor, *Proc. of the 10th International Workshop on Distributed Artificial Intelligence*. MCC Technical Report Nr. ACT-AI-355-90, 1990.

[Smi80] R.G. Smith. The contract net protocol: High level communication and control in a distributed problem solver. *IEEE Trans. on Computers*, 29:1104 – 1113, 1980.

[Ste95] D. Steiner. IMAGINE: an integrated framework for constructing multi-agent systems. In G. O'Hare and N. Jennings, editors, *Foundations of Distributed Artificial Intelligence*. Wiley Interscience, 1995.

[VHJL94] Gerd Völksen, Hans Haugeneder, Alex Jarczyk, and Peter Löffler. Cooperation-Ware: Integration of human collaboration with agent-based interaction. In S. M. Deen, editor, *Proceedings of the Second International Working Conference on Cooperating Knowledge Based Systems*, pages 307–330. DAKE Centre, University of Keele, 1994.

[Wer91] E. Werner. The Design of Multi-Agent Systems. In *Decentralized A.I. 3. Proc. Third European Workshop on Modelling Autonomous Agents in a Multi-Agent World*, pages 3–30, 1991.

Document-Centered Presentation of Computing Software: Compound Documents Are Better Workspaces

Wolfgang Weck

Institute of Scientific Computing, ETH Zürich, CH-8092 Zürich, Weck@inf.ethz.ch

1 Introduction

This extended abstract sketches some results from prototyping a document-centered component framework for mathematical software. An important aspect of the prototype was the separation of data management from computational software components. Compound documents have been used as data repositories.

2 Compound Documents as Workspaces

Interactive application programs must manage the data objects on which the user computes. The repository for these objects is called *the user's workspace* [3]. Compound documents, as introduced with OpenDoc and OLE [5], can be used to implement workspaces. Each data object is represented by a document part. This workspace implementation is particularly light-weight.

Traditional data repositories use name bindings. The disadvantage is the hidden state created by name bindings. Visual document parts avoid such hidden state. Direct manipulation of workspaces is possible. Further, document parts can display more information than names can. For instance, a mathematical formula can be typeset within a document part. Finally, document parts can react on user input. An in-place expression editor can be implemented that way. An example of such an editor is *MathType* within *Microsoft's Word 6.0* [4]. For a further discussion on document parts vs. name bindings see [8].

Maple Worksheets [2] or *Mathematica Notebooks* [6], are document-oriented front ends for workspaces using name bindings. In contrast, compound documents are workspaces themselves.

3 The Advantage of Document-Centered Software

The advantage of a document-centered software architecture is that many independent software components can share workspaces. In contrast, a process-centered program is a package deal, bundling data management and computational power. It is cumbersome to combine such programs.

Document-centered software allows for reuse of many existing software components. For instance, already available hypertext facilities pave the road towards interactive textbooks. For mathematical software, similar ideas were developed for *CaminoReal* [1] and *Euromath* [7]. The former extends an existing document editor, allowing for component reuse as mentioned above. The latter implements its own structure editor.

4 Enhancements

Some issues are not solved by *CaminoReal* and *Euromath*. One problem to be resolved has been posed in [1]: how can concurrent computation on a remote server be represented within an interactive document editor? We have solved this by using document parts as placeholders (promises) for upcoming results.

Another issue is scripting. One can use a scripting facility, as document editors typically provide, or program the script as a new software component to extend the editor. Usually, the former requires to compose the script from editing operations instead of mathematical operations. The latter may be too cumbersome for simple scripts.

In [8] a more general view of compound documents has been proposed. In particular, document parts within texts can be used as tokens of a formal language. Thus, a section of a document can be interpreted as a script. Document parts can be included to refer to objects from the workspace.

5 Conclusions

Using compound documents as workspaces has two advantages: separating data management from computation software is a prerequisite for decomposing mathematical software into components, and the direct manipulation of workspaces is an improvement over name bindings. Further, document parts can act as placeholders for upcoming results. Viewing document parts within a textual context allows scripting within documents.

References

1. Arnon, D., Beach, R., McIsaac, K., Waldspurger, C.: CaminoReal: An Interactive Mathematical Notebook. Proc. EP'88, Int. Conf. on Electronic Publishing, Document Manipulation, and Typography, 1988.
2. Char, B., Geddes, K., Gonnet, G., Leong, B., Monagan, M., Watt, S.: Maple V Language Reference Manual. Springer-Verlag, New York, 1991.
3. Gilman, L., Rose, A.: APL – An Interactive Approach. Wiley & Sons, N. Y., 1974.
4. Microsoft Corporation: Word User's Guide, version 6.0. Microsoft Press, 1993.
5. Orfali, R., Harkey, D., Edwards, J.: The essential distributed objects survival guide. Wiley& Sons, New York, 1996.
6. Soiffer, N.: Mathematical Typesetting in Mathematica. Proc. ISSAC'95, Montreal, Canada, pp. 140-149, ACM 1995.
7. von Sydow, B.: The design of the Euromath system. Euromath Bulletin 1, 1992.
8. Weck, W.: Putting Icons into (Con-) Text. Proc. of TOOLS EUROPE '94, Versailles, France, Prentice Hall, 1994.

Animating a Non-executable Formal Specification with a Distributed Symbolic Language

P.Ciancarini and S.Cimato

Dipartimento di Scienze dell'Informazione, Università di Bologna,
Piazza di Porta S. Donato 5, 40127 Bologna (Italy)
E-mail: {cianca,cimato}@cs.unibo.it

The introduction of formal methods in software engineering practice has been motivated by an increasing need of quality and reliability in the development of software systems. The use of a mathematical notation in the specification phase helps the specifier to give a more precise and rigorous description of the system being developed, reducing possible ambiguities and inconsistencies. The technique of *animating* a formal specification aims at overcoming the difficulties of using a non executable, highly abstract, declarative specification language, thus allowing the specifier to analyze and verify his formal specification document by observing its behavior. In this paper we show how we obtain a distributed program by automatic translation of an abstract specification written in Z [Spi92]. In our knowledge this is the first attempt to obtain a concurrent and distributed animation directly from a Z specification document.

The basic problem in building an animation system for Z consists of giving an operational semantics to a notation which is intrinsically declarative. The whole system can be totally described by predicates on the abstract state, while operations can be described by predicates which say how the state of the system can evolve. In Z, all this is modeled in schemas where either state components or operations are defined. In particular, operation schemas consist of a set of preconditions which specify the set of valid states of the system to which it is possible to apply that operation and a set of postconditions which define the set of states that can be reached after the application of the operation.

This basic semantics can be redefined using a concurrent semantic model [Eva94], namely we follow a *chemical* approach [CCM96]. An operation is enabled if the current state of the system satisfies its preconditions and two (or more) operations are concurrent if they are enabled at the same time. Once satisfied the preconditions of the schema considered, the defined operation can be executed. This implies that in an animation context, several operations can be executed at the same time.

In order to simplify the animation process we have defined a subset of Z, *Zel (Z elementary)*, restricting the number of admissible constructs [CCM96]. The limitations we have introduced concern types and use of variables, form of predicates, usability of quantifiers and logical operator (only $\wedge$ and $\neg$ can be used). No restrictions are imposed about schemas construction and connections.

Once a Zel specification has been obtained by manual refinement of the original Z document, the animation process continues using a translator which automatically produces a Shared Prolog program [CG96]. Shared Prolog is a coordination language based on the combination of the shared dataspace coordination model (as in Linda) with logic programming computing (as in Prolog). A SP program consists of a *tuple_space*, which is a multiset of logical tuples, and a set of modules, called *theories*, which are Prolog programs concurrently executed by active *agents* which share the tuple_space and are defined in the *initial goal.* The translator opportunely maps Zel schemas, variables and predicates into Sp theories and tuples.

The Shared Prolog program obtained from the transformation process has to be initialized before execution in order to satisfy the particular query we want to animate. After we have defined the initial goal and the initial tuple space corresponding to the query, the program is passed to the Sicstus Prolog compiler which produces the executable code. Having chosen SP as target language, our method produces, for a given specification, code that is executable by a distributed run-time system. A truly distributed execution of the animation has the remarkable advantage of highlighting concurrent aspects of system being built, that could not be observed using other sequential approaches. By inspecting the behavior of the animation, the specifier can increase the degree of trust in the system being developed, accelerating the validation of the specification with respect to the user requirements. Formal proofs of properties, that often are quite expensive, can be attempted after the specification has an acceptable behavior on the test cases.

With our approach, we have successfully animated several classic examples (eg., the library problem, the router problem, the vending machine problem), and many others, obtaining efficient animations. However, since a procedural style for writing specifications into Zel notation is requested, it is not always so immediate to refine a Z specification into a Zel document. Future developments of our method aim at extending both the Zel notation in order to augment the Z coverage, and the SP language with new constructs in order to increase its expressive power.

References

[CCM96] P. Ciancarini, S. Cimato, and C. Mascolo. Engineering formal requirements: an analysis and testing method for Z documents. Technical Report UBLCS-6, Dipartimento di Scienze dell'Informazione, Università di Bologna, Italy, 1996.

[CG96] P. Ciancarini and M. Gaspari. Rule Based Coordination of Logic Programs. *Computer Languages*, (to appear), 1996.

[Eva94] A. Evans. Specifying and Verifying Concurrent Systems Using Z. In M. Bertran, T. Denvir, and M. Naftalin, editors, *Proc. FME'94 Industrial Benefits of Formal Methods*, volume 873 of *Lecture Notes in Computer Science*, pages 366–380. Springer-Verlag, Berlin, 1994.

[Spi92] J. Spivey. *The Z Notation. A Reference Manual.* Prentice-Hall, 2 edition, 1992.

Uniform Representation of Basic Algebraic Structures in Computer Algebra*

Carla Limongelli+, Giuseppina Malerba+, Marco Temperini++

This work presents a methodological framework suitable for the design of innovative symbolic computation systems. A significant set of basic algebraic structures such as algebraic numbers, multivariate polynomials and square matrices, is defined and implemented by a uniform representation based on the algebraic structure of Truncated Power Series (TPS). By following this approach we can perform an operation between a number and a polynomial, as well as a polynomial and a matrix, without any need for explicit coercions.

We give a specification of the three basic algebraic structures cited above, by expressing their common algebraic nature of Ring and TPS structures. This specification is translated in the definition of a hierarchy of data structures, in which numbers, polynomials and matrices are uniformly represented as specializations of the TPS.

The idea of a uniform representation for numbers and polynomials has been already presented in [2]. Here we develop further the notion by providing an interpretation of matrices over ring elements, as TPS. Finally we have to model explicitly our matrix structure as a (specialization of) the algebraic structure of R-module [3].

Let N, P and M be symbols that represent algebraic numbers, multivariate polynomials and matrices resp. We deal with structures of "type" $\mathcal{T}$, by the following notation:

$$\mathcal{T} ::= N|P\mathcal{T}|M\mathcal{T}.$$

As a matter of fact algebraic numbers, polynomials with ring coefficients and matrices over ring elements are ring structures, hence $\mathcal{T}$ always denotes a ring structure. In order to reach a uniform representation for these structures, we need a suitable parametric structure, we will call $TPS_{h,r}$: in our view the algebraic numbers, polynomials and matrices will inherit their representation from $TPS_{h,r}$ and their algebraic properties from the Ring (R-module) structure.

Related to the algebraic operations we list the following observations:

1. The addition and the representation of additive unit are the same for numbers, polynomials and matrices but multiplication over matrices, does not follow the ***convolution rule*** like numbers and polynomials do.
2. In case of matrices, the ***multiplication by a scalar*** is defined (the scalar belongs to the matrix elements ring). So, the algebraic structure appropriate for representing matrices is R-module.

---

* Sponsored by Project MURST 40% "Calcolo Algebrico e Simbolico"; +Università degli Studi di Roma Tre, Dipartimento di Informatica, Via della Vasca Navale 84, I-00146, Roma Italy, e-mail: `limongel@inf.uniroma3.it`, `malerba@dis.uniroma1.it`; ++Università "La Sapienza", Dipartimento di Informatica e Sistemistica, Via Salaria 113, I-00198 Roma, Italy, e-mail: `marte@dis.uniroma1.it`

3. With respect to polynomials and numbers, the multiplicative unit has a different representation for matrices.

From these observations, it derives that we have to distinguish between two different rings: one for numbers and polynomials and another one for matrices. Nevertheless the concrete representation of these mathematical objects is always via a TPS.

We exploit this similarity, through the definition of $TPS_{h,r}$, where h is the base of the series and r is the truncation order. By this structure we can unify matrices, numbers and polynomials from a formal point of view. On the other hand we cannot consider $TPS_{h,r}$ as an algebraic structure itself, since it lacks algebraic operations. Such operations are then introduced in two different parametric structures, $TPS_{h,r}$-NP and $TPS_{h,r}$-M: they represent, respectively, polynomials (numbers) and matrices. Finally we want to stress that another methodological

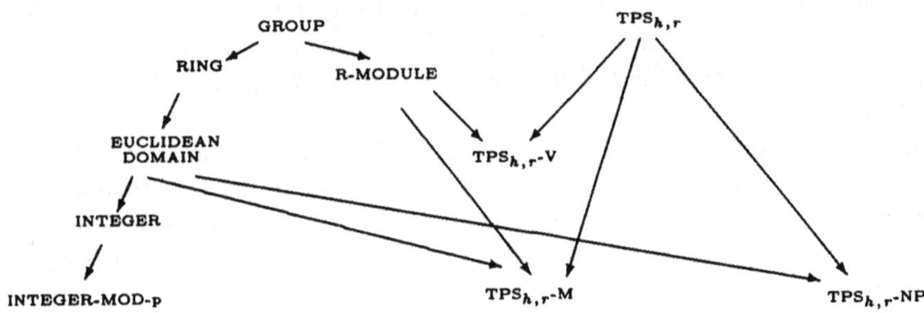

aspect of this work concerns the definition of basic mathematical structures under an object-oriented specification model. Such specification expresses and profits of the common algebraic nature of algebraic numbers, polynomials and matrices. The implementation is written in CLOS [1] where the derivation of a structure from another one is expressed by specialization inheritance.

We plan to develop further this work, aiming to define a more complete system where other algebraic structures can be defined in this style. Probably the choice of R-module structure is not yet general enough, but what we want to show by this work is mainly a methodology for the definition of new systems.

References

1. Sonya E. Keene. *Object Oriented Programming in Common Lisp.* Addison Wesley Publishing Company, 1989.
2. C. Limongelli and M. Temperini. On the uniform representation of mathematical data structures. In A. Miola, editor, *Design and Implementation of Symbolic Computation Systems, International (Symposium Disco '93)*, volume 722 of *LNCS*. Springer Verlag, 1993.
3. Henryk Minc and Marvin Marcus. *Introduction to Linear Algebra.* Macmillan, New York, 1965.

Integrating Computer Algebra with Proof Planning

Manfred Kerber[1], Michael Kohlhase*[2], Volker Sorge[2]

[1] School of Computer Science, The University of Birmingham
Birmingham B15 2TT, England
M.Kerber@cs.bham.ac.uk

[2] Fachbereich Informatik, Universität des Saarlandes
D-66141 Saarbrücken, Germany
{kohlhase|sorge}@cs.uni-sb.de

Abstract. Mechanised reasoning systems and computer algebra systems have apparently different objectives. Their integration is, however, highly desirable, since in many formal proofs both of the two different tasks, proving and calculating, have to be performed. In the context of producing reliable proofs, the question how to ensure correctness when integrating a computer algebra system into a mechanised reasoning system is crucial. In this contribution, we discuss the correctness problems that arise from such an integration and advocate an approach in which the calculations of the computer algebra system are checked at the calculus level of the mechanised reasoning system. We present an implementation which achieves this by adding a verbose mode to the computer algebra system which produces high-level protocol information that can be processed by an interface to derive proof plans. Such a proof plan in turn can be expanded to proofs at different levels of abstraction, so the approach is well-suited for producing a high-level verbalised explication as well as for a low-level (machine checkable) calculus-level proof.

1 Introduction

Mechanised reasoning systems (for short MRS in the following) may be built with various concrete purposes in mind. One goal is the construction of an autonomous theorem prover, whose strength achieves or even surpasses the ability of human mathematicians. Another may be to build a system where the user derives the proof, with the system guaranteeing its correctness. A third purpose might be the modelling of human problem-solving behaviour on a machine, that is, cognitive aspects are the focus. Advanced theorem proving systems often try to combine the different goals, since they can complement each other in an ideal way. However the ultimative motivation behind all of these systems is to give the human user assistance in formal (mathematical) reasoning tasks, and in particular theorem proving.

While all of the the approaches are in principle general enough to cope with any kind of proofs, they often neglect the fact that for many mathematical fields, everyday reasoning only partially consists in proving theorems. Calculation plays an equally important role, and mathematicians want to have support in both

* This work was supported by the Deutsche Forschungsgemeinschaft in SFB 314 (D2)

activities. More often than not the tasks of proving theorems and calculating simplifications of certain terms are interwoven and inseparable. In such cases traditional MRS will only provide rather poor support to a user, since they are very weak, when it comes to computation with mathematical objects.

In contrast, computer algebra systems (CAS in the following) manipulate highly optimised representations of the objects, which makes them very efficient. However they only provide answers to computations, but no proofs.

Although theoretically any computation can be reduced to theorem proving, this is not practical for non-trivial cases using traditional MRS, since the search spaces there are intractable. For many of these tasks, however, no search is necessary at all, since there are numerical or algebraic algorithms that can be used. If we think of Kowalski's equation "Algorithm = Logic + Control" [Kow79], general purpose MRS procedures do not (and can not) provide the control for doing the concrete computations that are explicitly encoded in CAS.

All of these facts point to the usefulness of an integration of CAS into MRS, where the former are used for solving subgoals in mathematical deduction. Consequently several experiments on combining CAS and MRS have been carried out recently. As pointed out by Buchberger [Buc96] the integration problem is still unsolved, but it can be expected that a successful combination of these systems will lead to "a drastic improvement of the intelligence level" of such support systems.

We will give an overview over these experiments in Sect. 3, advocate a particular approach built on top of the proof planning paradigm and describe the implementation of a prototype in Sect. 4. But let us first take a closer look at MRS in general and our Ω-MKRP system in particular.

2 Mechanised Reasoning and the Role of Proofs

Let us roughly divide existing theorem-proving systems into three categories: machine-oriented theorem provers, proof checkers, and human-oriented (plan-based) theorem provers. By *machine-oriented theorem provers* we mean theorem provers based on computational logic theories such as resolution, paramodulation, or the connection method. *Interactive proof checking* and *proof development systems* have been developed to carry out the meticulous final checking of proofs. And finally *human-oriented theorem-proving systems* incorporate some human problem solving behaviour. This is achieved for instance by using tactics, programs that manipulate proof-states by a series of calculus steps (first used in LCF [GMW79]) or by proof planning with so-called methods, which are tactics extended by specifications (e.g. CLAM [BvHHS90, BSvH+93]).

Normally all these systems do not exist in a pure form anymore, and in some systems like our own Ω-MKRP system [HKK+94] it is explicitly tried to combine the reasoning power of automated theorem provers as logic engines, the specialised problem solving knowledge of the proof planning mechanism, and the interactive support of tactic-based proof development environments.

MRS can also be classified with respect to the role proofs play in them. There are essentially two different views of proofs, the *realist's* (also called *Platonist*) and the *nominalist's* [Pel91]. A realist accepts abstract properties of proofs, in particular he/she is satisfied with the evidence of the existence of a proof in order to accept the truth of a theorem. A proof for a nominalist, on the other hand, makes only sense with respect to a particular calculus, hence he/she only accepts concrete proofs formulated in this calculus. The advantage of adopting the realist position is that reasoning systems can be built (and meta-theoretically extended) without bothering about the concrete construction of proofs. The advantage of the nominalist position is that it preserves the tradition that proofs can be communicated: The nominalist position guarantees the correctness of machine generated proofs without violating an essential of the traditional notion of proof, namely the possibility to communicate them. Furthermore explicit proofs can be checked by simple proof checkers and this seems currently to be the only way to ensure the correctness of proofs generated by large computing systems, which are inevitably error-prone.

In accordance with the two philosophical positions there are two possible ways to use an MRS: as trustworthy black box (trustworthy, for instance, since there are a lot of meta-arguments, why the system works properly) or as a system that produces communicable and checkable proofs.

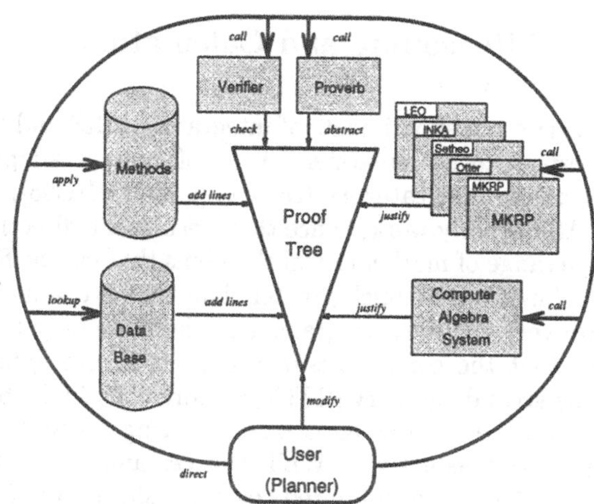

Fig. 1. Architecture of Ω-MKRP

In the Ω-MKRP-system to be described in the following, we advocate the stricter nominalist approach. The main reason for this is the reliability argument. Clearly, the sheer size of the systems involved prohibits to come up with provably correct MRS, but in particular makes it impossible to add different systems to an MRS like an external CAS *without* giving up any correctness requirement. In Ω-MKRP, a human user can apply different integrated tools to manipulate a proof tree, which stores the current partial natural deduction proof. In particular, new pieces of proof can be added by calling so-called methods, programs that store some proof information, by inserting facts from a data base, or by calling some external automated theorem prover (see Fig. 1). Furthermore the user can call the Proverb system for the generation of an abstract proof that can be verbalised. Finally, there is the possibility to call a computer algebra system as we will describe in the rest of the paper.

Since the correctness of the different components (in particular of the external ones) cannot be guaranteed, the final proof has to be checked by a simple verifier equipped with a fixed set of natural deduction rules. The soundness of the overall system only relies on the correctness of this checker and the correctness of the natural deduction rules. The price we have to pay for this is that in our approach each component must protocol its results in some universal format (in our case, a variant of Gentzen's calculus NK of natural deduction).

We are not going to present Ω-MKRP in great detail, most of its components are quite standard and are not important for the purpose of this paper. A main component, which is important for the integration of computer algebra into Ω-MKRP is its planning component. The entire process of theorem proving in Ω-MKRP can be viewed as an *interleaving process* of proof planning, method execution (the plan operators are called methods and essentially are tactics plus specifications) and verification. In particular, this model ascribes a problem-solver's reasoning competence to the existence of methods together with a planning mechanism that uses these methods for proof planning.

3 Syntheses of Reasoning and Calculation

We will now categorise the experiments of integrating MRS and CAS into three categories with respect to the treatment of proofs that is adopted. In the attempts belonging to the first category (see e.g. [CZ92, BHC95]) one essentially trusts that the CAS properly work, hence their results are directly incorporated into the proof. The range of mathematical theorems that can be formally proved by the system combinations is much greater than that provable by MRS alone. However, CAS are very complex programs and therefore only trustworthy to a limited extent, so that the correctness of proofs in such a hybrid system can be questioned. The second category [HT93] is more conscious about the role of proofs, and only uses the CAS as an oracle, receiving a result, whose correctness can then be checked deductively. While this certainly solves the correctness problem, this approach only has a limited coverage, since even checking the correctness of a calculation may be out of scope of most MRS, when they don't have additional information. A third approach of integrating computer algebra systems into a particular kind of mechanised reasoning system, consists in the meta-theoretic extension of the reasoning system as proposed for instance in [BM81, How88]. In this approach a constructive mechanised reasoning system is basically used as its own meta-system, the constructive features are exploited to construct a correct computer algebra system and due to bridge rules it can directly be used in the base system.

The main problem of integrating CAS into MRS without violating correctness requirements is that CAS are generally highly optimised towards maximal speed of computation but not towards generating explanations of the computations involved. Not only if one is interested in correctness, but in particular if explanations for theorems and calculations are required, this can turn out to be a major problem; in such a case explicit proofs (calling for a nominalist approach) are essential.

If we take the idea of generating explicit proofs and explanations seriously also for computations and do not want to give up the nominalist paradigm, we can neither just take existing CAS nor follow the approach of meta-theoretic extensions, since Ω-MKRP is a classical proof system and does not use constructive logic. But we cannot forgo using them either, if we want to tackle nontrivial computations inside proofs. For instance even the proof for the binomial formula $(x + y)^2 = x^2 + 2xy + y^2$ (a trivial problem for any computer algebra system) needs more then 70 single steps in the natural deduction calculus[3]. Thus using theorem provers or rewriting systems to find such proofs can produce unnecessarily large search spaces and thus absorb valuable resources. On the other hand such proofs show a remarkable resemblance to algebraic calculations themselves and suggest the use of a CAS not only to instantly compute the result of the given problem, but also to guide a proof in the way of exploiting the implicit knowledge of the algorithms. We propose to do this extraction of information not by trying to reconstruct the computation in the MRS after the result is generated—as we have seen, even in case of a trivial example for a CAS this may turn out to be a very hard task for an MRS—but rather to extend the CAS algorithm itself so that it produces some logically usable output alongside the actual computation.

The novel contribution of our approach is to use the mathematical knowledge implicit in the CAS to extract proof plans that correspond to the mathematical computation in the CAS. So essentially the output of a CAS should be transferable into a sequence of tactics, which presents a high-level description for the proof of correctness of the computation the CAS has performed. Note that this does not prove general correctness of the algorithms involved, instead it only gives a proof for a particular instance of computation. The high-level description can then be used to produce a readable explanation or evaluated to check the proof. If we want to check the whole derivation, these proof plans can then be expanded into detailed natural deduction proofs. The decision to extract proof plans rather than concrete proofs from the CAS is essential to the goal of being verbose without transmitting too much detail.

For our purpose, we need different modes, in which we can use the CAS. Normally, during a proof search, we are only interested in the result of a computation, since the assumption that the computation is correct is normally justified for established CAS. When we want to understand the computation—in particular, in a successful proof—we need a verbose mode of the CAS that gives enough information to generate a high-level description of the computation in terms of the mathematics involved. How this can be achieved is described in the next section in detail.

4 SAPPER – Integrating Computations into Proofs

The SAPPER system (System for Algorithmic Proof Plan Extraction and Reasoning), integrates a prototypical computer algebra system into a proof plan-

[3] Proofs of this length are among the hardest ever found by totally automatic theorem provers without domain-specific knowledge.

based mechanised reasoning system. The system is kept generic, but for the concrete integration we have used the Ω-MKRP-system as MRS and a self-written CAS, called μ-CAS. As mentioned in the previous section, for the intended integration it is necessary to augment the CAS with mathematical information for a *verbose mode* in order to achieve the proposed integration at the level of proofs. The μ-CAS-system is very simple and can at the moment only perform basic polynomial manipulations and differentiation, but it suffices for demonstrating the feasibility of our approach. Clearly, for a practical system for mathematical reasoning, a much more developed system like Maple [CGG+92], Reduce [Hea87], or Mathematica [Wol91] has to be integrated. Enriching such a large CAS with a corresponding verbose mode for producing additional protocol information, would of course require a considerable amount of work.

The SAPPER system can be seen as a generic interface for connecting Ω-MKRP with one or several CAS (see Fig. 2). An incorporated CAS is treated as a slave to Ω-MKRP which means that only the latter can call the first one and not vice versa. From the software engineering point of view, Ω-MKRP and the CAS are two independent processes while the interface is a process providing a bridge for communication. Its role is to automate the broadcasting of messages by transforming output of one system into data that can be processed by the other[4].

Unlike other approaches (see [GPT94, HC95] for example) we do not want to change the logic inside our prover. In the same line, we do not want to change the computational behaviour of the computer algebra algorithms. In order to achieve this goal the trace output of the algorithm is kept as short as possible. In fact most of the computations for constructing a proof plan is left to the interface. The proof plans can directly be imported into Ω-MKRP.

This makes the integration independent of the particular systems, and indeed all the results below are independent of the CAS employed and make only some general assumptions about the MRS (such as being proof plan-based). Moreover, the interface approach helps us to keep the CAS free of any logical computation, for which such a system is not intended anyway. Finally, the interface minimises the required changes to an existing CAS, while maintaining the possibility of using the CAS stand-alone. The only requirement we make for integrating a particular CAS is that it has to produce enough protocol information so that a proof plan can be generated from this information. The proof plan in turn can be expanded by the MRS into a proof verifying the concrete computation.

The interface itself can be roughly divided into two parts; the ***translation part***, and the ***plan-generator***. The first performs syntax translations between Ω-MKRP and a CAS in both directions while the latter only transforms verbose output of the CAS to Ω-MKRP proof plans. Clearly only the translation part depends on the particular CAS that is invoked.

For the translations a collection of data structures—called ***abstract CAS***—is provided each one referring to a particular connected CAS (or just parts of one). The main purpose of these structures is to specify function mappings, relating a particular function of Ω-MKRP to a corresponding CAS-function and the type of its arguments. Furthermore it provides functionality to convert the given arguments of the mapped Ω-MKRP function to CAS input. In the same fashion it

[4] This is an adaptation of the general approach on combining systems in [CMP91].

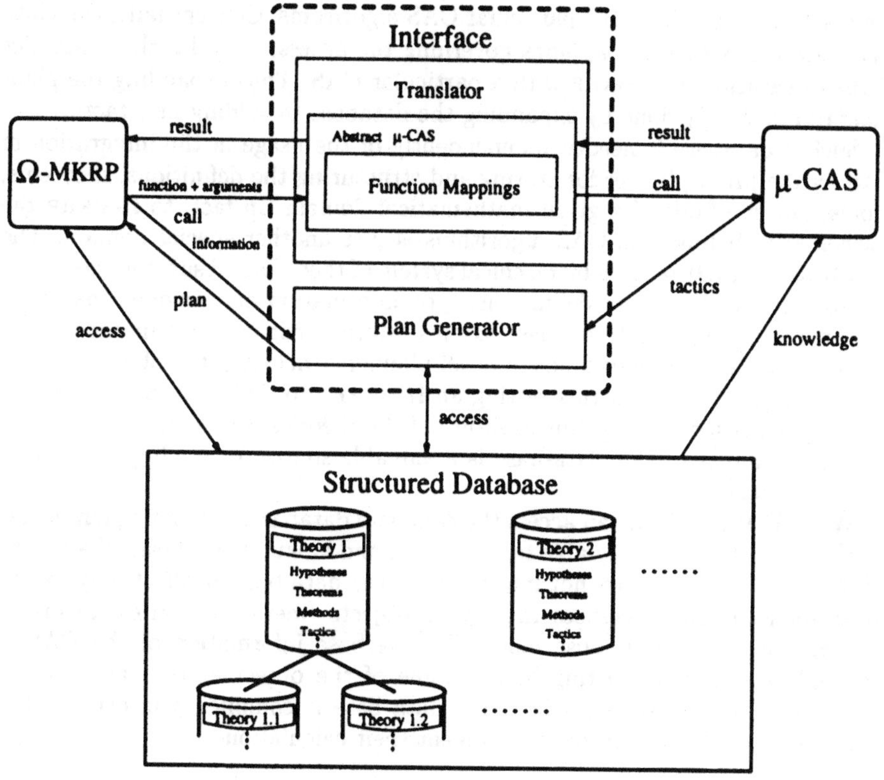

Fig. 2. Interface between Ω-MKRP and μ-CAS

transforms results of algebraic computations back into data that can be further processed by Ω-MKRP. These syntax transformations are implemented by two generic functions, one for each direction of data flow. The single methods for each function are selected according to the type of objects sent to and returned from the CAS. It is possible to expand the translation unit generically by simply adding new abstract CAS and methods for the two translation functions. The functionality in this part of our interface offers us the possibility of connecting any CAS as a black box system. For instance, we may want to use a very efficient system without verbose mode for proof search as black box system, and then another less efficient system with verbose mode for the actual proof construction, once it is clear what the proof should look like.

The plan-generator solely provides the machinery for our main goal, the proof plan extraction. Equipped with supplementary information on the proof by Ω-MKRP it records the output produced by the particular algebraic algorithm and converts it into a proof plan. Here the requirements of keeping the CAS side free of logical considerations and on the other hand of keeping the interface generic seem conflicting at the first glance. However this conflict can be solved by giving both sides of the interface access to a data base of mathematical facts formalising the mathematics behind the particular CAS algorithms. Conceptually, this data base together with the mappings governing the access, provides the semantics of the integration of Ω-MKRP with a particular CAS. Thus expanding the plan-generator is simply done by expanding the database by adding new tactics.

Such a database is needed independently of the usage in the integration of CAS by the proof planner for storing and structuring the definitions, theorems, proofs, and methods of a given mathematical domain. In fact, to Ω-MKRP the mathematics behind the CAS algorithms is just another special domain. The data-base is structured in a hierarchical system of theories and sub-theories. Such a theory is a collection of definitions, type information and axioms, theorems, and lemmata derivable from these axioms. Moreover it contains methods, tactics, etc. usable by various kinds of proof planners (including the plan-generator for a particular CAS). In this setting an Ω-MKRP proof plan obtains a natural hierarchy corresponding to the structure of the theories. In particular, the hierarchical structure of this data-base is a valuable source for guiding the search for proof plans.

While Ω-MKRP itself can access the complete database, SAPPER's plan-generator in the interface is only able to use tactics and lookup hypotheses of a theory (cf. Fig. 2). The CAS does not interact with the data base at all: it only has to know about it and references the logical objects (methods, tactics, theorems or definitions) in the verbose mode. This verbose information of the CAS is returned as strings consisting of the name of the object and its appropriate arguments. Thus knowledge about the data base is compiled a priori into the algebraic algorithms in order to document their calculations.

5 An Example of the Integration

As an example of the integration we consider the case of extracting proof plans from a recursive algorithm for adding polynomials. Even though this algorithm is quite trivial from the CAS point of view, we present it here, since it sheds some light on the spirit of the integration.

Let us now take a look at the different representations of a polynomial $p(x_1, \ldots, x_r) = \sum_{i=1}^{n} \alpha_i x_1^{e_{1_i}} \cdots x_r^{e_{r_i}}$ in the variables $x_1, \ldots, x_r$. The logical language of Ω-MKRP is a variant of the simply typed λ-calculus, so the polynomials are represented as λ-expression where the formal parameters $x_1, \ldots, x_r$ are λ-abstracted (mathematically, p is a function of r arguments):

$$p: \; \lambda x_1 \cdots \lambda x_r.(+\,(*\alpha_n\,(*\,(\uparrow x_1\, e_{1_n})\cdots))\cdots(*\alpha_1\,(*\,(\uparrow x_1\, e_{1_1})\cdots)),$$

For the notation, we use a prefix notation; the symbols $+$, $*$ and $\uparrow$ denote binary functions for addition, multiplication and exponentiation on the rationals. In this representation, we can use β-reduction for the evaluation of polynomials, but we have to define a special function for polynomial addition for any arity r.

In μ-CAS, we use a variable dense, expanded representation as an internal data-structure for polynomials (as described in [Zip93] for instance). Thus every monomial is represented as a list containing its coefficient together with the exponents of each variable. p is represented as $((\alpha_n \; e_{1_n} \cdots e_{r_n}) \cdots (\alpha_1 \; e_{1_1} \cdots e_{r_1}))$ for instance.

Let us now turn to the actual μ-CAS algorithm for polynomial addition. This simple algorithm adds polynomials p and q by a case analysis on the exponents with recursive calls to itself. So let $p = \sum_{i=1}^{n} \alpha_i x_1^{e_{1_i}} \cdots x_r^{e_{r_i}}$ and $q = \sum_{i=1}^{m} \beta_i x_1^{f_{1_i}} \cdots x_r^{f_{r_i}}$. We have presented the algorithm in the jth component of p and the kth component of q in a LISP-like pseudo-code in Fig. 3. Intuitively, the algorithm proceeds by ordering the monomials, advancing the leading monomial either of the first or the second arguments; in the case of equal exponents, the coefficients of the monomials are added.

(poly-add (p q)
  $(= (e_{1_j} \cdots e_{r_j})(f_{1_k} \cdots f_{r_k}))$
    (tactic mono-add)
    (cons-poly $(\alpha_j + \beta_k) x_1^{e_{1_j}} \cdots x_r^{e_{r_j}}$
      (poly-add $\sum_{i=j+1}^{n} \alpha_i x_1^{e_{1_i}} \cdots x_r^{e_{r_i}} \ \sum_{i=k+1}^{m} \beta_i x_1^{f_{1_i}} \cdots x_r^{f_{r_i}}$**))**
  $(> (e_{1_j} \cdots e_{r_j})(f_{1_k} \cdots f_{r_k}))$
    (tactic pop-first)
    (cons-poly $\alpha_j x_1^{e_{1_j}} \cdots x_r^{e_{r_j}}$
      (poly-add $\sum_{i=j+1}^{n} \alpha_i x_1^{e_{1_i}} \cdots x_r^{e_{r_i}} \ \sum_{i=k}^{m} \beta_i x_1^{f_{1_i}} \cdots x_r^{f_{r_i}}$**))**
  $(< (e_{1_j} \cdots e_{r_j})(f_{1_k} \cdots f_{r_k}))$
    (tactic pop-second)
    (cons-poly $\beta_k x_1^{e_{1_k}} \cdots x_r^{e_{r_k}}$
      (poly-add $\sum_{i=j}^{n} \alpha_i x_1^{e_{1_i}} \cdots x_r^{e_{r_i}} \ \sum_{i=k+1}^{m} \beta_i x_1^{f_{1_i}} \cdots x_r^{f_{r_i}}$**)))**

Fig. 3. Polynomial addition in μ-CAS.

Obviously, the only expansions of the original algorithm needed for the verbose mode are the additional `(tactic ...)` statements[5]. They just produce the additional output by returning keywords of tactic names to the tactic generator and do not have any side effects. In particular, the computational behaviour of the algorithm does not have to be changed at all. If we apply the algorithm to the simple polynomials $p := 3x^2 + 2x + 5$ and $q := 5x^3 + x + 2$ we obtain the following verbose output: `(pop-second, pop-first, mono-add, mono-add)`.

First the cubic monomial from q (the second argument) and then the quadratic one from p (the first argument) are raised, since they only appear in one argument, and finally the remaining monomials are summed up.

[5] Observe that in this case, the called tactics do not need any additional arguments, since our plan-generator in the interface keeps track of the position in the proof and thus knows on which monomials the algorithm works when returning a tactic. This way we need not be concerned what form a monomial actually has during the course of the algorithm.

In this simple case, each of the verbose keywords directly corresponds to a tactic with the same name, so the verbose output directly represents a proof plan for polynomial addition of the concrete polynomials p and q.

Let us now take a look at the pop-second tactic to understand its logical content. The tactic itself describes a reordering in a sum that looks in the general case as follows:

$$(a + (b + c)) = (b + (a + c)) \tag{1}$$

For the current example we can view a and c as arbitrary polynomials and b as a monomial of rank greater than that of the polynomial a. It is now obvious that the behaviour of pop-second is determined by the pattern of the sum it is applied to. If in (1) the polynomial c does not exist, pop-second is equivalent to a single application of the law of commutativity. Otherwise, like in our example, the tactic performs a series of commutativity and associativity steps. In the example the tactic can be expanded in two steps. First we have a one step inference with pop-second corresponding to two lines in our proof. (In the actual proofs, the terms are of course embedded in contexts.)

$((3x^2 + 2x + 5) + (5x^3 + x + 2))$
$(5x^3 + ((3x^2 + 2x + 5) + (x + 2)))$ (pop-second)

Expanding this plan adds two intermediate lines to the proof which then reflects the single step applications of the laws of commutativity and associativity.

$((3x^2 + 2x + 5) + (5x^3 + x + 2))$
$(((3x^2 + 2x + 5) + 5x^3) + (x + 2))$ (associativity)
$((5x^3 + (3x^2 + 2x + 5)) + (x + 2))$ (commutativity)
$(5x^3 + ((3x^2 + 2x + 5) + (x + 2)))$ (associativity)

Finally to receive the complete calculus level proof for the computation step described by the pop-second tactic we further expand all three justifications of the above lines. This leads to a sequence of eliminations of universally quantified variables in the corresponding hypothesis, the axioms of commutativity and associativity. In our example the commutativity axiom would be transformed in the following fashion:

$\forall a \forall b. (a + b) = (b + a)$ (HYP)
$\forall b. ((3x^2 + 2x + 5) + b) = (b + (3x^2 + 2x + 5))$ ($\forall$E $(3x^2 + 2x + 5)$)
$((3x^2 + 2x + 5) + 5x^3) = (5x^3 + (3x^2 + 2x + 5))$ ($\forall$E $5x^3$)

Altogether this single application of the pop-second-tactic is equivalent to a calculus-level proof of 11 inference steps. The overall proof of this trivial polynomial addition has a length of 47 single step.

6 Conclusion

In this work we reported on an experiment of integrating a computer algebra system into the interactive proof development environment Ω-MKRP, not only at the *system* level, but also at the level of *proofs*. The motivation for such an integration is the need for a support of a human user when his/her proofs contain non-trivial computations. Unfortunately, we could not use a standard CAS for the integration, since such a system provides answers, but no justifications, which turned out to be essential in an environment that is built to construct communicable and checkable proofs.

In order to achieve a solution that is compatible with such a strong requirement, we have adopted a generic approach, where the only requirement for the CAS is that it provides a verbose mode for the generation of communicable and checkable proofs. Since we want to achieve two goals simultaneously, namely to have high-level descriptions of the calculations of the CAS for communicating them to human users as well as low-level ones for a mechanical checking, we represent the protocol information in form of high-level hierarchical proof plans which can be expanded to the desired detail. Fully expanded proof plans are natural deduction proofs which can be mechanically checked. In the case that the CAS has made a mistake in the computation, the proof checker can detect it on this level. Thus our approach provides a paradigm of verification of calculations that is much simpler than that of verifying the correctness of the CAS itself. The usefulness of the integration for proof development can already be seen in the case of our simple μ-CAS. After the integration we are able to prove optimisation problems which were out of reach without such a support.

We have tested proof plan extraction from simple recursive and iterative CAS algorithms, where it works quite well. However, more complicated schemes like divide-and-conquer algorithms (e.g. the polynomial multiplication of Karatsuba and Ofman [KO63]) cannot be adapted to our approach so easily. Highly elaborated and efficient algorithms in systems like *Mathematica* [Wol91] or *Maple* [CGG+92] might be hard to augment with verbose modes. However, even if it proves impossible to extract verbose information that is valuable at the conceptual, mathematical level, it is always possible to reserve these elaborated techniques for the quiet mode used in proof discovery, and use more basic algorithms for the proof extraction phase. It may even be, that algorithms using elaborated data-structures and calculation schemes can contribute internal information that improves informed runs of simpler algorithms. It furthermore would be desirable to make use of sophisticated type systems and axiom specifications like those in *Axiom* [Dav92] or *MuPAD* [Fuc96] to return domain specific tactics in the verbose output.

References

[BHC95] C. Ballarin, K. Homann, and J. Calmet. Theorems and algorithms: An interface between Isabelle and Maple. In A. H. M. Levelt, editor, *Proceedings of International Symposium on Symbolic and Algebraic Computation (ISSAC'95)*, pages 150–157. ACM Press, 1995.

[BM81] R. S. Boyer and J S. Moore. Metafunctions. In R. S. Boyer and J Strother Moore, editors, *The Correctness Problem in Computer Science*, pages 103–184. Academic Press, 1981.

[BSvH+93] Alan Bundy, Andrew Stevens, Frank van Harmelen, Andrew Ireland, and Alan Smaill. Rippling: A heuristic for guiding inductive proofs. *AI*, **62**:185–253, 1993.

[Buc96] Bruno Buchberger. Mathematische Software-Systeme: Drastische Erweiterung des "Intelligenzniveaus" entsprechender Programme erwartet. *Informatik Spektrum*, 19/2:100–101, 1996.

[BvHHS90] Alan Bundy, Frank van Harmelen, Christian Horn, and Alan Smaill. The OYSTER-CLAM system. In Mark E. Stickel, editor, *Proceedings of the 10th CADE*, pages 647–648, Kaiserslautern, Germany, 1990. Springer-Verlag.

[CGG+92] Bruce W. Char, Keith O. Geddes, Gaston H. Gonnet, Benton L. Leong, Michael B. Monagan, and Stephen M. Watt. *First leaves: a tutorial introduction to Maple V.* Springer-Verlag, 1992.

[CMP91] D. Clément, F. Montagnac, and V. Prunet. Integrated software components: A paradigm for control integration. In *Proceedings of the European Symposium on Software Development Environments and CASE Technology*, Springer-Verlag, LNCS 509, 1991.

[CZ92] Edmund Clarke and Xudong Zhao. Analytica – A theorem prover in mathematica. In *Proceedings of the 11th CADE*, pages 761–763, Saratoga Springs, New York, 1992, Springer-Verlag,

[Dav92] J. H. Davenport. The AXIOM system. AXIOM Technical Report TR5/92 (ATR/3) (NP2492), Numerical Algorithms Group, Inc., Downer's Grove, IL, USA and Oxford, UK, 1992.

[Fuc96] Benno Fuchssteiner et al. (The MuPAD Group). *MuPAD User's Manual.* John Wiley and Sons, Chichester, first edition, 1996.

[GMW79] Michael Gordon, Robin Milner, and Christopher Wadsworth. *Edinburgh LCF: A Mechanized Logic of Computation.* LNCS 78. Springer-Verlag, 1979.

[GPT94] Fausto Giunchiglia, Paolo Pecchiari, and Carolyn Talcott. Reasoning theories – towards an architecture for open mechanized reasoning systems. IRST-Technical Report 9409-15, IRST (Istituto per la Ricerca Scientifica e Tecnologica), Trento, Italy, June 1994.

[HC95] K. Homann and J. Calmet. An open environment for doing mathematics. In M. Wester, S. Steinberg, and M. Jahn, editors, *Proceedings of 1st International IMACS Conference on Applications of Computer Algebra*, Albuquerque, USA, 1995.

[Hea87] A. C. Hearn. Reduce user's manual: Version 3.3. Technical Report, Rand Corporation, Santa Monica, CA, USA, 1987.

[HKK+94] Xiaorong Huang, Manfred Kerber, Michael Kohlhase, Erica Melis, Dan Nesmith, Jörn Richts, and Jörg Siekmann. Ω-MKRP: A proof development environment. In Alan Bundy, editor, *Proceedings of the 12th CADE*, pages 788–792, Nancy, 1994. Springer-Verlag, LNAI 814.

[How88] D. J. Howe. Computational Metatheory in Nuprl. In Lusk and Overbeek, editors, *Proceedings of CADE 9*, pages 238–257, 1988. Springer-Verlag.

[HT93] J. Harrison and L. Théry. Extending the HOL theorem prover with a computer algebra system to reason about the reals. In C.-J. H. Seger J. J. Joyce, editor, *Higher Order Logic Theorem Proving and its Applications (HUG '93)*, pages 174–184, 1993. Springer-Verlag, LNCS 780.

[KO63] A. Karatsuba and Y. Ofman. *Multiplication of Multidigit Numbers by Automata.* Soviet Physics-Doklady, 1963.

[Kow79] Robert Kowalski. Algorithm = Logic + Control. *CACM*, 1979.

[Pel91] Francis Jeffry Pelletier. The philosophy of automated theorem proving. In John Mylopoulos and Ray Reiter, editors, *Proceedings of the 12th IJCAI*, pages 538–543, Sydney, 1991. Morgan Kaufmann.

[Wol91] S. Wolfram. *Mathematica: A System for Doing Mathematics by Computer.* Addison-Wesley, 1991.

[Zip93] Richard Zippel. *Effective Polynomial Computation.* Kluwer Academic Press, 1993.

Structures for Symbolic Mathematical Reasoning and Computation

Karsten Homann and Jacques Calmet

Universität Karlsruhe
Institut für Algorithmen und Kognitive Systeme
Am Fasanengarten 5 · 76131 Karlsruhe · Germany
{homann,calmet}@ira.uka.de

Abstract. Recent research towards integrating symbolic mathematical reasoning and computation has led to prototypes of interfaces and environments. This paper introduces computation theories and structures to represent mathematical objects and applications of algorithms occuring in algorithmic services. The composition of reasoning and computation theories and structures provide a formal framework for the specification of symbolic mathematical problem solving by cooperation of algorithms and theorems.

Keywords: Integration of computing and reasoning paradigms, Interfaces

1 Introduction

The combination of systems performing any kind of mathematical computation is a young and active research field. We call such systems *mathematical services* which include computer algebra systems (CAS), theorem provers and proof checkers (TPS), mathematical knowledge representation systems, tools for visualization and editing, ... A major requirement to qualify as mathematical service is the ability to cooperate by incremental and restartable computation and deduction.

There has been work to integrate homogeneous services, i.e. combining CAS in CAS/π [13] and OPENMATH [1], combining TPS in OMRS [9] and many others. The integration of CAS and TPS in a common environment has not yet led to powerful systems. However, some prototypes were developed which prove the advantages of such a combination, e.g. ANALYTICA [6], interfaces between HOL and MAPLE [10], ISABELLE and MAPLE [2], and NUPRL and WEYL [12].

We classify communication and cooperation methods for such environments in [5]. However, we believe that the lack of a general formal framework is one reason why nowadays prototypes do not qualify as mathematical services. [1] introduces a semantics of mathematical objects and interfaces and [9] initiates the formal specification of reasoning theories and reasoning structures for combining logical services. However, a formal framework to combine CAS and TPS is not given.

Another reason is that nowadays CAS behave like black boxes. They are not designed to allow provisional or restartable computation and do not provide access to their context. To implement contexts in mathematical reasoning and computation has been initiated in [8] by proposing extensions to the interactive mathematical proof system IMPS [7]. Again, there is no semantics of the integration.

The long term goal of this work is to provide a methodology for constructing complex systems by composing mathematical services while the actual goal of this paper is to provide a formal method for their specification. Such a methodology has to consist of both a formal theory and structures for its representation. We introduce ***computation structures*** to represent objects and applications of algorithms appearing in algorithmic services by extending and modifying the concept of reasoning structures given in [9]. The notions, definitions and theorems given here are intentively kept very close to reasoning theories to illustrate their similarities and to construct natural combinations of deduction and computation. Because of the restricted length of the paper we left some notions informal or without examples and give references. The composition of reasoning and computation theories and structures is subject of ongoing research.

The paper is organized as follows. Mathematical services are defined in section 2. They include interactive and automated open systems performing any kind of mathematical computation. The formal specification of reasoning structures is sketched by some examples of IMPS in section 3. Section 4 and 5 introduce computation structures and their construction and manipulation. Finally, section 6 illustrates examples of combining structures for symbolic mathematical reasoning and computation.

2 Specification of Mathematical Services

A formal description of mathematical services has to ensure interaction capabilities to combine several systems, to contain meta information and justifications on functions, it should allow easy extension by subpackages and interaction of existing systems. Such a description should consist of several levels such as the communication level for dynamic distribution of messages and events among the services and the abstract functionality of the interface (see [17] and [16] for examples).

The formal approach given in this paper provides a description of theories for reasoning and computation respectively. [9] introduces the concept of *Open Mechanized Reasoning Systems* which we call *logical services* $\mathcal{LS}$. These systems are based on reasoning theories and structures. By defining computation theories for *algorithmic services* $\mathcal{AS}$ and composition of theories we introduce a framework for the structures given in the next section.

Definition 1.
A ***mathematical service*** ($\mathcal{MS}$) is an implementation of a mathematical computation or processing and an interaction component.

In the rest of the paper we restrict $\mathcal{MS}$ to reasoning and computation systems.[1]

$$
\begin{aligned}
\mathcal{MS} &= \mathcal{LS} \mid \mathcal{AS} \\
\mathcal{LS} &= \textit{Reasoning System} + \textit{Interaction} \\
\mathcal{AS} &= \textit{Computation System} + \textit{Interaction} \\
\textit{Reasoning System} &= \textit{Reasoning Theory} + \textit{Control} \\
\textit{Computation System} &= \textit{Computation Theory} + \textit{Control} \\
\textit{Reasoning Theory} &= \textit{Sequents} + \textit{Rules} \\
\textit{Computation Theory} &= \textit{Objects} + \textit{Algorithms}
\end{aligned}
$$

Reasoning theories are defined in [9] to consist of a sequent system $S_{sys} = \langle S, C, \models, I, \_[\_]\rangle$, a set of identifiers Id and a rule set $\tilde{r} \in Rset[S_{sys}, Id]$. Sequents allow the use of schematic variables and can be instantiated. A rule is a relation on tuples consisting of a non-empty sequence of sequents and a finite set of constraints.

Definition 2.
An ***object system*** is a structure

$$O_{Sys} = \langle O, C, \models, I, \_[\_]\rangle$$

with a set of ***objects*** O, a set of ***constraints*** C, a ***constraint satisfaction mechanism*** $\models \subseteq (P_\omega(C) \times C)$, a set of ***instantiations*** I, and an ***application of instantiations*** $\_[\_] : [O \times I \to O]$ and $\_[\_] : [C \times I \to C]$.

Objects in CAS include polynomials, numbers, matrices, equations, sequences, sets, expressions and many others. Conditions in terms of constraints are provided as local context of objects, e.g. type constraints. We allow objects and constraints to be schematic and they can be instantiated.

Constraints are also introduced to guarantee certain properties when applying algorithms. Such an algorithm is a labelled relation on tupels of objects, input parameters and a unique result, which is closed under instantiation, with Id a set of identifiers.

Definition 3.
A ***set of algorithms over an object system*** $A \in Algs[O_{Sys}, Id]$ is such that

$$Algs[O_{Sys}, Id] = [Id \longrightarrow \{A \subseteq (O^* \times O \times P_\omega(C)) \mid (\forall \langle \bar{o}, o, \tilde{c}\rangle \in A)(\forall \iota \in I) \\ (\langle \bar{o}, o, \tilde{c}\rangle[\iota] \in A)\}]$$

A computation theory specifies a set of objects and a set of algorithms. They are designed to provide a methodology for the design and implementation of future CAS according to the ideas given in [3]. The intermediate computations are represented by computation structures which are introduced in the next section.

[1] $\mathcal{LS}$ corresponds to OMRS in [9], $\mathcal{AS}$ to OMCS and $\mathcal{MS}$ to OMME in [4].

Definition 4.
Let O_{Sys} be an object system, Id a set of identifiers and $\tilde{a} \in Algs[O_{Sys}, Id]$. A ***computation theory*** CT is a structure

$$CT = \langle O_{Sys}, Id, \tilde{a} \rangle \ .$$

To define a disjoint composition of theories for composing proofs and computations is subject of ongoing research. It is done by gluing together seperate reasoning and computation theories using additional rules called *bridges*. Such bridges may include syntax transformations or instructions for rigorous systems how to verify results of external theories. Section 6 sketches two examples of bridges.

3 Reasoning Structures

Let $RT = \langle S_{Sys}, Id, \tilde{r} \rangle$ be an arbitrary but fixed reasoning theory. Reasoning structures [9] were designed to represent the structures appearing in the construction of derivations and proofs. They can be illustrated as labelled graphs with edges and two kinds of nodes: sequent nodes and link nodes. These nodes are labelled by their corresponding sequents and justifications respectively. To enable vertical flexibility the justifications may contain nested reasoning structures together with an instantiation as well as premiss and conclusion nodes.

Definition 5.
Basic reasoning structures are elements of $RS_0[RT, SN, LN]$, which is the set of structures $rs = \langle Sn, Ln, g, sg, sL, lL \rangle$ such that

1. $Sn \in P_\omega(SN)$ set of sequent nodes of rs,
2. $Ln \in P_\omega(LN)$ set of link nodes of rs,
3. $g : [Ln \rightarrow Sn]$ maps to associated goal sequent node,
4. $sg : [Ln \rightarrow Sn^*]$ maps to possibly empty set of subgoal sequent nodes,
5. $sL : [Sn \rightarrow S]$ sequent node labelling map;
6. $lL : [Ln \rightarrow [Id \times P_\omega(C)]]$ link node labelling map such that $\forall ln \in Ln$:

$$lL(ln) = \langle id, \tilde{c} \rangle \wedge \bar{s} = sL(sg(ln)) \wedge s = sL(g(ln)) \Rightarrow \exists \tilde{c}' : \tilde{c} \models \tilde{c}' \wedge \langle \bar{s}, s, \tilde{c}' \rangle \in \tilde{r}(id)$$

For example, a basic reasoning structure rs_I in the proof construction of theorem 6 is illustrated in figure 1 and expanded in figure 2. The labelling of the sequent nodes was omitted because of readability.

Theorem 6.
$\forall x, c \in \mathbf{R}$ *and partial functions* $f : \mathbf{R} \rightarrow \mathbf{R}$ *such that* $\mathcal{D}(f)(x) \downarrow$:

$$\mathcal{D}(c * f(x)) = c * \mathcal{D}(f)(c)$$

Vertical flexibility allows to nest structures within others to achieve better presentations and readability of proofs.

```
((#{IMPS-sqn 1}
  (FORCE-SUBSTITUTION
   (#{IMPS-sqn 2}
    (PRODUCT-RULE
     (#{IMPS-sqn 5})
      (#{IMPS-sqn 6} GROUNDED)))
   (#{IMPS-sqn 3} GROUNDED))))
```

Fig. 1. Illustration of a deduction graph in IMPS

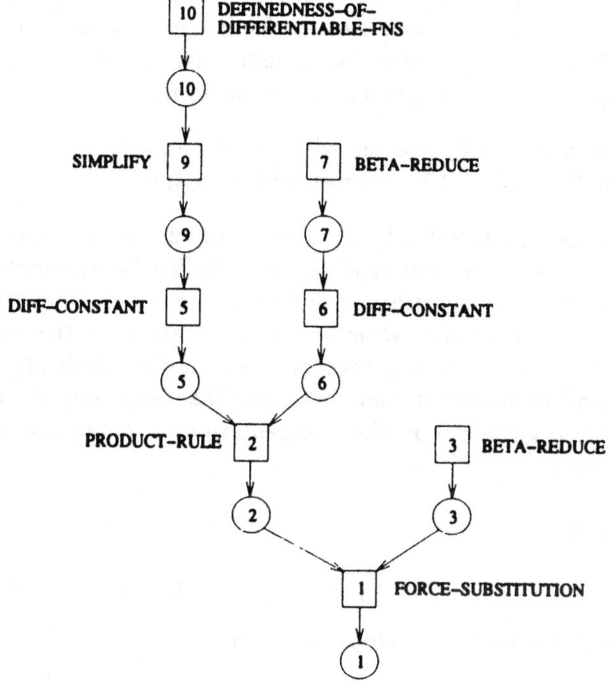

Fig. 2. Reasoning structure rs_I

Definition 7.
A ***reasoning structure*** is an element of

$$RS[RT, SN, LN] = \bigcup_{n \in \mathbb{N}} RS_n[RT, SN, LN]$$

with $RS_n[RT, SN, LN]$ the set of ***reasoning structures of level n***. RS_{n+1} is recursively defined as the set of structures $rs = \langle Sn, Ln, g, sg, sL, lL\rangle$ satisfying (1.-5.) of definition 5 and

6_{n+1}. $vB : [Vk \to [Id \times P_\omega(C)] + [P_\omega(C) \times I \times [Sk^*, Sk] \times DS_n[DT, SK, VK]]]$ is a natural extension of the link labelling function.

Reasoning structures can be instantiated and can represent derivations and proofs. Such proofs are derivations without open assumptions, e.g. figure 2. Theorems about reachability, instantiation of derivation, elimination of nesting and derivation of trees are given in [9].

4 Computation Structures

Let $CT = \langle O_{Sys}, Id, \tilde{a}\rangle$ be an arbitrary but fixed computation theory. Corresponding to reasoning structures we define computation structures to represent graphs of computations which are constructed by the application of algorithms. The graphs consist of edges and two kinds of nodes:

- *object nodes*, ON, represented by circles and
- *algorithm nodes*, AN, represented by squares.

Object nodes are labelled by $\Gamma|o$ consisting of a local context Γ of constraints and an object o. Algorithm nodes are labelled by *explanations* which consist of the name of an algorithm with its parameters or a quadrupel $\langle C, \iota, [on], cs\rangle$ of a *nested computation structure* cs such that C is the set of constraints, ι instantiation, and $[on]$ sequence of object nodes. Each algorithm node has a unique link to its result and there are links from object nodes to algorithm nodes. These graphs allow the representation of contexts for symbolic algebraic computation.

Definition 8.

$$cs = \langle On, An, r, p, oL, aL\rangle \in CS_0[CT, ON, AN]$$

is a ***basic computation structure*** with

1. $On \in P_\omega(ON)$ set of object nodes of cs,
2. $An \in P_\omega(AN)$ set of algorithm nodes of cs,
3. $r : [An \to On]$ maps to resulting object node,
4. $p : [An \to On^*]$ maps to possibly empty set of input parameter objects,
5. $oL : [On \to O]$ object node labelling map,
6. $aL : [An \to [Id \times P_\omega(C)]]$ algorithm node labelling map such that $\forall an \in An$:

$$aL(an) = \langle id, \tilde{c}\rangle \wedge \bar{o} = oL(p(an)) \wedge o = oL(r(an)) \Rightarrow \exists \tilde{c}' : \tilde{c} \models \tilde{c}' \wedge \langle \bar{o}, o, \tilde{c}'\rangle \in \tilde{a}(id)$$

Definition 9.
The ***labelled graph*** $Graph(cs)$ of a basic computation structure $cs = \langle On, An, r, p, oL, aL\rangle$ is the graph with nodes $On \cup An$ and edges

$$\{(an, r(an)) \mid an \in An\} \cup \{(on, an) \mid an \in An \wedge on \in e(an))\}$$

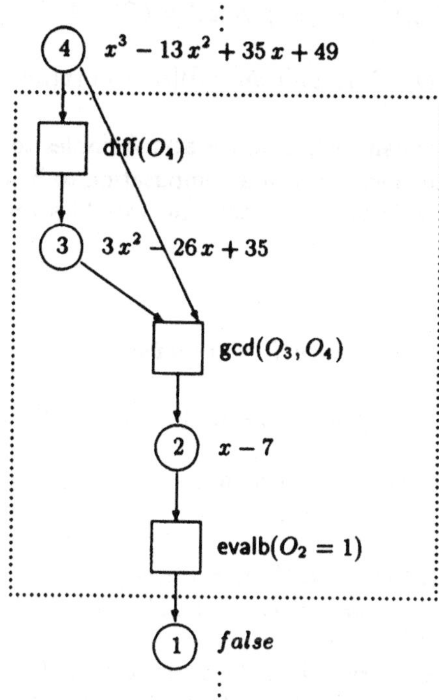

Fig. 3. Parts of a computation structure cs_I

Relying on basic structures we introduce vertical flexibility for algorithmic services by nesting structures. Computation structures are defined by their depth and recursive elements. Figure 3 illustrates the graph of a basic computation structure without constraints.

To increase the readability of computations and to group several steps we introduce vertical flexibility by nesting computation structures. The nested structure usually includes schematic variables and one can assign a unique identifier, e.g. **Squarefree**. Nesting computation structures correspond to the concept of subroutines in symbolic computation systems. To allow for cooperation within the computation of the squarefree property a CAS must handle structures for subroutines.

Definition 10.

$$cs = \langle On, An, r, p, oL, aL \rangle \in CS[CT, ON, AN] = \bigcup_{n \in \mathbb{N}} CS_n[CT, ON, AN]$$

is a ***computation structure***. The set CS_{n+1} of ***computation structures of depth n*** is recursively defined as the set of structures such that (1.-5.) in definition 8 holds and

6_{n+1}. $aL : [An \rightarrow [Id \times P_\omega(C)] + [P_\omega(C) \times I \times [On^*, On] \times CS_n[CT, ON, AN]]]$
the algorithm nodes labelling map such that (6.) holds for basic structures

and if $aL(an) = \langle \tilde{c}, \iota, [\bar{on}, on], cs' \rangle \;\wedge\; cs' = \langle On', An', r', p', oL', aL' \rangle$ then

$$[\bar{on}, on] \in (On')^* \;\wedge\; oL'([\bar{on}, on])[\iota] = [oL(p(an)), oL(r(an))]$$

Computation structures can include schematic variables to allow for schematic computations. It was defined for such a computation to be closed under instantiation. Examples for objects with schematic variables are $gcd(n, 2n) = n$ or $\int x^n\, dx = \frac{x^{n+1}}{n+1}$.

Definition 11.
An ***instantiation of a computation structure***

$$cs = \langle On, An, r, p, oL, aL \rangle \in CS[CT, ON, AN] \text{ and } \iota \in I$$

is $cs[\iota] = \langle On, An, r, p, oL[\iota], aL' \rangle$ such that

$\forall an \in An$
(ra) $\quad aL(an) = \langle id, \tilde{c} \rangle \;\Rightarrow aL'(an) = \langle id, \tilde{c}[\iota] \rangle$ and
(nest) $\quad aL(an) = \langle \tilde{c}, \iota_1, [\bar{ok}, ok], cs_1 \rangle \;\Rightarrow aL'(an) = \langle \tilde{c}[\iota], \iota \circ \iota_1, [\bar{ok}, ok], cs_1 \rangle$.

Computation structures represent the computation of a resulting object from a set of given input parameters if no algorithm application has unsolved constraints. The following lemma states that vertical flexibility can be eliminated by vertical unfolding.

Lemma 12. *Elimination of nesting*
Let $cs \in CS[CT, ON, AN]$ be a computation structure with result o and input parameters $\tilde{o}$ then there exists a computation structure $cs_0 \in CS[CT, ON, AK]$ of depth 0 with result s and input parameters $\tilde{o}$.

5 Construction and Manipulation of Structures

Operations $O_{rs} = \{addS, linkR, solveC, linkN\}$ for constructing and manipulating reasoning structures are given in [9], including the empty reasoning structure *emptyRS*, introduction of sequents, link by forward and backward application of rules, constraint solving, and nesting. This section introduces the corresponding operations on computation structures.

Definition 13.

$$emptyCS = \langle \emptyset, \emptyset, \emptyset, \emptyset, \emptyset, \emptyset \rangle$$

is the empty computation structure.

Since $emptyCS = emptyRS$ we denote empty structures as *empty*.

Definition 14.
Objects $o \in O$ are introduced by *addO*

$$addO(cs, o) = \langle On \cup \{on\}, An, r, p, oL\{on \mapsto o\}, aL\rangle$$

with new object node $on \in ON \setminus On$.

Definition 15.
Let $on \in On, \bar{on} \in On^*, a = \langle id, \bar{o}, o, \tilde{c}\rangle \in \tilde{a}, \models \{\bar{o} \sim oL(\bar{on}), o \sim oL(on)\}$. Links by application of algorithms are defined as

$$linkA(cs, \bar{on}, on, a) = \langle On, An \cup \{an\}, \\ r\{an \mapsto on\}, p\{an \mapsto \bar{on}\}, oL, aL\{an \mapsto \langle id, \tilde{c}\rangle\}\rangle$$

with new algorithm node $an \in AN \setminus An$.

Definitions of operations for solving constraints, nesting links, extending operations to nested computation structures and additional operations to combine structures by introducing bridge rules correspond to [9] and are given in [11]. One can easily proof that the operations of $\mathcal{O}_{cs} = \{addO, linkA, solveC, linkN\}$ are sound, independent, and complete generators of any computation structure out of *empty*.

Let $a = \langle id, \bar{o}, o_0, \tilde{c}\rangle$ be an algorithm such that $\bar{o} = [o_1, \ldots, o_n]$ and $\models \tilde{c}$. We define the application of an algorithm to be a computation structure as a sequence of operations from $\mathcal{O}_{cs}$.

Definition 16.
Let $cs = \langle On, An, r, p, oL, aL\rangle$ be a computation structure such that $\bar{on} = [on_1, \ldots, on_n] \in On^* \land \forall\, 1 \le j \le n : oL(on_j) = o_j$. Let $on_0 \notin On \land an_0 \notin An$. $applyA(cs, a, \bar{on})$ is the sequence of operations

- $cs_0 = addO(cs, o_0)$ by introducing on_0;
- $cs_1 = linkA(cs_0, \bar{on}, on_0, a)$ by introducing an_0;
- $cs' = solveC(cs_1, vk_0, \emptyset)$.

6 Examples

6.1 Isabelle and Maple

We designed and implemented an interface between the tactical theorem prover Isabelle [15] and Maple [14] by extending the simplifier of Isabelle ([2], figure 4). The simplifier is extended by new kinds of rules to call external functions of the CAS. The interactive proof involves computation structures with only one algorithm application. As an example, the inductive proof of

$$\forall n \in \mathbf{N} : 5 \le n \Longrightarrow n^5 \le 5^n.$$

expands all of the products in the induction step. $x+1$ is an object in the sequent of a reasoning structure which serves as input node to the **expand** algorithm of Maple.

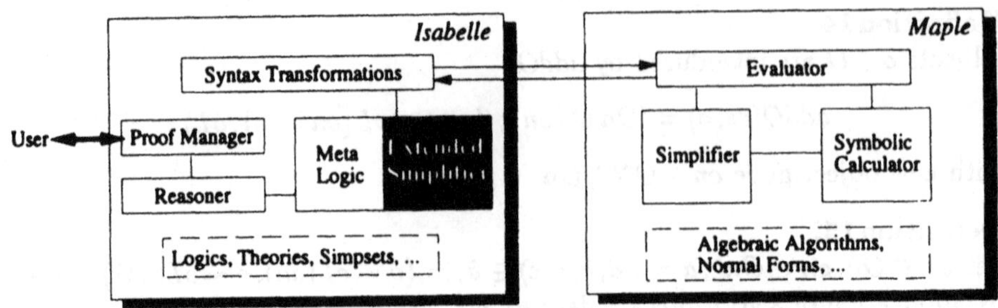

Fig. 4. Schematic Link between Isabelle and Maple

```
- by (res_inst_tac [("P",
= "%y.expand((x + 1) ^ 5) <= y")] expandE 1);
n ^ 5 <= 5 ^ n
 1. !!x. [| x : Nat; 5 <= x; x ^ 5 <= 5 ^ x |] ==>
         expand((x + 1) ^ 5) <= expand(5 ^ (x + 1))
- by (asm_simp_tac Nat_simplify_ss 1);
n ^ 5 <= 5 ^ n
 1. !!x. [| x : Nat; 5 <= x; x ^ 5 <= 5 ^ x |] ==>
         x ^ 5 + 5 * x ^ 4 + 10 * x ^ 3 +
         10 * x ^ 2 + 5 * x + 1 <= 5 * 5 ^ x
```

The interface was implemented without explicit construction of computation structures. The semantic of the interaction is given in [2].

6.2 Imps and Calvin

To introduce external function calls in a rigorous theorem prover is difficult. Either algorithms must be proven to be correct or the results of algorithms must be fully proven by the proof system because rigorous systems can not trust external tools. [10] introduces the concept of trust in cooperated reasoning and computation. Such cooperations can be used to guide proofs where verification is easier as computation by theorems. Since a huge subset of mathematical algorithmic computation can be easily verified by simple equations we started to implement an interface between IMPS [7] and MAPLE along the concept of external macetes introduced by [8]. External computation and verification of results could be successfully implemented, e.g., for factorisation of polynomials, anti-derivatives, gcd computations, Hensel lifting, and chinese remainder. Problems with completeness of results occur by applications of the `solve` operator, i.e. solving linear and differential equation systems, recurrence relations and equational simplification. The verification of numerical or "direct" computations, however, is of same complexity.

A link with an existing CAS allows only to apply computation structures with just one algorithm node. To implement a CAS capable to manage contexts is one of our long term research projects. The system, called CALVIN, allows restartable and incremental application of algorithms. These contexts can be used to guarantee correctness of computations, e.g. requesting the context of n when integrating x^n.

7 Conclusion and Further Research

This paper introduces computation theories and structures which serve as a formal framework and tool for representing mathematical objects and applications of algorithms appearing in algorithmic services. The composition of reasoning and computation theories and structures will provide a theoretical and technical framework for the specification and implementation of symbolic mathematical problem solving by cooperation of algorithms and theorems.

Nowadays CAS behave like black boxes. Cooperation with such systems result in computation structures with just one algorithm node. Although the benifit of the given theory still is a formal description of the cooperation between provers as well as several CAS and horizontal flexibility, major redesign of CAS is required to achieve high level cooperation. We have shown how computation structures naturally correspond to reasoning structures and give access to intermediate knowledge. To develop the CAS CALVIN along the concepts presented in this paper is part of an ongoing research project.

Cooperating services must provide some kind of access to reasoning and computation structures. The functionality of interfaces for distributed reasoning and computation are given in [16] and [11] respectively. Although they might not explicitly be implemented cooperations among mathematical services are based on manipulation of structures for symbolic mathematical reasoning and computation.

References

1. J. ABBOTT, A. VAN LEEUWEN, A. STROTMANN
   Objectives of OpenMath. Submitted to Journal of Symbolic Computation, 1995.
2. C. BALLARIN, K. HOMANN, J. CALMET
   Theorems and Algorithms: An Interface between Isabelle and Maple. In A.H.M. LEVELT (Ed.), Proceedings of International Symposium on Symbolic and Algebraic Computation (ISSAC'95), pp. 150–157, ACM press, 1995.
3. B. BUCHBERGER
   Mathematica: Doing Mathematics by Computer? In A. MIOLA (Ed.), Advances in the Design of Symbolic Computation Systems, Texts and Monographs in Symbolic Computation, Springer, 1996.
4. J. CALMET, K. HOMANN
   Distributed Mathematical Problem Solving. In Proceedings of 4th Bar-Ilan Symposium on Foundations of Artificial Intelligence (BISFAI'95), pp. 222-230, 1995.
5. J. CALMET, K. HOMANN
   Classification of Communication and Cooperation Mechanisms for Logical and Symbolic Computation Systems. In F. BAADER, K.U. SCHULZ (Eds.), Proceedings of First International Workshop on Frontiers of Combining Systems (FroCoS'96), pp. 133–146, Kluwer Series on Applied Logic, 1996.

6. E. Clarke, X. Zhao
Combining Symbolic Computation and Theorem Proving: Some Problems of Ramanujan. In A. Bundy (Ed.), Automated Deduction (CADE-12), pp. 758–763, LNAI 814, Springer, 1994.
7. W.M. Farmer, J.D. Guttman, F.J. Thayer
IMPS: An Interactive Mathematical Proof System. In Journal of Automated Reasoning 11:213–248, 1993.
8. W.M. Farmer, J.D. Guttman, F.J. Thayer
Contexts in Mathematical Reasoning and Computation. In Journal of Symbolic Computation 19:201–216, 1995.
9. F. Giunchiglia, P. Pecchiari, C. Talcott
Reasoning Theories - Towards an Architecture for Open Mechanized Reasoning Systems. In F. Baader, K.U. Schulz (Eds.), Proceedings of First International Workshop on Frontiers of Combining Systems (FroCoS'96), pp. 97–114, Kluwer Series on Applied Logic, 1996.
10. J. Harrison, L. Théry
Extending the HOL Theorem Prover with a Computer Algebra System to Reason About the Reals. In J.J. Joyce, C.-J.H. Seger (Eds.), Higher Order Logic Theorem Proving and its Applications (HUG'93), pp. 174–184, LNCS 780, Springer, 1993.
11. K. Homann
Symbolic Mathematical Problem Solving by Cooperation of Algorithmic and Logical Services, (in german), Dissertation, University of Karlsruhe, 1996.
12. P. Jackson
Exploring Abstract Algebra in Constructive Type Theory. In A. Bundy (Ed.), Automated Deduction (CADE-12), pp. 590–604, LNAI 814, Springer, 1994.
13. N. Kajler
CAS/PI: a Portable and Extensible Interface for Computer Algebra Systems. In Proceedings of ISSAC'92, pp. 376–386, ACM press, 1992.
14. M.B. Monagan, K.O. Geddes, K.M. Heal, G. Labahn, S. Vorkoetter
Maple V Programming Guide, Springer, 1996.
15. L.C. Paulson
Isabelle — A Generic Theorem Prover, LNCS 828, Springer, 1994.
16. C. Talcott
Handles - Specialized logical services, Working Draft, 1994.
17. N. Venkatasubramanian, C. Talcott
Reasoning about Meta Level Activities in Open Distributed Systems, to appear in PODC, 1995.

Implementing FS_0 in Isabelle: Adding Structure at the Metalevel

Seán Matthews

Max-Planck-Inst., Im Stadtwald
66123 Saarbrücken, Germany
sean@mpi-sb.mpg.de

Abstract. Often the theoretical virtue of simplicity in a theory does not fit with the practical necessities of use. An example of this is Feferman's FS_0, a theory of inductive definitions which is very simple, but seems to lack all practical facilities. We present an implementation in the Isabelle generic theorem prover. We show that we can use the facilities available there to provide all the complex structuring facilities we need without compromising the simplicity of the original theory. The result is a thoroughly practical implementation. We further argue that it is unlikely that a custom implementation would be as effective.

1 Introduction

A great many logics have been proposed in computer science for formal, and even machine checked reasoning. However, if we try to implement these, we find that what has been proposed by theoreticians as a practical tool has to be augmented in all sorts of ways before it really becomes a practical tool: the basics of structured programming and many other facilities need to be imported. Unfortunately, this means that a deduction system which could be summarised on a page or so grows to fill a manual, and is augmented with decision procedures and other extras which can be difficult to verify.

A suggested solution to some of this problem is what has become known as a *Logical framework*: a system designed to be suitable for implementing a wide range of different logics easily, so that they can be presented in a uniform manner to users, allowing the same theorem proving facilities to be reused across a wide range of implemented logics, instead of having to be rebuilt from scratch each time. The basic idea of a logical framework is that logics should be implementable via an equation

$$\text{Syntax} + \text{Axioms} + \text{Rules} = \text{Theorem prover}.$$

But really we get more than this: because a logical framework based system is generic, its designers can afford to invest in much more powerful facilities, since they are likely to be reused in a range of different contexts. In fact, since the designers have to provide these facilities prior to the implementation of any particular theory, they must, of necessity, make them powerful and general to improve that chance that they will be usable in unimagined contexts. Thus given a theory, a logical framework based system should not only be able to provide an implementation much more quickly, but also be able to provide some, or all of the structuring facilities that are needed for practical proof development in a way that is independent of the theory. If this is so, then the theory itself does not have to be extended, and can be practically implemented in a way that is closer to the original proposal.

Feferman's FS_0 is a theory perfect for illustrating this point. It has been proposed as a suitable vehicle for machine checked proof theory but, while it is 'simple', it so primitive that it is not at all clear that it is remotely practically usable (especially as Feferman gives it). It comes with none of the structuring facilities upon which we depend when developing large systems, so these have to be provided somehow in an implementation; but there are good reasons why we should not modify the theory itself in any way.

We describe an implementation we have built for the Isabelle system, exploiting the combination of higher-order logic, proof-construction by higher order resolution and the ML programming language available there to provide effective versions of the facilities we need purely in the metalevel. This in turn illustrates how general the type-theoretic logical framework view of theorem provers is. We also claim that the result is probably more effective than a custom implementation, since the design of so many of the powerful facilities we provide is inspired directly by the facilities that a generic system has make available, and we suspect unlikely to occur to the 'naïve' designer of a bespoke system, who would anyway have fewer resources available to implement them. Finally, we supply evidence that FS_0 is a plausible theory for real computer supported theory development (answering, in passing, a question raised by Pollack), by presenting the first practical implementation, So far as we, or Feferman, know.

2 The theory FS_0

FS_0 (the reader is referred to [4] for more details than we have room for here) is in the tradition of Gödel's incompleteness results: a 'rational reconstruction' of the results of Gödel's preliminary development in arithmetic (i.e. tools for doing 'Gödel-numbering'), and a conservative extension of primitive recursive arithmetic. A first impression is that it is very simple, and Feferman argues strongly for not adulterating that simplicity, so we should implement it pretty much as it stands: which is as a three-sorted classical first order, finitely axiomatisable theory of s-expressions, primitive recursive functions and recursively enumerable classes, that resembles Pure Lisp. We have:

- A sort S of s-expressions, containing a leaf object nil, and closed under a pairing function $(\_,\_)$; equality is defined in the obvious way over s-expressions.
- A sort F of functions, of type $S \to S$ with function application denoted by '; i.e. if f is of type F and t of type S then $f`t$ is a function application of type S. We have a small set of basic functions: Id (identity), car, cdr, and

Dec (decide). Most of these should be well known, apart from *Dec*, for which

$$Dec`(((a,b),c),d) = \begin{cases} a = b \rightarrow c \\ a \neq b \rightarrow d \end{cases}$$

There are also constant functions $K(a)$ where $K(a)`b = a$.
There are three combinators over functions of type $(F \times F) \rightarrow F$: composition, $\_ * \_$, where $(f * g)`t = f`(g`t)$; pairing, $[\_,\_]$, where $[f,g]`t = (f`t, g`t)$; and structural recursion, $Rec(\_,\_)$, where, if $h = Rec(f,g)$,

$$\begin{aligned} h`nil &= nil \\ h`(a, nil) &= f`a \\ h`(a,(b,c)) &= g`((((a,b),c), h`(a,b)), h`(a,c)) \end{aligned}$$

Equality on functions is defined extensionally.
- A sort C of classes; We have the class containing only *nil*, i.e. $\{nil\}$ and can take the intersection and union $\cup$ and $\cap$ of classes, as well as the inverse image of a class c under a function f, $f^{-1}c$ where $t \in f^{-1}c \leftrightarrow f`t \in c$. More complicated, we can also build recursively enumerable classes $I(a,b)$, containing a and closed under the rule $t_1, t_2/t$, where $((t,t_1),t_2) \in b$. Equality and the subset relation on classes are defined extensionally.
- Finally, we have induction; over the universe,

$$nil \in c \rightarrow \forall a \forall b(a, b \in c \rightarrow (a,b) \in c) \rightarrow \forall x(x \in c)$$

and over inductively defined sets,

$$y \subset x \rightarrow \forall a \forall b \forall c(b, c \in x \rightarrow ((a,b),c) \in z \rightarrow a \in x) \rightarrow I(y,z) \subset x.$$

FS_0 is essentially a language for defining recursively enumerable (r.e.) classes and, less often, recursive functions. Development in it consists of constructing these, and then proving properties about them. When we try actually to do this, however, it soon becomes clear that it is very difficult: the structure of definitions is not intuitive, and when we try to check that what we have defined has at least the preliminary properties that we want, we find, almost invariably, that it is full of hard to correct errors. We do, however, have a theorem that characterises the classes we are able, theoretically, to define, in the form of a comprehension principle: If we define the class of $\exists^+$-formulae to be the class containing equalities and inequalities between S sorted terms, and set membership, and closed under disjunction, conjunction and existential quantification of S sorted variables, then we have a comprehension principle:

> Given a $\exists^+$-formula $P(x)$ with one free variable x of type S, there exists a class c such that $FS_0 \vdash x \in c \leftrightarrow P(x)$.

The class of $\exists^+$-formulae provides us with a readable specification language for r.e. classes. Note, however, that this result is a meta-theorem; it is neither an axiom schema nor a theorem of FS_0, and while we could add it as an axiom to the theory directly we have already said that we want to avoid such extensions. How we provide comprehension is in fact one of the main facilities that we document.

3 Isabelle

The Isabelle generic theorem prover (the reader is referred to Paulson [7] for more details) comes as a collection of extensions for the ML programming language. It cannot accurately be described as a program that just happens to be written in ML; the relationship between the two is much closer than that: we work with Isabelle directly through the ML interpreter command line, meaning that we also have direct access to ML to program extensions, or as a tactic language; this is a powerful, safe facility—the strong typing of the language is an effective prophylaxis against accidental damage to an implemented system.

Isabelle is based on the observation, dating back to the Automath [3] project, that an effective generic proof system can be built from the $\forall/\rightarrow$ fragment of a weak higher-order logic (called the *metalogic*), a language which is suitable for both describing natural deduction style deduction systems and binding operations (a traditional problem for implementations). In fact Isabelle is best thought of as a collection of tools for deriving and manipulating terms in a higher-order logic in which we can encode the language, formulae and rules of *object-logics*. Since the metalogic makes no distinction between rules and theorems, the best way to regard theorems is simply as degenerate rules with no premises and no schematic variables. Further, Isabelle is not based on rewriting: we combine terms of the metalogic using higher-order resolution, which provides some interesting effects; for instance we can use the same rule for both forward and backwards proof (since resolution is 'bidirectional'); we can even have 'metavariables' in formulae we want to prove, allowing these to be instantiated as we build a proof, picking up information from rules as we apply them.

Some of the details of how Isabelle allows structuring of collections of definitions in ML are as follows:

- A new data-type `theory` for packaging the definitions of a declared theory.
- Functions for combining and extending `theorys`; i.e. if we have `theory` of first-order logic called `Pred`, we can extend it to a `theory` of arithmetic `Nat`. Or combine `Nat` with, `List` to get a theory of lists of numbers and lists inherits all the theorems that have been proven for either.
- A new data-type `thm` for the terms (i.e. theorems and rules) derived using `theorys`; importantly (c.f. §6) Isabelle makes no distinction between derived and basic theorems of a theory, they are all just `thms`.

Along with the basic system, we get some tools for building such things as rewrite systems. We also get a few predefined logics, including sorted classical first-order logic; and using this we can immediately build a naïve implementation of FS_0: all we have to do is declare sorts S, F, C then type in the axioms Feferman gives, to get a `theory` we call `FS0`.

4 The basic implementation

We have already argued that a naïve implementation of FS_0 is not usable, but we can assess what it does make available. This amounts (in Isabelle) mostly to an effective method for modularising development: developments in FS_0 consist of definitional extensions and associated theorems. In Isabelle there is an effective way to include definitional extensions to a theory simply by adding metalevel equations to a **theory** defining a new name to be syntactically equivalent to some term. We can thus take over the general Isabelle theory handling tools directly to handle FS_0 theories, treating definitional extensions of FS_0 as definitional extensions, in Isabelle, of FS0. Further, since equality on functions and classes is purely existential, if we then prove suitable a collection of theorems showing these definitions have the properties we actually want, we can completely ignore in future the implementation work hidden behind the definitions, and just use the theorems about them as if they were axioms *defining* the behaviour of the newly added names. (Remember that Isabelle does not distinguish between the derived theorems and basic axioms of a declared **theory**.) Then when we combine such theories in the usual way in Isabelle, these theorems get inherited immediately; thus we have a simple but effective tool for structuring the development of large theories as collections of small ones, and for hiding the implementational details.

5 A new interface

Now we can consider in more detail the other, new, facilities we need to get a usable implementation. We have already said that the three stage process of constructing, by hand, large FS_0 terms, assigning them to new names, and finally proving that they correspond to the classes and functions we think they do, is not practical. But the comprehension principle we give above says that there is a close relationship between FS_0 classes and a certain class of logical formulae, and we have said that this relationship is central to how we use FS_0, so it would seem that, if possible, we would like appeal to it directly, using $\exists^+$-formulae as specifications, and avoiding the three stage process altogether.

Unfortunately we cannot do this and keep the standard Isabelle interface, which is designed explicitly for working with theories which do fit the three stage approach. Thus we develop a new interface, by defining a new function **extend_FS0_theory** (by analogy with Isabelle's **extend_theory**, part of the standard interface). This takes an FS_0 theory and a list of definitions in a suitable specification language and returns a package of a new theory and a collection of useful theorems that we have been able to generate automatically. In fact the interface we provide in **extend_FS0_theory** is very general: not only does it allow us to define classes directly using comprehension, but we can also define new functions (or any other terms) with respect to logical formulae in the same way (if not quite so elegantly). We also take the opportunity to add a facility for 'top down' development.

Currently four sorts of definitions are possible, all with the same basic form: we write

```
sort(name,args,...)
```

to define **name** using the style of **sort**; the **args** parameter is a device which we can use to 'schematize' or parameterise a definition over a list of arguments. We discuss each of the possibilities in turn.

Simple constants The first sort of definition is just the 'hand built' style, where we directly enter the term we want a name to abbreviate. This sort of definition makes sense if we want to encode, e.g., the natural numbers as lists of *nils*; i.e. $0 \triangleq nil$, $s(0) \triangleq (nil, nil)$, $s(s(0)) \triangleq (nil, (nil, nil))$ etc., which we can define as:

```
def("zero", [], "nil"),
def("s", [], "[const(nil),Id]")
def("suc", ["x"], "(nil,x)")
```

resulting in the defined theory containing the new name *zero* for *nil* the new function *s*, where $s`t = (nil, t)$, which can also write '$suc(t)$. Note that the equation for *s* is a theorem that we have to prove, **def** declarations are so general that it is not possible to extract any extra information from them; however we can see the beginning of an abstract interface here: we prove a set of theorems describing the abstract behaviour of *suc* and *zero* in the natural numbers, without regard to their implementation, and we can use these instead of the concrete details in future.

Comprehension More interesting than **def** is comprehension. The secret to making this available is to examine the proof (on paper) of the comprehension theorem. This shows, by induction on the structure of $\exists^+$-formulae, that it is always possible to construct an equivalent class. Most of the cases are easy; for instance there is an obvious equivalence between $\cap/\cup$, and $\wedge/\vee$. The trickiest is existential quantification ($\exists$): we have to show that given $x \in c \leftrightarrow P(car`x, cdr`x)$, there is some construction $Ex(c)$ such that $x \in Ex(c)$ iff $\exists y.P(x, y)$. The construction of Ex is a bit tedious but we only have to do it once (see Feferman for details).

The type theory of Isabelle is too weak to formalise all of this argument, since it does not support induction. This is the only thing that cannot be formalised though—all the step cases are provable, i.e. we can derive rules for each possible reduction step needed by the proof. For instance we can derive the rule

$$\frac{\forall x(x \in z \leftrightarrow P(car`x, cdr`x))}{\forall x(x \in Ex(z) \leftrightarrow \exists a(P(a, x)))}$$

(given some z such that ..., then there is some $Ex(z)$ such that ...).

But since we build proofs in Isabelle by resolution (it may help here to think of Isabelle as a sophisticated version of Prolog), we can get the equivalent of the induction we need by backchaining through these derived rules. For instance imagine we want to define the class corresponding to the 'less than' relation. We might start with a goal of the form

$$\forall x(x \in ?c \leftrightarrow \exists w \exists y \exists z(x = (y, z) \wedge plus`(x, w) = z)) \qquad (*)$$

where $?c$ is a *metavariable*, or 'hole'; that is $?c$ is not an variable in the language of FS_0, and what we are trying to prove is not a formula in the language of FS_0, but a schema. This makes no difference to Isabelle though, which we can now use to find a 'solution' for $?c$ such that we have a theorem.

Thus we first apply the rule for the existential case. The conclusion of this resolves with (*) leaving us to prove

$$\forall x(x \in ?d \leftrightarrow \exists y \exists z(cdr`x = (y, z) \wedge plus`(cdr`x, car`x) = z))$$

where $?c$ has been instantiated with $Ex(?d)$. We can repeat this step twice more, then we have to change to the rule for $\wedge$, and so on. Eventually we have only goals $\forall x(x \in ?e \leftrightarrow x \in d)$ which we can dispose of simply by resolving with the theorem $\forall x(x \in y \leftrightarrow x \in y)$.

Consider what we have done: We have *not* proved (*), rather what we have done is to to find an instantiation of $?c$ for which (*) is a theorem, providing us simultaneously with two things: a term which we *know* corresponds to the comprehension of $\exists w \exists y \exists z(x = (y, z) \wedge plus`(x, w) = z)$, and a ready made machine checked theorem stating that correspondence. We only have to attach a name to the term and record the theorem.[1]

Finally, we have implemented this so that a definition can be given directly in the form

```
comp("ltC", [], "(y,z)", "EX w. plus'(y,w)=z")
```

(the extra existential quantifiers are generated automatically), to define the new name `ltC` and generate the associated theorem.

General synthesis For comprehension we have a uniform procedure which when given theorem-with-a-hole of a particular form, can fill out that hole. However there are other sorts of objects, e.g. functions, that we will want to define, for which no uniform procedure exists, but we still want to avoid the three stage process of piecing them together by hand out of primitives, verifying them, etc. We do this by extending the idea that we have just used to implement the comprehension theorem to a much larger class of synthesis problems.

As an example, consider the 'less than' ordering again, only this time we want to define it as a function, not as a relation. We adopt the same method of synthesis, starting with a formula-with-a-hole that we try to prove, as follows:

$$\begin{aligned} &?lt`nil = false \wedge ?lt`(a, zero) = false \wedge \\ &\quad ?lt`(a, s`a) = true \wedge (a \neq b \rightarrow ?lt`(a, s`b) = ?lt`(a, b)) \end{aligned}$$

We can read this as the specification of a function $?lt$, which we can use Isabelle to solve. The difference this time is that, instead of being able to

[1] This technique is similar to the idea suggested in, e.g., [1] as a general method for formal program synthesis.

use a uniform procedure, we have to do this interactively, ourselves. We allow **extend_FS0_theory** thus to take a definition of the form

```
sch_def("lt", [],
   "?lt'nil=true & ?lt'(a,zero)=false &
    ?lt'(a,s'a)=true) & (a~=b --> ?lt'(a,s'b)=?lt'(a,b))"
   ltsynthtac)
```

This is like comp except that it has one more parameter, which we use to supply the procedure (in this case ltsynth) needed to solve the specification. Then, assuming that ltsynthtac doesn't fail, the result is exactly like before: the new function is synthesised and attached to a new constant, and the theorem that has been proved is returned, with the new abbreviation replacing the synthesised term. We could even have defined ltC this way, with the definition

```
sch_def("ltC",
        "x:?ltC <-> EX w y z. x = (y,z) & plus'(y,w)=z",
        comprehension_tac)
```

Top down specification The facilities we have described for constructing classes and functions are very effective for synthesising terms, attaching them to names, and keeping track of their properties. However they are also 'bottom up', since they build objects out of basic components. This is very safe, since it ensures that everything we define *is* a definitional extension of FS_0, but it also means that 'top down' development is not possible. But there are times when we might want to postpone work on some part of a development, or do several things in parallel. Thus we also provide a way to add, temporarily, new constants and axioms to a theory directly, instead of just definitional extensions. This looks very like sch_def, but has one less parameter:

```
abstract("lt", [],
   "lt'nil=true & lt'(a,zero)=false &
    lt'(a,s'a)=true) & (a~=b --> lt'(a,s'b)=lt'(a,b))")
```

(notice that the formula here has no 'holes'). With comp and sch_def a definitional extension of the given theory is generated and a theorem about it is proven. With abstract this process is short-circuited: no effort is made to try to generate a suitable definition, or prove a theorem. The result is a new constant lt, which has an associated new *axiom* defining its behaviour. This is dangerous because this extension is not definitional and the associated theorem is not a theorem of FS_0 (unlike a theorem, nothing stops us adding a false axiom), but abstract is supposed to be used only as a temporary, stop-gap, measure; we would expect all instances of abstract to be removed from a development before the end.

If we consider what we have developed in this section we see a single idea, applied in a variety of ways. We have tried constantly to hide the details of FS_0

definitions behind abbreviations, which we treat as new objects, with new defining axioms, added to the theory. And in general, we try to build these indirectly using unification, from logical specifications, rather than directly. Really FS_0 is used only as an underlying foundation to the extension we define. But it is nevertheless important that, if we really need to, we can strip these levels of abstraction away, leaving the unadorned theory, since everything is, in the end, just a definitional extension. We extend this theme in the next section, when we talk about how Isabelle allows us to implement rewriting.

6 In use

So far we have described the tools we have built in Isabelle to define objects in FS_0. We now consider how practical it is to work with those objects once we have defined them. We will consider two important problems: induction and rewriting.

Induction Clearly a lot of the work of using FS_0 is in defining classes. But we encounter the problem of classes corresponding to predicates not only when we are building theories, but also regularly when we are working with them. The commonest method of proof in FS_0 is induction, of one sort or another, and the only induction we have is over classes, in spite of the fact that in practice we almost never want to do this.

By the comprehension principle we know that we can find an equivalent class to any $\exists^+$-formula, so we have induction over this class (which is enough in for most practical things). However the metatheory of Isabelle does not allow us to prove derived rules of this form; i.e. we cannot prove a single metatheorem corresponding to the rule

$$\frac{P(nil) \quad \forall x \forall y (P(x) \rightarrow P(y) \rightarrow P((x,y)))}{\forall x P(x)}$$

(where P is a $\exists^+$-formula). We are, however able to prove something almost identical in the rule

$$\frac{\exists c \forall x (x \in c \leftrightarrow P(x)) \quad P(nil) \quad \forall x \forall y (P(x) \rightarrow P(y) \rightarrow P((x,y)))}{\forall x P(x)}.$$

This moves the side condition (that P be $\exists^+$) into the rule as an extra side goal. But we already have a tool available to dealing with this side goal: the uniform proof procedure which we use to implement comp above, so we can always dispose of this goal immediately. Again we have been able to abstract away from the concrete details of FS_0.

Rewriting The other large and ubiquitous problem-in-use for FS_0 is term simplification since many theorems are proven basically by selecting a suitable induction then simplifying the resulting terms. It maybe not immediately clear why this is a problem: the functions provided by FS_0 are a well defined class with for which we have a complete set of normalising rewriting rules (cf. §2); surely we need only implement these and arrange for names to be unfolded to their definitions as necessary?

But consider as an example a function to number the elements of a list with their positions. We can specify this as follows:

$$\begin{aligned} \mathit{label}`l &= \mathit{labeld}`(\mathit{zero}, l) \\ \mathit{labeld}`(n, \mathit{nil}) &= \mathit{nil} \\ \mathit{labeld}`(n, (f, r)) &= ((n, f), \mathit{labeld}`(s`n, r)) \end{aligned}$$

and (with some effort) we can find a solution for this in terms of FS_0 primitives.

However if we try to evaluate the resulting function term by following the basic rewriting rules, we discover that its operational behaviour is nothing like our specification leads us to expect, and extremely 'bad'. Not only is its computational complexity ***not*** linear w.r.t. length, but the normalisation does not follow the path that we expect it should: the normalisation algorithm is usually forced to expand definitions that do not seem to have anything to do with the calculation of the result. Consider simply the step case of ***labeld***; we might imagine that given a solution to ***labeld***, we could prove this equation by unfolding the function definitions into their basic components, then reducing the two sides to normal form. But if we try this, we find that it does not work: the result of normalisation is an unreadable, partially normalised mess of s-expressions and functions. In fact this technique can be used only to prove equivalence for ground instances; we need to use induction to prove the general case. This problem seems to be unavoidable: the primitive recursive nature of functions in FS_0 seems to impose these awkward functions on us. Further, it is a serious enough problem that Pollack [8, §5.2] mentions it as a possibly insuperable obstacle to the practical use of FS_0 as a framework.

What is needed is a rewriting system respecting the abstract properties of what we have defined, not the concrete details of the implementation. We have already mentioned that Isabelle does not distinguish between abbreviations and constants, or between basic axioms and derived theorems; we can exploit this in combination with the provided general rewrite package (**`term_simp_tac`**) to avoid the problems we have pointed outa. **`term_simp_tac`** takes, as a parameter, a set of equational theorems to be used as rewrite rules. The important thing is that these theorems need not describe the actual normalisation path that an expression would take—in fact the package will happily accept a diverging set of rules. Thus after we have verified (inductively) that, e.g., ***label*** and ***labeld*** behave as we want according their specification, we can use the specification, rather than the implementation, in the rewriter. In practice, these equations are often immediately available if we use the **`sch_def`** facility we have described.

Thus we have continued, and finished, the theme of abstracting the system as the user sees it from the primitive details of FS_0, complementing the term abstraction facilities for defining theories described in the last section with abstract induction and rewriting which we can use to work in theories.

7 Experience

FS_0 is designed for doing proof theory; one specific intended application for it is to build a machine checked, structured, proof of Gödel's second incompleteness theorem. This, however, is very much a long term goal which can only be built on a foundation of general facilities for proof-theory, which are of independent interest in themselves. We have concentrated on a developing a range of such facilities.

One ubiquitous and notorious problem with formalising proof theory is the formal treatment of binding and substitution, but there is no facility for doing this in FS_0 (unlike in other, type-theoretic, proposed logical frameworks). We have, therefore, implemented a version of Talcott's algebra of binding [9]; even if this work were not important for future use of FS_0 as a framework, the theory itself is notably complex, and a formal development of it from scratch is a demanding test of how our implementation handles a real application. In fact, we have found that the facilities available in Isabelle again substantially and positively contributed to course of the work development and the final presentation [5]. The other facility we need for a general framework for doing metatheory is a notation and associated tools for treating consequence-style proof systems (i.e., rather Hilbert style, or axiomatic systems), this is current work in progress. In support of them, we have also developed a range of theories for, e.g., the natural numbers, lists, finite functions. The development scripts for some of these are also available [6]. Independently of this work we have also used our implementation to investigate how effective FS_0 is for structuring metalevel development. This work is described in [2].

8 Conclusions

At the centre of any formal proof development system are two things: an underlying logic which can be used to build and work in various theories, and a way to make that building and working as easy as possible, by allowing users to impose various sorts of structure on the development. This might mean (among other things): modularity of development, top down development and abstraction.

We have shown that in a system like Isabelle, which provides sophisticated structuring and development facilities of the sort we have listed, and which are provided prior to, and therefore independent of, any theory we might implement, we can take a simple (i.e. primitive) theory, in this case FS_0, lacking any such facilities and provide them purely at the metalevel, while preserving the simplicity of the theory itself. We have been able to provide a very practical and usable system (we have performed several complex developments in it) which is implemented using exactly the axioms in Feferman's original paper; further all development in the system can easily be unwound to that level.

We even believe it would be difficult to build a custom theorem prover as effective as what we have produced, since many of the more useful facilities are inspired directly by our experience of what Isabelle provides, and these facilities are so much more powerful than we can imagine trying to program 'from scratch'.

We could also see this implementation as a feasibility study for a program of practical 'minimal mathematics' in computer science. It is interesting to compare what we describe with other theorem proving systems where developments of the sort discussed in §7 might be undertaken. Almost all of these are for very powerful foundational logics such as set theory, or predicative or impredicative type theories. But in practice this formal power is never needed, it is only there to provide access to structuring facilities. This paper shows that such structuring does not need to be supplied in the logic, it can instead be imposed externally, allowing us to use much simpler finitary systems as foundations.

References

1. D. Basin, A. Bundy, I. Kraan, and S. Matthews. A framework for program development based on schematic proof. In *Proc. 7th Int. Workshop on Software Specification and Design*. IEEE, 1993.
2. D. Basin and S. Matthews. Structuring metatheory on inductive definitions. In J. Slaney, editor, *Proc. CADE-13*. Springer, Berlin, 1996.
3. N. G. de Bruijn. A survey of the project Automath. In J. R. Hindley and J. P. Seldin, editors, *To H. B. Curry: Essays in Combinatory Logic, Lambda Calculus and Formalism*, pages 579–606. Academic Press, New York, 1980.
4. S. Feferman. Finitary inductive systems. In *Logic Colloquium '88*. North-Holland, Amsterdam, 1990.
5. S. Matthews. A general binding facility in FS_0. Available from `http://www.-mpi-sb.mpg.de/~sean/`.
6. S. Matthews. Worked examples in FS_0. Available from `http://www.mpi-sb.-mpg.de/~sean/FSexamples.html`).
7. L. C. Paulson. *Isabelle: A generic theorem prover*. Springer, Berlin, 1994.
8. R. Pollack. On extensibility of proof checkers. In P. Dybjer et al., editors, *Types for Proofs and Programs*, pages 140–161. Springer, Berlin, 1995.
9. C. Talcott. A theory of binding structures, and applications to rewriting. *Theoret. Comput. Sci.*, 112:99–143, 1993.

An Approach to Class Reasoning in Symbolic Computation *

Gianna Cioni**, Attilio Colagrossi***, Marco Temperini†

Abstract. A methodology, following the object-oriented paradigm, for modeling mathematical structure by axiomatization is presented. A suitable programming language is defined in terms of its class construct and inheritance. The class construct allows for expressing also the properties of the structure represented by the class. By reasoning about the properties defined in classes (*class reasoning*) logical relations among the related mathematical structures can be evaluated. We show how class reasoning can be applied in order to rearrange and/or modify a hierarchy of (classes representing) mathematical structures.

1 Introduction

The research about the use of mathematical logic into symbolic computation systems has surged ahead during last years. In this paper we define a methodology for modeling mathematical structures by axiomatization. We support this methodology by a suitable object-oriented programming language and by the integration of an automated deduction tool into the language. The definition of hierarchies of mathematical structures is obtained by implementing classes through the language. We define such a language in terms of its class construct and its inheritance mechanism. The class construct allows for expressing also the properties of the structure represented by the class. By reasoning about the properties defined in classes (*class reasoning*) logical relations among the related mathematical structures can be evaluated. We show how class reasoning can be applied in order to rearrange and/or modify a hierarchy of (classes representing) mathematical structure. The rearranging operation is shown to be equivalence preserving, in the sense of *h*-equivalence, we define. The topics described in this paper are implemented in a prototype object-oriented programming environment, built on top of the CLOS language.

---

* This research has been partially supported by the Italian MURST Project: "Calcolo Algebrico, Sistemi di Manipolazione Algebrica".

** Istituto di Analisi dei Sistemi ed Informatica, C.N.R., Viale Manzoni, 30 - 00185 Roma, Italy, e-mail: cioni@iasi.rm.cnr.it

*** Presidenza del Consiglio dei Ministri, Dipartimento per l'Informatica e la Statistica, Via della Stamperia, 7 - 00187 Roma, Italy, e-mail: colagross@iasi.rm.cnr.it

† Dipartimento Informatica e Sistemistica - Università "La Sapienza", Via Salaria 113, I-00198 Roma, Italy, e-mail marte@dis.uniroma1.it

1.1 Axiomatization and hierarchization

It is well known that a mathematical system can be completely defined by a triple of elements: a nonempty set E of elements, a set O of relations and operations on E and a set A of axioms concerning the element of $E \cup O$. A mathematical system is classified as ***abstract*** if only the elements of A are defined, i.e. if properties of the system are defined while "structure" of the elements and operations on them might be undefined ([Des64] pag. 24). On the other hand a ***concrete system*** is a system, where all sets A, E and O are completely specified (properties and methods). In this paper we are interested in reasoning on such systems, specially on abstract systems. We consider a system defined by the process of axiomatization previously sketched. By reasoning we intend to derive new properties from an axiomatization, and to evaluate relationships holding among systems (mainly derivation and equivalence). Since we expect that both specification of systems and reasoning about systems must be performed by a software tool, we define a class construct allowing for the specification of a system. Moreover, we define a specialization inheritance discipline, to allow for expressing the derivation of a system from another system, by inheritance between the classes corresponding to the systems. A mathematical structure can be defined by ***axiomatization*** by giving the underlying sets, operations and axioms. An example of collection of mathematical structures defined by axiomatization is shown in Fig. 1.

Semigroup	**Monoid**	**Group**	**Cyclic Group**	**Abelian Group**
$, op	$, op	$, op	$, op	$, op
Axioms:	*Axioms:*	*Axioms:*	*Axioms:*	*Axioms:*
associativity	associativity	associativity	associativity	associativity
	unit	unit	unit	unit
		inverse	inverse	inverse
			xx=unit	commutativity

Fig. 1. A collection of algebraic structures

The same collection can be arranged in a hierarchy as shown in Fig. 2. In that figure, the derivation of a system from another is expressed by an intuitive notion of inheritance, such that the specification of a derived structure needs only the new features to be stated, assuming that everything else is inherited by the "ancestor"[1].

As a matter of facts, a programming language suitable for representing mathematical structures must support axiomatization. In particular, Fig.1 and Fig.2 show two possible requirements for the definition of such languages: in the first

[1] These specifications are given very informally, at the moment: syntax of the formulas (axioms) is explained in section 4. Let us assume that $ is a specification of elements of a set, giving the sort of interest for the system; *op* is a binary operator on $; by ***associative law***, the associative property of *op* over $ is intended; other properties, occurring in the examples, refer to *op* as well.

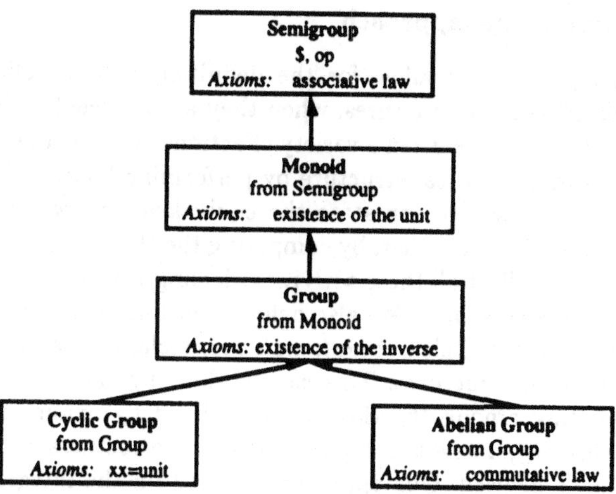

Fig. 2. An algebraic hierarchy

figure, the representation of mathematical structures requires a language supporting data abstraction via Abstract Data Types (ADT); in the second figure, an object-oriented language is required for expressing inheritance relationships. A variant of the latter approach is discussed in [Lit92], where basic abstract structures of formal algebra are defined incrementally via a suitable specialization inheritance mechanism. In Sec. 2 we define the basic features of a programming environment for the axiomatic specification of mathematical structures. This environment must allow to capture the natural hierarchization of mathematical systems. One of the main features must be the capability of reasoning about the properties derivable by the axiomatization and about the relationships among structures defined in a hierarchy.

There are several symbolic computation systems which allow for treating mathematical objects by means of ADT approach and use of hierarchization. In AXIOM [Jen92], specification of structures and methods at a high level of abstraction is allowed through the concepts of Category and Domain. Views [Abd86] exploits the same concepts of object-oriented paradigm, basically with the aim of ruling the use of some polymorphic functionality over compatible domains. In these systems some well known benefits of object-oriented methodology are accessible. Also the software environment of TASSO project [Lmt91] is based on the object-oriented programming approach. It aims to provide a tool suitable for the definition of systems for computer mathematics, by programming the axiomatization and manipulation of mathematical data structures. All those systems allow for modeling a mathematical structure via a class construct. Unfortunately the sets of axioms characterizing a given structure can not be fully exploited for reasoning.

1.2 Description of the approach

The aim of this paper is to allow for the definition and the effective use of properties of mathematical structures, when they are modeled by hierarchies of classes. By "effective use" we mean a variety of actions: we can get new properties of the underlined mathematical structure by performing logical inferences from axioms defined in a class; we can state the equivalence of two systems or the logical derivation one from another, by comparing the theories defined by related axiomatizations. We collect all these activities, dealing with properties of classes, under the name *class reasoning*. We also define a peculiar use of class reasoning, as the possibility of making clearer (*rationalizing*) a class hierarchy that models a set of mathematical structures. This can be done by checking possible logic relationships existing among the theories underlined by axioms specified into the classes. In this way, a new tool for verifying the correctness of mathematical software systems is devised. Our approach is based on the integration of logic features into an object-oriented language suitable for symbolic computation. We define the language in terms of its class construct and its inheritance mechanism. The integration of logic capabilities is obtained by class reasoning. Thus, our approach differs from those that aim to build a single language for achieving both logic programming and object-oriented features, such as, for example, the one proposed in [Dem95].

2 Basic definitions of the programming language

In this section we show a class construct suitable to support class reasoning in an object-oriented language. Through such class construct, and the related inheritance, we aim to provide an expressive mechanism for the definition of hierarchies of mathematical systems by specialization.

The class construct is based on a generic object-oriented language, but refers to the specialization inheritance mechanism called ESI (*Enhanced Strict Inheritance*, [Dbt95]). Here we enlarge the classical notion of class and the ESI itself, to deal with properties collected within the class definition. We call ESI_P the new inheritance we define.

In the following definition of class structure, *inheritance list* stands for the set of direct superclasses; the declaration $v_j : C_j$ denotes an instance variable v_j of class C_j; $m_i\ (p_i : P_i) : R_i$ denotes a method signature composed of an argument p_i of class P_i and a result class R_i; $A_i; \cdots; A_k$ stand for the axioms defining properties of the instances of the class (they are just symbols, here, representing formulas of the first order logic: further details will be given in section 4).

Definition 1 class structure.
<u>class</u> *name* {inheritance list}
$v_1 : C_1; \cdots; v_m : C_m;$
$m_1(p_1 : P_1) : R_1$ <u>is</u> $\{m_1$ body$\}$; ... $m_n(p_n : P_n) : R_n$ <u>is</u> $\{m_n$ body$\}$;
$A_1; \cdots; A_k;$
<u>endclass</u>

We want to define the specialization inheritance for managing the defined class construct; before doing that, we have to state and clarify the usage we preview for axioms.

In general a property can be represented as a 4-tuple: $< ID, VL, FL, B >$ where: ID is the identifier of the property (property name); VL is the list of all the typed variables which the property is concerning with; FL is the list of all the functions (corresponding to method invocation) called in the subterms of the body B; B, the body of property, is a wff in a suitable logic language.

In a hierarchy of abstract classes, each property name is unique, that is there are not two properties with the same property name.

Properties can be inherited through classes, as well as other attributes do. So we can distinguish between two kinds of properties in the axiomatization given by a class: we call *axiom* property, each formula explicitly included in the class definition, and we call *inherited* properties the ones defined in a superclass of the class[2]. Finally we represent by *pr*(C) = *ax*(C)∪*inh*(C) the set of all properties involved in the definition of C, where *ax*(C) stands for the set of all axioms defined in C and *inh*(C) stands for the set of all properties inherited by C, since defined in its superclasses. Let us note that *pr*(C) is a suitable representation for the whole theory built up by the properties holding for the mathematical system defined by the class C.

ESI [Dbt95] is equal to strict inheritance plus covariant redefinition. So we have to present the notion of covariant redefinition before of stating the definition of ESI_P. Notice that by *redefinition* we intend the possibility for some instance variables or methods defined in a class of having another explicit definition in a subclass. Note that redefinition is not allowed for the axiom properties. The essence of covariance is that an inherited attribute can have a redefinition, provided that the new definition narrows the older one[3]. So the multiple inheritance case is considered as stated in [Dbt95].

In the following definition, the notation C' *is subclass* of C is used to mean that either $C' = C$ or C' is defined by inheriting from C ($C' \leq_{ESI_P} C$).

Definition 2 covariant redefinition. If C' is subclass of C:

1. the redefinition $v : C'_v$ in C', of the variable $v : C_v$ defined in C, is a covariant redefinition iff C'_v is subclass of C_v.
2. the redefinition $m(p : P') : R'$ in C', of the method $m(p : P) : R$ defined in C, is a covariant redefinition iff P' is subclass of P and R' is subclass of R.

The following relation, $\leq_{ESI_P}$, extends the ESI definition, by introducing the conditions 4. and 5. It allows subclassing only in those cases when the subclass is a specialization of the superclass also in the sense that the properties of subclass instances imply the properties of instances of the superclass. Actually the test for correctness of subclassing consists in ensuring that no inconsistency is introduced by the new axioms w.r.t. the inherited properties.

---

[2] So, inherited properties join the subclass axioms by inheritance.

[3] In principle, by starting from a class, we can define a specialization by adding attributes and/or narrowing (specializing by redefinition) attributes.

Definition 3 $\leq_{ESI_P}$. Given a class C' and a set of classes $\{C_1,\ldots,C_n\}$, with $n \geq 1$,

$C' \succeq_{ESI_P} \{C_1,\ldots,C_n\}$ iff either one of following cases 1. or 2. or 3. does hold,

1. $C' \leq_{ESI} C'$;
2. if n=1 and $C' \neq C_1$, then $C' \leq_{ESI} C_1$ iff C' inherits every attribute of C_1, possibly redefining some of them by covariant redefinition;
3. if n>1, then $C' \leq_{ESI} \{C_1,\ldots,C_n\}$ iff C' inherits every method of $C_1,\ldots,C_n$ possibly redefining some of them by covariant redefinition w.r.t C_i $\forall i \in I$, and C' inherits all the variable attributes defined in the classes $C_1,\ldots,C_n$ ensuring the following constraint: if v is an attribute belonging to the interface of more than one class, say the $\{C_j\}$ with $j \in J \subset I$, then either v is redefined in C' ensuring the covariant redefinition w.r.t. C_j $\forall j \in J$, or a variable $v : V_k$ (for some $k \in J$) is inherited such that $\forall j \in J$ $V_k \leq_{ESI} V_j$ (note that V_j is the class declared for v in C_j).

and the following conditions 4. and 5. are verified

4. $(\forall i \quad \forall p \in pr(C_i)) \quad \bigcup_i pr(C_i) \not\vdash \neg p$
5. $(\forall p \in ax(C')) \quad inh(C') \not\vdash \neg p$ [4]

If one of the first three conditions stated in Definition 3, together with the 4. and 5. do hold, then the class C' owns the properties of C besides those defined in it, and C' can correctly be considered more specialized than C. This is expressed by the following theorem.

Theorem 4 specialization. *If $C' \leq_{ESI_P} C$, then C' is a specialization of C; in particular $pr(C') \Rightarrow pr(C)$.*

To prove the theorem, we see that C' has to contain all the instance variables and methods (possibly specialized by covariant redefinition) defined in C, since $C' \leq_{ESI_P} C$. Regarding properties, we can observe that $pr(C') \Rightarrow pr(C)$, since $pr(C') = inh(C') \cup ax(C')$ and $inh(C') = pr(C)$. Then $pr(C')$ strengthens $pr(C)$ with the axioms $ax(C')$. So we can call C' a structural and behavioural specialization of C.

Let us note that variable attribute is not relevant in our considerations, because only abstract classes are considered. But they are included in our formalization for future development of this work, in the direction of considering instances of abstract classes at the run-time. In an object-oriented programming language the relation between classes $\leq_{ESI_P}$ must be verified by type checking. Type checking must be enhanced by a deduction tool. Such a tool will verify, for instance, that the commutative law defined in Abelian Group is not inconsistent with all the properties inherited from the superclasses of Abelian Group. We preview this check be performed by a deduction tool called automatically within the language during the specification phase. We meet this aspect in section 4.

[4] That is, inconsistencies between the axiom properties of C', $ax(C')$, and the inherited ones, $inh(C')$, must be prevented.

3 Static class reasoning

Every class C belonging to a hierarchy H models a mathematical structure whose theory is given by $pr(C)$. Of course, the same structure can be modeled by another class, C', included in a different hierarchy, H', provided that $pr(C) \equiv pr(C')$. We call *h-equivalent* two hierarchies such that for each class C in H at least a class C' in H' does exist, such that $pr(C) \equiv pr(C')$. Then, class reasoning is the process that allows for detecting the relationships between properties of abstract classes, providing techniques for rearranging the hierarchy to which the classes belong. Since class reasoning takes into account only the properties of the classes, the operators and the theorem we will present can be applied whenever the classes involved are homogeneous with respect to the definition of data and method attributes.

Different hierarchies might be differently preferable, depending on the characteristics of the specific application: bigger expressivity can be particularly crucial in tutorial applications, while in other cases a specific readability must be pursued. So, in some situations a hierarchy which expresses all the possible relationships among different classes could be preferably represented by a graph with many arcs connecting the classes. In other cases a linearized hierarchy could be more adequate to model the specific application. These considerations lead us to evaluate as an important accomplishment to rearrange a given hierarchy in order to make it suitable in some respects to the needs of the user.

Given a hierarchy H the following four operators can be applied to its classes, in order to obtain another hierarchy H' which is h-equivalent to H.

Definition 5 rearranging operators.

Elimination Operator ε

Let H be a hierarchy, and A and B two classes in H such that:

- A and B are homogeneous with respect to data and method attributes;
- A is an ancestor of B;
- $inh(B) \cap ax(B) \equiv ax(B)$;
- $pr(A) \equiv pr(B)$;

then, the application $\varepsilon(H, A, B)$ of the operator ε to the hierarchy H and its classes A and B, returns a new hierarchy H' obtained from H eliminating the class B and attaching all the direct descendants of B to the direct ancestors of B.

Simplification Operator σ

Let H be a hierarchy, and A and B two classes in H such that:

- A and B are homogeneous with respect to data and method attributes;
- A is an ancestor of B;
- $ax(A) \cap ax(B) \neq \emptyset$;
- $pr(A) \not\equiv pr(B)$;

then, the application $\sigma(H, A, B)$ of the operator σ to the hierarchy H and its classes A and B, returns a new hierarchy H' obtained from H by replacing the set $ax(B)$ of the class B with $ax(B) - ax(A) \cap ax(B)$.

Collapsing Operator γ

Let H be a hierarchy and A and B two classes in H such that:

- A and B are homogeneous with respect to data and method attributes;
- $pr(A) \equiv pr(B)$;

then, the application $\gamma(H, A, B)$ of the operator γ to the hierarchy H and its classes A and B, returns a new hierarchy H' obtained from H by collapsing the class A and B in a new class C such that:

1. A and B are homogeneous with respect to data and method attributes;
2. $ax(C) = ax(A) \cup ax(B)$.

Linearizing Operator λ

Let H be a hierarchy and A and B two classes in H such that:

- C is homogeneous with A and B with respect to data and method attributes;
- A and B have a same direct ancestor
- $pr(A) \Rightarrow pr(B)$

then, the application $\lambda(H, A, B)$ of the operator λ to the hierarchy H and its classes A and B, returns a new hierarchy H' which differs from H only because A and B are arranged in a linear structure, where B comes first A.

Theorem 6 rearrangement of the hierarchies. *Let H be a hierarchy and let A and B be two of its classes. Let H' be the hierarchy obtained by applying whatever of the operators $\varepsilon, \sigma, \gamma, \lambda$ to H, A and B. The hierarchies H and H' are* h-*equivalent.*

In order to prove the theorem, we consider the application of each operator separately.

- ε The hierarchies H and H' are h-equivalent because H' has all the classes of H except the class B, and by hypothesis $pr(B) \equiv pr(A)$.
- γ The hierarchies H and H' differs for having collapsed A and B of H in a new class C of H', and by hypothesis $pr(A) \equiv pr(C)$ and $pr(B) \equiv pr(C)$, so H and H' are h-equivalent.
- σ The application of σ produces a hierarchy H' which is the same as H except for B has been replaced by a new class B' such that $pr(B') \equiv pr(B)$, and then H and H' are h-equivalent.
- λ Let C be the direct ancestor of A and B in H; then $\lambda(H, A, B)$ returns a hierarchy H' obtained from H, where C is the direct ancestor of B and B is the direct ancestor of A. Let us rename A and B in H' as A' and B', respectively. For H and H' to be h-equivalent $pr(A) \equiv pr(A')$ and $pr(B) \equiv pr(B')$ must hold. It is $pr(B') \equiv pr(B)$, since B' a direct descendant of C, as well as B in H. Moreover, by definition, it is $pr(A) = Pr(C) \cup ax(A)$ and $pr(A') = Pr(B') \cup ax(A')$, where $pr(B') = Pr(B) = Pr(C) \cup ax(B)$. So we can write $pr(A') = Pr(C) \cup ax(B) \cup ax(A') = Pr(A) \cup ax(B) \equiv Pr(A)$ (the equivalence id due to the hypothesis of λ-applicability: $pr(A) \Rightarrow pr(B)$).

In the following we show two examples of manipulation of hierarchies of abstract classes by application of some of the defined operators. The examples come from elementary algebra; the class names are self-explanatory.

Example 1

The Collapsing Operator γ is used through Fig. 3. It can be applied to the hierarchy in part (a) ($pr(Group - left) \equiv pr(Group - right)$ does hold); so the classes **Group-left** and **Group-right** collapse in a new class **Group** and the hierarchy in part (b) comes out.

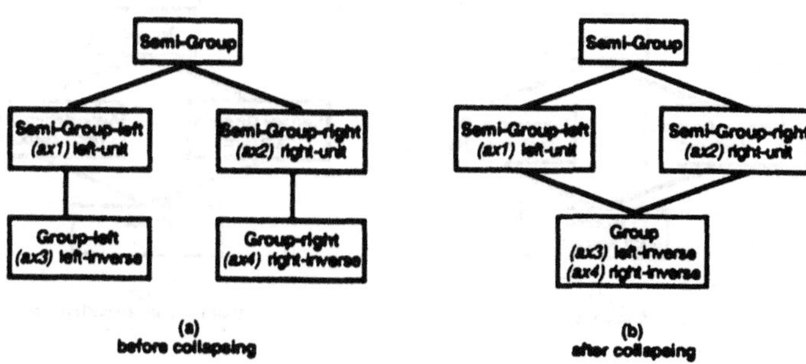

Fig. 3. Collapsing

Example 2

Here we show the collapsing operator γ and the simplification operator σ working in combination. Let be the hierarchy of Fig. 4(a) given. The conditions for application of γ are verified, so the hierarchy is modified to become as in Fig. 4(b), where the collapsed class is the **Monoid**.

The new hierarchy verifies the conditions for the application of the Simplification Operator σ to the class **Monoid** and one out of **Unitary-Groupoid** and **Associative-Groupoid**. Once we first apply σ to **Monoid** and **Unitary-Groupoid** we get the modification of class **Monoid** as in Fig. 4(c). Then we can apply σ once more to the class **Monoid**, so far modified, and the class **Associative-Groupoid**: this operation modifies the class **Monoid** by eliminating the remaining properties (the last, in fact), producing the hierarchy in Fig. 4(d).

4 Defining and manipulating properties

Many different languages can be suitable for representing classes including properties. In this paper we consider a prototypal implementation of an object-oriented language we are developing starting from the CLOS [Bdg88]. This CLOS extension features the class structure described in Sec. 2, managing the ESI_P. It is integrated by an automated deduction tool based on sequent calculus. A preprocessor which builds for each class C the sets $inh(C)$ and $ax(C)$ and, from them, a sequent to be processed by the sequent calculus is also provided.

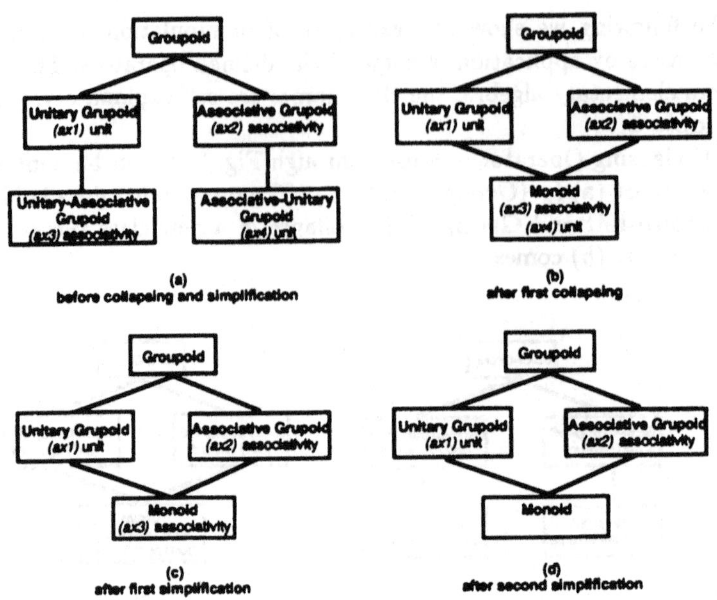

Fig. 4. Collapsing and Simplification

In order to clarify the elements of the language concerning the definition and manipulation of the properties, we outline that the bodies of the axioms are formulas of First Order Logic with equality, containing references to typed variables and methods of the class. So, for example, $\forall x, y : \$ \; op(x, y) = op(y, x)$ is a way to say the operator $op : \; \$ \times \$ \longrightarrow \$$ is commutative. The bodies of the properties of the classes, being part of the class attribute, have the following characteristics:

- variable symbols represent typed variables and the type checking of the language verifies whether they are correct w.r.t. the ESI_P;
- properties contain method invocations; the type consistency of such invocations and of the occurrences of variable symbols is checked by the type checking;
- properties contain relational operators such as $=$, $<$, $>$.

Each property is tested by the type checking of the language, together with the other attributes of the class. Moreover, in order to guarantee the correctness of the whole specification a deduction tool is necessary. This tool has to check the different cases of the Def. 3. In particular it has to verify that the commutative law is *not inconsistent* with all properties inherited by the class Abelian Group, that is those specified in the class Group, in Monoid and in Semigroup. Every deduction tools defined for First Order Logic can be used for this purpose. In particular natural deduction tools are specially suitable for this purpose, since they have the characteristics of being flexible, self explanatory and friendly. In the prototypal implementation of our language we adopt the sequent calculus described in [Ccm95]. We have an implementation of such calculus, built in plain CLOS, whose proof process can be monitored and controlled by the user. Moreover, the deduction tree generated during the process of verification by sequent calculus allows for easy distinguishing between the cases of inconsistency of $ax(C')$ of Def. 3 and $pr(C')$. In order to apply the mentioned deduction tool, all the properties of the classes, written in the previously mentioned language, are transformed in a sequent form by a preprocessor, putting in the Antecedent of the sequent the conjunction of the inherited properties and in the Succedent the negation of the properties defined in the class.

Let us consider as an example the hierarchy given in Fig. 2. The properties of its classes are expressed in the following form:

$class\ Semigroup(\$)\quad \forall x,y,z:\$\ op(x,op(y,z))=op(op(x,y),z)$

$class\ Monoid\ inherit\ Semigroup\quad \exists u:\$\ \forall x:\$\quad op(x,u)=x \quad\wedge\quad op(u,x)=x$

$class\ Group\ inherit\ Monoid\quad \forall x:\$\ \exists xi:\$\quad op(xi,x)=u \quad\wedge\quad op(x,xi)=u$

$class\ Cyclic-Group\ inherit\ Group\quad \forall x:\$\quad op(x,x)=u$

$class\ Abelian-Group\ inherit\ Group\quad \forall x,y:\$\quad op(x,y)=op(y,x)$

The preprocessor transforms such properties in a sequent that can be manipulated by the automatic proof system based on sequent calculus [Ccm95]. Our language prototype provides two functions ax and inh available to the user: given a class C, they return the sets $ax(C)$ and $inh(C)$. In the previous example, the following sequent is built: $pr(Cyclic-Group) \Rightarrow pr(Abelian-Group)$; then it can be submitted to the proof system; it is valid, so the hierarchy of Fig. 2 becomes the following linearized one:

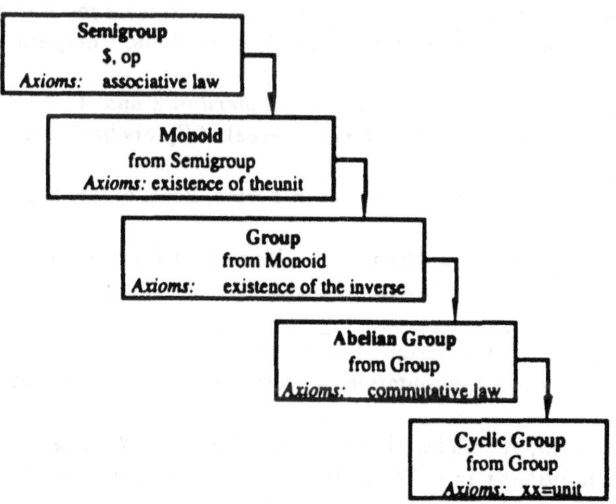

Fig. 5. A linearized collection of algebraic structures

5 Conclusions

In this paper we have discussed some features a language for symbolic computation should have in order to deal with logic properties of mathematical data structures.

We have shown how axiomatization of mathematical structures can be supported by an object-oriented language equipped with suitable class construct and inheritance. In particular a class construct including the property attributes, has been presented together with the definition of ESI_P. By means of such inheritance mechanism the properties belonging to a given class are defined as the union of two sets of properties (*axiom* and *inherited* properties. All such properties are expressed by a first order language whose terms include references to typed variables and method invocations. We have just sketched the prototypal implementation of the programming environment we are working on.

We have focussed on class reasoning, i.e. the logical activity performed on properties defined into classes. We showed one out of the possible applications of class reasoning: transforming class hierarchies into equivalent ones. In future work we expect to further clarify and enlarge the applicability of class reasoning.

6 References

[Abd86] S.K. Abdali, G.W. Cherry, N. Soiffer, *An Object-Oriented Approach to Algebra System Design*, in B.W. Char (Ed.) Proc. of ACM SYMSAC '86, 1986.

[Bdg88] D. G. Bobrow, L. G. DeMichiel, R. P. Gabriel, S. E. Keeneand, G. Kiczales, D. A. Moon, *Common Lisp Object System Specification*, Techn. Rep. X3J13 doc. 88-002R, 1988.

[Ccm95] G. Cioni, A. Colagrossi, A. Miola, *A Sequent Calculus for Automated Reasoning in Symbolic Computation Systems*, Journal of Symbolic Computation, **19**, (1-3), 175-199, 1995.

[Dbt95] P. Di Blasio, M. Temperini, *Subtyping Inheritance and Its Application in Languages for Symbolic Computation Systems*, Journal of Symbolic Computation, **19**, (1-3), 39-63, 1995.

[Dem95] G. Delzanno, M. Martelli, *Objects in Forum*, Proc. of Int. Logic Programming Symposium, 1995.

[Des64] W.E. Deskins, *Abstract Algebra*, The Macmillan Co. New York, 1964.

[Jen92] R.D. Jenks, R.S. Sutor, *Axiom, the scientific computation*, Springer Verlag, New York, 1992.

[Lit92] C. Limongelli, M. Temperini, *Abstract Specification of Structures and Methods in Symbolic Mathematical Computation*, Theoretical Computer Science, 104, pp. 89-107, 1992.

[Lmt91] C. Limongelli, A. Miola, M. Temperini, *Design and Implementation of Symbolic Computation Systems*, in P. W. Gaffney, E.Houstis (Eds) Proc. IFIP TC2/WG2.5 Working Conference on Programming Environments for High Level Scientific Problem Solving, Elsevier Scientific Publ., 1991.

An Intelligent Interface to Numerical Routines*

Brian J. Dupée and James H. Davenport **

School of Mathematical Sciences University of Bath
Claverton Down Bath BA2 7AY United Kingdom

Abstract. Links from Computer Algebra Systems to Numerical Libraries have been increasingly made available. However, they remain, like the numerical routines which comprise these libraries, difficult to use by a novice and there is little help in choosing the appropriate routine for any given problem, should there be a choice.
Computer Algebra Systems use generic names for each problem area. For example, 'integrate' (or 'int') is used for integration of a function, whatever method the code may use. Numeric interfaces still use different names for each method together with a variety of extra parameters, some of which may be optional. Ideally, we should extend the generic name structure to cover numerical routines. This would then, necessarily, require algorithms for making an assessment of the efficacy of different methods where such a choice exists.
This paper considers the link to the NAG Fortran Library from version 2.0 of Axiom and shows how we can build on this to extend and simplify the interface using an expert system for choosing and using the numerical routines.

1 Introduction

There exist, increasingly, links from many Computer Algebra packages to numerical libraries enabling the use of such an invaluable wealth of experience and expertise encapsulated within the numerical routines by users either not familiar with Fortran or preferring the more modern GUIs of CA systems. The success of Mathematica, Maple, Macsyma, Axiom and so on has been largely due to the provision of an interactive environment where the methodologies employed by the system, which may embody complex mathematical techniques, are entirely hidden from the user.

To perform an integration, Axiom and Macsyma have the function '*integrate*'; Maple, Mathematica and Reduce use '*int*', which are generic names to a variety of functions all of which can do integration on a subset of problems. But this is not the case when we look at the interfaces to the library routines. In general, the numerical routine is called using the same name as the library name. So, for

* The project "More Intelligent Delivery of Numerical Analysis to a Wider Audience" is funded by the UK Govt. Joint Information Systems Committee under their New Technologies Initiative (NTI-24)

** Email: {bjd,jhd}@maths.bath.ac.uk

example, the NAG numerical integration routine **D01AJF** is called using the name 'd01ajf' by all of the systems with links to the NAG Library. To use a different routine, one must change the name of the called external function, and perhaps supply different parameters, or a different order to the parameters. The method is thus no longer transparent to the user. Indeed the user will, and must, know a great deal about the implementation and inner workings of such a routine in order to understand both the input parameters and the results.

However, given that there exists a choice of routines, deciding amongst the several approaches may, or may not, be an easy task for the user. Different attributes of the problem must be identified and their possible interactions weighed up before a final decision can be made. Most of this analysis may be achieved using the computer algebra system itself.

So what we wish to achieve is a generic interface to all routines of the same problem domain, preferably consistent with the equivalent within the CA system. For example, given an integral to calculate, and the end points of integration, the system should make the choice of whichever routine would best be used to achieve a desired accuracy.

2 Axiom & the NAG Link

The fundamental operation of the Axiom-NAG[3] link is described by Dewar (1994). From the (curious) user's point of view, a command such as

```
d01ajf(-1 ,1 ,0.0 ,1.0e-4 ,800 ,200 ,-1 ,(4/(1+X^2)::Expression Float)
       ::ASP1(F))
```

performs the following tasks.

1. Suitable values are given to all the non-data[4] parameters that have not been specified — in this case just the *output only* parameters **RESULT, ABSERR, W** and **IW** (NAG 1993). The HyperDoc interface goes a little further in this respect by automatic selection of **LIW**. (see Fig. 1)
2. The formula $4/(1+X^2)$ is converted into a Fortran subroutine which will be an argument to **d01ajf**, and this ASP (Argument SubProgram) must therefore be *of the kind required by* **d01ajf** — see Davenport et al. (1991) for a detailed description of the difficulties here.

[3] Bath and NAG Ltd. are grateful for the support provided by the UK's Teaching Company Scheme for the development of this link.

[4] Axiom follows the IRENA description in Dewar (1991) of NAG routine parameters as being: **data** — those that describe the actual problem to be solved (for example, it makes no sense to say "integrate" numerically without specifying the integrand and the range); **algorithmic control** such as error tolerances, the treatment of errors (hard or soft fail), bounds on the number of iterations etc. and **housekeeping** such as the dimensions of arrays, parameters required by the exigencies of Fortran, and so on.

3. Any additional ASPs representing, for example, derivatives of functions, or Jacobians of systems, are written automatically, using symbolic differentiation.
4. An appropriate Fortran main program is written, which may need knowledge of the machine precision etc. of the target machine.
5. This program and the ASP are passed to a program called the NAG daemon, potentially on a remote machine with a different architecture[5], which:
   - compiles them (if there are no suitable compiled versions from a previous attempt);
   - links them with a suitable main program in C and the NAG Fortran library;
   - passes the results, or an error indication if that happened (we are using C's error-trapping to detect hard errors in the Fortran) back to Axiom.
6. Converts the results into an appropriate Axiom data structure, which will contain not only the "mathematical" result, i.e. the integral in this case, but also any indication of numerical accuracy, and other information.

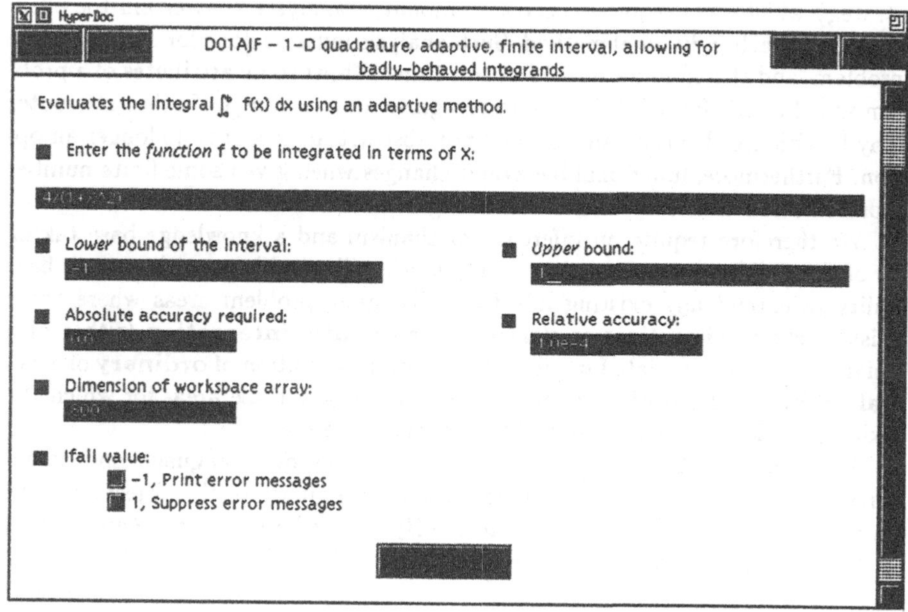

Fig. 1. NAG Routine D01AJF HyperDoc Input Page

It is clear that this technology has done a lot for us: we no longer need to write and debug Fortran; machine dependencies have been coped with (XDR is

[5] The Naglink allows the numerical code to be run on a remote machine which, if parallel, would automatically use parallel processing

used for transferring results); any differentiation required is performed for us; a somewhat impenetrable 12-volume manual has been replaced by defaulting and command-line keywords (and indeed also by Axiom's HyperDoc pages, which provide a "fill in the boxes" interface); and the results are presented interactively. In fact one more thing has been done: the Axiom interface, like IRENA (Davenport et al. 1991) does a great deal to regularise the parameterisation and naming of the parameters of the NAG routines: for example absolute errors are always called **abserr** and rectangles are represented consistently, including the treatment of unbounded rectangles. Furthermore, the functionality of the link is somewhat greater than that of the individual subroutines of the NAG library: where multiple routines need to be used to perform a mathematically useful result, a Fortran program (called a 'jacket') has been written to perform this linking (Richardson 1996).

3 The Expert System – ANNA

The existence of a choice of numerical methods may necessitate some overall strategy which then requires a certain amount of analysis to ascertain the effectiveness of each individual method. Such analysis must be done for each specified problem, and therefore performed at run-time. Whilst some attributes of a problem may be easy for a CA system to recognise, measurement of other attributes may be either difficult or impossible since abstract argument is no longer an option. Furthermore, functional behaviour changes when given some finite number representation necessary for numerical algorithms.

We therefore require an inference mechanism and a knowledge base taking all of these difficulties into account and which will provide us with enough flexibility to extend any existing interface. The main problem areas where there exists a choice of routines or strategies are **definite integration** (either of a univariate or multivariate function), the numerical solution of **ordinary** or **partial differential equations** and **optimization**. It is these areas for which the Axiom/NAG Numerical Analyst (ANNA) has been created.

The code for ANNA is written using Axiom's own language and structures – Categories, Domains and Packages using as much of the object-oriented paradigm as we require. Extensions to the Hypertext based interface are written in "boot" code, an interface to the underlying LISP.

3.1 Knowledge Base

A table of available routines and their application areas has been created. This also includes such details as an indication whereby, if a routine has already been found which has a very good likelihood of being able to be used efficiently, a lazy evaluation mechanism is triggered. This table can also include details of the **IFAIL** values and indications of preferable fall-back strategies.

Further tables are used which contain dynamic information such as that knowledge gained about the attributes of the problem currently under investigation. This will be used as part of the inference machine and also as part of the explanation mechanism.

3.2 Inference Machine

The inference machine must be able to consider all of the available methods within its knowledge base which might be addressed to the current problem. In doing so it analyses the input problem to find out about any attributes that could affect the ability of the methods under consideration to perform effectively.

We would consider the most effective routine as one which would perform the task to the required accuracy using the least resources i.e. least computational expense. Since some of this analysis may, in itself, be expensive, we might under certain circumstances waive this explicit requirement in favour of one whereby we require a routine which is *likely* to be inexpensive.

An Axiom domain is created for each method or strategy for solving the problem. These method domains each implement two functions with a uniform (method independent) interface:

measure: A function which calculates an estimate of suitability of this particular method to the problem if there is a possibility that the method under consideration is more appropriate than one already investigated.
If it may be possible to improve on the current favourite method, the function will call computational agents to analyse the problem for specific features and calculate the measure from the results these agents return.

implementation: A function which may be one of two distinct kinds. The first kind uses the interface to the NAG Library to call a particular routine with the required parameters. Some of the parameters may need to be calculated from the data provided before the external function call.
The other kind will apply a "high level" strategy to try to solve the problem e.g. a transformation of an expression from one that is difficult to solve to one which is easier, or a splitting of the problem into several more easily solvable parts. This may thus enforce some recursion on the *measure* function.

Computational Agents Computational agents investigate the attributes of the input function or functions, such as *Is this function continuous?* and will usually be functions or programs written completely in the Axiom language and implemented using computer algebra. Some agents will be common to more than one problem domain whereas others will be specific to a single domain. They may also vary greatly in their complexity. It will be a simple task to return details about the range of a function since this information will have been included in the problem specification. It is a different order of complexity to return details of its singularity structure.

Each of these computational agents which may be called by a number of method domains retain their output in a dynamic hash-table, so speeding the process and retaining efficiency.

Let us take, for example, the search for the singularity structure of some function. There are two types of test for continuity required reflecting the definitions of continuity within a given range and at a point. The first is a general search for all singularities within (interior to) the range of definition of the input function. A workable strategy for dealing with functions with multiple singularities is to split the function at those points and use each segment separately. In this case we need to know not only that such singularities exist, but also where in the range of definition they are.

For the most part we use a set of rules to recursively split the function and a look-up table.

Example 1. Let a, b be elementary functions, i.e. real functions defined using a finite number of algebraic, polynomial, logarithmic, exponential and trigonometric operations, and Ψ, Φ be operations such that $\Psi(\alpha) =$ the list of singularities of the elementary function α and $\Phi(\gamma)$ = the list of zeros of the elementary function γ.

$$\begin{aligned}
\Psi(ab) &\subseteq \Psi(a) \cup \Psi(b) \\
\Psi(\frac{1}{a}) &= \Phi(a) \\
\Psi(e^a) &= \Psi(a) \\
\Psi(a+b) &\subseteq \Psi(a) \cup \Psi(b) \\
\Psi(\log a) &= \Phi(a) \\
\Phi(ab) &\subseteq \Phi(a) \cup \Phi(b) \cup \Psi(ab) \\
\Phi(\frac{1}{a}) &= \Psi(a)
\end{aligned}$$

These form a set of production rules which can be used in the search for singularities as well as those of more specific nature such as trigonometric expressions for which the look-up table is used. Since we are being conservative, we can replace the $\subseteq$ by $=$ above. This may overestimate the number of singularities, including the possibility of indeterminate forms, but such points can be investigated using series methods should the need arise.

There remains the significant problem of finding the zeros of expressions of the form $a + b$. If a is constant, the system reduces so we only need to find solutions of the form $-b/a = 1$ and either the look-up table or inverse functions can be used.

Even easier is the case where $a + b$ is polynomial for which there exist algorithms using, say, Sturm sequences (Collins & Loos 1983; Mignotte 1992). The difficulty arises when we have non-trivial expressions a and b. To find the zeros of expressions $a + b$ we will revert to power series methods, in particular the Padé Approximant, and solve the numerator polynomial (Baker & Graves-Morris 1981).

The second need for a test of continuity is allied to the search for algebraic or logarithmic end-point singularities for which routines such as the NAG routine

D01APF have been designed i.e. singularities of the form

$$f(x) = (x-a)^c(b-x)^d \log(x-a)^m \log(b-x)^n g(x) \quad s.t. \quad x \in [a,b].$$

Here we use power series methods. If the function is differentiable, it is necessarily continuous, so if a Taylor Series exists at the particular end-point, continuity is assured. If we can find a Laurent Series or a Puiseux Series (a power series with powers of fractional degree) and the leading term is negative, the function is not continuous at that point. For essential singularities, we use exponential series and test the limits as we approach the end-point.

Measure Functions We have taken for our model the system recommended by Lucks & Gladwell (1992) which uses a system of measurement of compatibility allowing for interaction and conflict between a number of attributes. All of these processes may not be required if the choice is clear-cut e.g. we have an integral to calculate and it has a particular singularity structure for which one particular method has been specifically created. However, for more difficult cases a composite picture should be built up to calculate a true measurement.

How the compatibility functions interpret the measurements of various attributes is up to them and may vary between differing methods. It is this area that takes as its basis the *judgement* of Numerical Analysis "experts" whether that be from the documentation, which may be deficient in certain respects, or from alternative sources[6]. However, its assessment of the suitability or otherwise of a particular method to a particular problem is reflected in a single normalised value facilitating the direct comparison of the suitability of a number of possible methods.

Composite Algorithms Sometimes we can be faced with the problem that either symbolic algebra or numerical methods on their own cannot efficiently give an answer to a particular analysis question as put by a computational agent. Here we can consider algorithms which contain aspects of both paradigms.

For example, one of the major complications of computing solutions of a system of ordinary differential equations is the problem of stiffness. Each system of differential equations has an associated Jacobian matrix which can be evaluated at, or near, the initial values. The eigenvalues of such a matrix give an indication of the stiffness of such a frozen system (Lambert 1973, pp 228–236). It is assumed in many models of dynamic systems that the stiffness ratios are constant throughout the range. Even where this is not the case and the Jacobian is non-linear, some indication of the degree of stiffness is given by this method.[7] In more arbitrary systems one might have to use alternative methods (Decker & Verwer 1984, pp 10–12).

---

[6] The IRENA project is grateful to the staff of NAG's Numerical Library Division for their assistance in the analysis of alternative methods

[7] This could be alleviated by the addition of a modified LSODE method to the NAG Foundation Library together with the appropriate link

We therefore have a computational agent *StiffnessFactor* which calculates symbolically the Jacobian matrix and its real eigenvalues. Since symbolic methods can be expensive in finding complex eigenvalues, the agent uses a technique that would seem perverse if performed manually. Should the number of real eigenvalues not equal the dimension of the Jacobian, the algorithm calls on the appropriate NAG Fortran Library routine to calculate them numerically. This can be considered since, once the link to the NAG Library is in place, the computational expense is not as high as would be either to create the code manually or to continue to use symbolic methods. This, in one way, can be thought of as a bizarre form of recursivity — creating a Fortran program to find out which form of Fortran program should be created to solve the problem. However, it is a useful addition to the program's armoury.

At the same time as investigating the stiffness of the ODE, it calculates a stability factor which is the proximity of this complex eigenvalue to the imaginary axis. The effect of this proximity is a system with a large sine or cosine factor in its solution. This can also have a detrimental effect on certain routines, in particular those implementing the BDF method and, to a smaller extent, the Adams method.

4 The Generic Interface

Once these mechanisms are in place the command line interface can be made generic, i.e. the integration

$$\int_0^2 \frac{\log(2-x)\ \log x}{(2-x)^{\frac{2}{3}}\ \sqrt{x}}\ dx$$

is input as

```
integrate((log(2-x)*log(x))/((2-x)^(2/3)*sqrt(x)), 0.0..2.0, 1.0e-6)
```

or

```
integrate((log(2-x)*log(x))/((2-x)^(2/3)*sqrt(x)), 0.0..2.0)
```

and the system will automatically choose the correct numerical method to achieve the required result. The corresponding HyperDoc input page is shown in Fig.2. To obtain solutions to sets of ODE's or PDE's the command is **solve** and minimization of a function or set of functions is achieved using the single command **optimize**.

4.1 Default Parameters

Some of the algorithmic control parameters for the NAG routines and all of the housekeeping parameters can be given default values. These include array dimensions (which can be calculated from the input arrays), workspace (which is often a function of the size of the problem), iteration limits and parameters controlling output messages. Derivatives and Jacobians can also be calculated symbolically and input automatically.

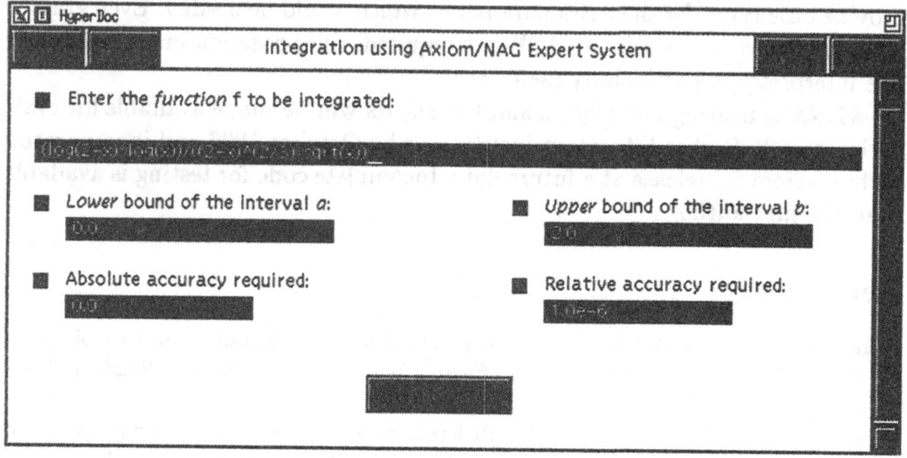

Fig. 2. ANNA Integrate HyperDoc Input Page

4.2 Optional Parameters

Other parameters can be made optional, such as tolerances, whereby if the error tolerance is included on the command line, that value is used. Otherwise some default is used. Some routines allow for absolute error tolerances as well as relative error tolerances. One or both of these could be set by default.

Where we would consider that a parameter, or group of parameters, should be optional, we create a number of overloaded functions. Those that set these parameters by default use either table look-up or explicit assignment.

4.3 Automatic Coercion of ASPs

The Fortran code and subsequently the ASPs (see §2) are automatically created within the Axiom Domains from the input parameters for use by the link to the NAG library (Keady & Nolan 1994).

5 Conclusion

The new interface has two major benefits — one obvious and the other perhaps a little surprising. It is evident that the interface is easier to use, both in the lack of extraneous parameter requirements and the generic name for a number of possible methods. This is particularly noticeable in the optimization chapter where the routine **E04UCF** requires 43 different parameters and ANNA reduces this to just 5 in order to solve the identical problem (and doesn't require a particular ordering of the constraint functions).

Possibly unexpected is that it is significantly quicker for Axiom to analyse various aspects of the input problem than for the input parser to interpret and provide base types for the extra parameters which would be needed. Even greater is the saving where coercion to ASPs are required, since coercion in the command line interpreter is particularly inefficient.

ANNA is nearing completion and the results will be made available for evaluation by UK Higher Education Institutions by October 1996 and incorporated within Axiom for release at a future date. Incomplete code for testing is available from the first author.

References

Baker, G.A., Graves-Morris, P.R.: *Encyclopaedia of Mathematics and its Applications: Padé Approximants, Part 1: Basic Theory.* Addison Wesley, Reading, Massachusetts, 1981.

Broughan, K.A., Keady, G., Robb, T., Richardson, M.G., Dewar,M.C.: Some symbolic computing links to the NAG numeric library. *SIGSAM Bulletin*, **25**:28–37, July 1991.

Buchberger, B., Collins, G.E., Loos, R., with Albrecht, R., editors: *Computer Algebra: Symbolic and Algebraic Computation.* Springer-Verlag, Wien, 1983.

Collins, G.E., Loos, R.: Real zeros of polynomials. In Buchberger et al. 1983, pp 83–94.

Davenport, J.H., Siret, Y., Tournier, E.: *Computer Algebra: Systems and Algorithms for Algebraic Computation.* Academic Press, London, 1988.

Davenport, J.H., Dewar, M.C., Richardson, M.G.: Symbolic and numeric computation: The IRENA project. In: *Proceedings of the Workshop on Symbolic and Numeric Computation (Helsinki, 1991)*, pp 1–18, Computing Centre of Helsinki University, 1991.

Dekker, K., Verwer, J.D.: *Stability of Runge-Kutte Methods for Stiff Nonlinear Differential Equations.* North Holland, Amsterdam, 1984.

Dewar, M.C.: *Interfacing Algebraic and Numeric Computation.* PhD thesis, University of Bath, 1991.

Dewar, M.C.: Manipulating Fortran code in Axiom and the Axiom-NAG link. In: Apiola, H., Laine, M., Valkeila, E., editors: *Proceedings of the Workshop on Symbolic and Numeric Computation (Helsinki, 1993)*, pp 1–12, 1994. Research Report B10, Rolf Nevanlinna Institute, Helsinki.

Dupée, B.J., Davenport, J.H.: Using computer algebra to choose and apply numerical routines. *AXIS*, **2(3)**:31–41, Sep 1995.

Hall, G., Watt, J.M., editors: *Modern Methods for Ordinary Differential Equations.* Clarendon Press, Oxford, 1976.

Hawkes, E., Keady, G.: Two more links to NAG numerics involving CA systems. In: *Proceedings of the IMACS Conference on Applications of Computer Algebra (University of New Mexico)*, May 1995.

Hopkins, T., Phillips, C.: *Numerical Methods in Practice: Using the NAG Library.* Addison-Wesley, Wokingham, England, 1988.

Houstis, E.N., Rice, J.R., Vichnevetsky, R., editors: *Expert Systems for Scientific Computing.* North-Holland, Amsterdam, 1992. Proceedings of the Second IMACS International Conference on Expert Systems for Numerical Computing, Purdue University, April 1992.

Kamel, M.S., Ma, K.S., Enright, W.H.: ODEXPERT: An expert system to select numerical solvers for initial value ODE systems. *ACM Transactions on Mathematical Software*, **19(1)**:44–62, March 1993.

Keady, G., Nolan, G.: Production of argument subprograms in the Axiom-NAG link: examples involving nonlinear systems. In: Apiola, H., Laine, M., Valkeila, E., editors: *Proceedings of the Workshop on Symbolic and Numeric Computation, Helsinki, May 1993*, pp 13–32, 1994. Research Report B10, Rolf Nevanlinna Institute, Helsinki.

Lambert, J.D.: *Computational Methods in Ordinary Differential Equations*. Wiley, London, 1973.

Lucks, M., Gladwell, I.: Automated selection of mathematical software. *ACM Transactions on Mathematical Software*, 18(1):11-34, March 1992. Also published in Houstis *et. al.* 1992, pp 421–459, as 'A Functional Representation for Software Selection Expertise'.

Mignotte, M.: *Mathematics for Computer Algebra*. Springer-Verlag, New York, 1992.

NAG: *Fortran Library Manual - Mark 16*. NAG Ltd, Oxford, UK, 1993. NAG Publication Code NP2478.

Prothero, A.: Introduction to stiff problems. In: Hall, G., Watt, J.M. 1976, pp 123–135.

Richardson, M.G.: *User Interfaces for Numeric Computation in Symbolic Environments*. PhD thesis, University of Bath, To be presented in 1996.

Computer Algebra and the World Wide Web

Anthony C. Hearn

RAND, Santa Monica CA 90407-2138, E-mail: hearn@rand.org

Abstract. We discuss various ways in which the World Wide Web can be used to improve algebraic computation research and development. In particular, dynamic updating of programs becomes possible, enabling geographically distributed researchers to use the same version of a program. This can be accomplished by using platform-independent pseudocode for execution. We also consider other ways in which the Web can improve collaboration in this area.

1 Introduction

User acceptance of computer algebra as a calculational tool is at an all-time high. Mass distribution of programs such as Derive, Maple, and Mathematica has brought sophisticated mathematical computational techniques to a rapidly growing audience.

One reason for this explosion is the ever-increasing power and availability of the personal computer, which well exemplifies the dream of a personal algebra machine I expressed in 1980 [1]. In addition, we are beginning to see algebraic engines, particularly Maple, embedded in other products.

In spite of this success, I have some concerns. In particular, neither the number of people working on the *development* of computer algebra algorithms and systems, nor the number of centers advertising such research as a speciality, seem to be growing. Furthermore, it is rare to find more than two or three people at any one center working on further development, as opposed to applying existing systems to scientific problems. Attendance at key meetings such as ISSAC and DISCO has also remained fairly static, suggesting a limit to the number of people working in this area. Finally, because algebraic computation sits at the juncture between computer science and mathematics, it is often hard to get reviewers from these different fields to agree on the worthiness of a research proposal, leading to its declination. This problem also affects university faculty in the algebraic computation area seeking promotion and tenure.

Of those persons who do describe themselves as working in the field, we find a significant number of these doing product development rather than research. In particular, much effort on the part of system developers during the past few years has been expended on better interface design. While this is important for user acceptance and convenience, it does nothing to enhance the range of algebraic capabilities that an algebra engine offers and is rarely seen by those doing algorithmic development as particularly important. Where algebraic development *is* undertaken, it is often in response to the need to match the capabilities of a competitive system, which allows the designer to claim that he or she too supports that capability, even if the implementation is rather shallow, and the range of problems capable of solution is fairly limited. In other words, there has been an emphasis on breadth as opposed to depth in such development.

These tendencies may in fact reflect the fact that all the "easy" algorithmic and system problems have been solved, and that it will take a lot of hard work to make further progress. However, we cannot afford as a society to neglect this challenge. We still face problem-solving challenges that cry out for better algorithms and corresponding system support. For example, in the area of nonlinear polynomial equation-solving, the euphoria over the possibilities inherent in the Buchberger algorithm has given way to a realization that there remain many problems for which this technique is inadequate. This is not to say that the Buchberger algorithm is not useful; indeed, there have been major successes reported in many fields (e.g., [2]), and alternative techniques such as sparse resultant evaluation [3] may help. However, many of the recent algorithmic improvements have enabled us to handle existing solved problems faster than to increase the range of problems solved. We therefore need more breakthroughs in this area to enlarge the solution boundary.

The issue I therefore wish to address in this paper is: How can we develop the best possible infrastructure for algorithmic and system development? In particular, given the sparse distribution of people working in this area, how can we build an appropriate collaboratory environment that best facilitates their work? The answer, I believe, lies in the use of the World Wide Web, hence the title of my talk.

2 Collaboration in Computer Algebra

Long before the advent of the World Wide Web, the computer algebra community had recognized the importance of networking as a means of alleviating the isolation produced by the sparse distribution of researchers. E-mail has long played an important role in this. In the 1970s, the Macsyma system was initially offered only by remote access, providing support to a widely distributed community from a computer at MIT. In the particular case of REDUCE, many aspects of its development recognized the need for distributed use. Since it is used on many different computing platforms, portability has always been important. Source code portability was achieved by defining a standard programming language based on Lisp [4]. Later, a virtual machine [5] for this language was defined in terms of a set of machine independent macros, which could then be expanded into assembly language for a target machine, thus simplifying compilation.

One requirement for effective collaboration in a software-based field is that each person work with the same software. System maintenance has therefore always been an important concern of the REDUCE community. A full update was provided to relevant users on tape or other media every one or two years or so. However, in between there were numerous changes to the code reflecting additional features, new packages, and inevitably, bug fixes. This is one area where networking has proved to be very important in general, and essential for collaborative system development. Even though ftp had been available from the earliest days of the Arpanet for file distribution, most of the REDUCE user community could not use this until relatively recently, since many were not directly connected to the Internet or the Arpanet before this. As a result, in the early days changes were distributed where possible by e-mail. However, the advent of the Netlib system [6] about a decade ago enabled the REDUCE community to obtain updates by e-mail when they wanted them, rather than depending on automatic mailings that are hard to track, or waiting for a response to a manual request. Those in the Maple community followed soon after with their own Netlib server. E-mail servers can also be used in support of computation, as for example the CONVODE[1] system of Alain Moussiaux for solving differential equations.

The advent of gopher servers on the Internet provided an even more convenient means of access than Netlib, since it was much easier to browse the available material and download any needed items, and again like most other systems, the REDUCE community quickly made use of this capability[2].

In spite of these advances, system maintenance was still a fairly difficult task. A user picking up a new or updated package had to install it by running a script. In addition, it was not possible to put complete versions of the core system packages on a publicly accessible server and still maintain control over the distribution of the software. To resolve this problem, we developed a few years ago a "patch" protocol that enabled individual functions from anywhere in the system to be redefined in a single file. Again, it is necessary for the user to run an update process to install these changes.

3 Collaboration and the World Wide Web

The widespread acceptance of the World Wide Web as a means of information dissemination has revolutionized the way that users obtain information about algebra systems and relevant code. All major computer algebra systems operate Web sites, and REDUCE[3] is no exception. This particular site has online copies of all the documentation distributed with the system, downloadable demonstration copies of the software, and a pointer to the gopher-based compendium of patches and updated packages. Another site[4] also offers access to a full version of REDUCE for calculations.

All these capabilities can be offered on a gopher server, except for the improved presentation offered by HTML. However, the Web offers much more. A few sites offer form-based access to the software itself. For example, Eberhard Schruefer is offering a REDUCE server[5] at GMD in St. Augustin, Germany. One can enter an expression on a form and receive back the results of the computation. Such an approach works quite well for specific capabilities as exemplified by Richard Fateman and his collaborators at Berkeley who are providing a table-based symbolic integrator, TILU[6] [7], on the Web. In this case, a running process listens at a socket for requests. In the REDUCE case, a separate process is spawned by each request. We are also looking at using a running process. However, there are so many possible side-effects arising from a given calculation (e.g., the change in the setting of a switch) that it is difficult to ensure that the system has been set back to its previous state after a given calculation. One could imagine a whole set of such servers providing users with specific capabilities. However, except for the nature of the interface, the services provided by such systems are little more than that provided

[1] See: URL http://www.physique.fundp.ac.be/physdpt/administration/convode.html

[2] See: URL gopher://info.rand.org:70/11/software/reduce

[3] See: URL http://www.rrz.uni-koeln.de/REDUCE/

[4] See: URL http://www.zib-berlin.de/Symbolik/reduce/

[5] See: URL http://cartan.gmd.de/cgi-bin/REDUCE.pl

[6] See: URL http://http.cs.berkeley.edu/~fateman/htest.html

by the Macsyma computer in the1970s. To go further, we need to allow client programs to query such servers and return the results to a user's workspace. The Berkeley group has proposed this possibility. However, for this to succeed, we need better standards for the interchange of mathematical objects such as those under development by the Open Math[7] community [8].

All these examples so far consider the Web as an information resource, albeit one that includes computation as a component. However, Web developers have long been concerned about *collaboration* as well. As pointed out last year in a statement[8] by Berners-Lee, although the Web was initially designed as an interactive hypertext system for communication through shared knowledge, there are benefits to be gained by raising its functionality to allow interaction. These include:

1. Allowing groups to use the Web at an informal level, mixing formal and informal hypertext, as a powerful tool for solving problems and working in teams; allowing authorized people to update and comment on extant information, reducing the amount of out-of-date information and hence raising the overall quality of the Web;

2. Reducing the administrative overhead of those whose part in the production cycle is largely mechanical;

3. Providing through human interaction a more lively, stimulating, and enriching environment for all Web activities including education, research, and commerce.

Several groups are currently working on the necessary extensions to provide for such interaction. For example, the Networkplace project [9] at the National Center for Supercomputing Applications (NCSA) is working on changing the Web paradigm from a fairly static information resource to a place where work is conducted collaboratively. The goal is to allow members of an online work team to work together, regardless of the geographic location, time frame, and computing platform of individual team members. This collaborative environment on the World Wide Web offers a viable way of "telecommuting," or working remotely from one's physical office.

Let us now consider some ways in which the Web can be used in computer algebra to address Berners-Lee's second point above. As mentioned above, portability was achieved in the REDUCE distribution by defining a standard source programming language based on Lisp and a virtual machine for this language to simplify compilation. However, different compiled code was still required for each machine. So using the standard Web support, updating requires downloading a patch file or new module and running a script to compile it on the client machine. Alternatively, one could provide compiled versions of the necessary code on the server, but since many architectures are supported, this would require considerable work to keep things current.

As a further step in easing program maintenance, an ANSI C-based Codemist Standard Lisp (CSL) compiler [10] was recently developed that produces efficient machine-independent pseudocode as output. In other words, any package in the system can now be compiled into this pseudocode, which can then be executed by a running REDUCE

[7] See: URL http://www.can.nl/~abbott/OpenMath/index.html

[8] See: URL http://www.w3.org/pub/WWW/Collaboration/Activity.html

process on any platform. Client support comes in the form of an ANSI C-based interpreter for the pseudocode, plus a function library to support standard operations such as extended precision arithmetic, input-output, and the like.

To see how such pseudocode can be more effectively used, let us look in more detail at the REDUCE development model. Developers send source code to a central repository, maintained at RAND, which is mirrored in several countries . Users obtain updates from an appropriate repository, compile this code on their client machine, build a new executable image if core code was updated, and then run this image. To improve this process, we need to exploit the possibility the Web offers for dynamic interactions between the client and server. One system for this purpose, Oil Change, has recently been announced by Cybermedia[9]. This product will automatically update software on a client machine. Oil Change scans your hard disk to determine which software programs you have and then checks at the Web site of the relevant software vendor for upgrades and bug fixes. These are then routed back to your machine and installed. In this model, you could arrange to have your algebraic computation software updated daily, weekly, or even hourly without any action on your part. However, this approach is still batch-oriented, in the sense that updates do not normally occur when a user needs them, but by a predetermined schedule.

A model we are now developing goes further than this in that it uses the dynamic possibilities inherent in the Web to update an executing process. In this model, developers maintain their own code on their local server rather than sending it to a central repository. A user starts a REDUCE job on a client machine. If the calculation requires code that has been marked for possible updating, the executing process checks to see if a more recent version is available on the developer's server. If it is, pseudocode for that package will be downloaded and used (and stored for later use). If the server is unresponsive, the existing package on the client machine is used.

For such a scheme to work efficiently, it is important that the updated code segments be as small as possible. Fortunately, the version of REDUCE used for code development is already set up to support this. In the distributed version, source code is distributed in files that define *packages*. These encompass a relatively complete capability, such as integration or the solution of various classes of equations. Such files contain smaller units called *modules* that define subtasks within that capability. In the development version, these modules are themselves independent files. To facilitate their development, each module is defined to be independently compilable. Necessary declarations, macros and so on that apply across the whole package are contained in a module with the same name as the package. For example, the *solve* package contains eight modules. As I have just mentioned, the first has the same name (*solve*) as the package. The other seven modules (*solve1, ppsoln, solvelnr, glsolve, solvealg, solvetab,* and *quartic*) each take care of a particular case in the overall solving process. In general, such modules are a few hundred lines of code, and compile into the order of a hundred kilobytes of pseudocode. As a result, even on a loaded network these days, it is a fairly efficient process to download code for an individual module from a server.

To provide as robust an environment as possible, we must keep older versions of the code around in case an error occurs during downloading, or the developer informs the

[9] See: URL http://www.cybermedia.com/

cooperating clients that the latest version downloaded version has more serious bugs than the previous version! As a result, the server must keep a table of those sites that download code from them, in addition to the table kept by the client associating a host server with each downloadable module.

Since executable code is being downloaded, there are obvious security concerns. Initially, we are limiting the user community to a small group of trusted contributors using the development version of REDUCE as the basis. However, in the long run, it will be necessary to face the possibility of a virus or other malicious code being inserted in one of the executable packages as we extend these capabilities to the standard distributed version. We shall closely follow the experience of the Java [11] community in this regard.

In fact, the model we have described has obvious similarities to Java. The code being downloaded is platform-independent, portable, distributed, high-performance, and dynamic. On the other hand, it is neither object-oriented nor secure to the same extent. A new version of REDUCE under development [12], however, *is* object-oriented, and it would be possible to modify the pseudocode to reflect this more directly. Given all the current hype about Java, it is reasonable to ask why we are not using it instead of our own pseudocode. In response, it is important to realize that Java is not yet proven as a programming language in our environment, whereas we have had considerable experience with our CSL pseudocode. This is not to say that Java will not be useful to the algebraic computation community, and in this regard Fitch and Norman [13] describe some possibilities for its use. Their point is that Java can be used as a mechanism for coordinating the transfer of information between REDUCE and other software components such as browsers, graphical viewers, and so on. Rather than replace the REDUCE code by Java, Fitch and Norman investigate the steps needed to make it possible to mix the two codes as freely as possible.

As long as we are talking about browser support, it is worth pointing out that to use the full potential of the Web for collaboration, it will be necessary to use Web-compatible languages for documents. LaTeX has been enormously successfully for scientific document preparation. However, in order to share documents in a Web-based environment, especially if several distributed collaborators are working on the same document on a shared whiteboard, it will be necessary to be able to dynamically update that document. In that environment, a distributed WYSIWYG editor is needed, and the batch-oriented model used by LaTeX, even if backed up by Adobe Acrobat output (which several Web browsers can now easily handle) does not work well. One possible approach might be to use a product such as Scientific Word, which could presumably be extended to output HTML rather than LaTeX. Another possibility is to use HTML for direct support. For example, this paper was first produced using the WYSIWYG editing capabilities of Netscape Navigator Gold, then translated into LaTeX using F.J. Faase's HTML to LaTeX[10] program. However, this paper has no mathematical formulas, and one problem with HTML at the present time is its poor support for such formulas. One can use Java applets for this purpose as in Robert Miner's WebEQ[11] system. In the long run, though, let us hope that groups such as Open Math will propose suitable HTML extensions for

10 See: URL http://wwwis.cs.utwente.nl:8080/~faase/H/htmltools.html

11 See: URL http://www.geom.umn.edu:80/~rminer/jmath/

this purpose, or that more flexible editors for Acrobat documents will be created.

The previous paragraph also reminds us that Web collaboration involves more than just software maintenance. In particular, the "whiteboard" model raises the question as to whether two or more distributed collaborators should be able to input data to the same program, let alone write to the same document. I am somewhat skeptical about this possibility for program development. Normally, one discusses a problem, identifies a bug, or proposes an extension. Someone then has to work on implementing the changes. This is not something that lends itself easily to collaboration, although of course the final results need to be quickly shared with others. The model I have proposed above addresses that issue.

4 Conclusion

In this paper, I have pointed to a vision of the future in which networking expands our capacity for cooperative work, and software updating in particular. Not everyone believes in this vision. For some, distributed cooperative problem-solving and code development cannot take the place of face-to-face interactions. For others, the need to have the very latest software is secondary to the need to have stable, familiar software, even if it is somewhat out-of-date. However, it seems inevitable that networked computing will become dominant in the years ahead, and in that environment, the things I have talked about here will become so common as to be transparent, and, as such, expected by most people. However, it is very hard to predict just what will happen in, say, ten years from now. The World Wide Web has burst on the scene and established itself as an essential part of the computing infrastructure in an amazingly short time. In fact, hardly any of the ideas in this paper would have made sense as little as two or three years ago, when the notion of the World Wide Web was limited to a small group of people pursuing their own research goals. The information revolution is moving at a very fast pace these days, with things happening in months that used to take years. One can therefore only wonder at the possibilities that will become available to us in the years ahead. It should be an exciting time for all of us.

5 Acknowledgments

I am grateful to Arthur Norman and Eberhard Schruefer for many discussions on these topics.

6 References

1. A.C. Hearn. The Personal Algebra Machine. *Information Processing 80*, 621-628, North-Holland, 1980.
2. H. Melenk. Practical Application of Groebner Bases for the Solution of Polynomial Equation Systems. *IV. International Conference on Computer Algebra in Physical Research, 1990*, D. V. Shirkov, V. A. Tostovtsev, V. P. Gerdt, eds. (1991), 230-235, World Scientific, Singapore.

3. J. Canny and I. Emiris. An Efficient Algorithm for the Sparse Mixed Resultant. *Proc. Intern. Symp. Applied Algebra, Algebraic Algorithms and Error-Correcting Codes, Lect. Notes in Comp. Science 236,* Springer Verlag, 1993, 89-104.
4. A.C. Hearn. REDUCE - A User Oriented Interactive System for Algebraic Simplification. *Interactive Systems for Experimental Applied Mathematics,* (edited by M. Klerer and J.Reinfelds), Academic Press, New York, 79-90, 1968.
5. M.L. Griss and A.C. Hearn. A Portable LISP Compiler. *Software Practice and Experience* 11, 541-605, 1981.
6. J.J. Dongarra and E. Grosse. Distribution of Mathematical Software via Electronic Mail. *Comm. ACM* (1987) 30, 403-407.
7. T.H. Einwohner and R.J. Fateman. Searching Techniques for Integral Tables. *Proc. ISSAC '95,* ACM Press, 1995, 133-139.
8. J. Abbott, A. van Leeuwen and A. Strotmann. Objectives of OpenMath. *RIACA Technical Report* 12 (June 1996).
9. L. Smarr. Private communication.
10. A.C. Norman. Compact Delivery Support for REDUCE. *Journ. Symbolic Computation* 19, 133-143, 1995.
11. D. Flanagan. Java in a Nutshell. O'Reilly, 1996.
12. A.C. Hearn and E. Schruefer. A Computer Algebra System based on Order-sorted Algebra. *J. Symb. Comp.* 19 (1995) 65-77.
13. J.P. Fitch and A.C. Norman. Interfacing REDUCE to Java. These proceedings.

Interfacing REDUCE to Java

Arthur Norman[1] and John Fitch[2]

[1] University of Cambridge Computer Laboratory
Cambridge CB2 3QG, England, *and* Codemist Ltd.
[2] School of Mathematical Sciences, University of Bath
Bath BA2 7AY, England *and* Codemist Ltd.

Abstract. For some time it has been clear that algebra systems ought not to exist as isolated software packages, but should be viewed more as components in a more general scientific problem-solving environment. A so-called "software bus" would then link perhaps several algebra engines, each with special areas of strength, to separate tools to support numerical calculation, visualisation, domain-specific scripting or other capabilities. The term "bus" as used here derives from the same word used to describe a mode of transport, and as such is an abbreviation for omnibus — stressing the fact that to be useful it must be universally available. The language and system Java[5] presents itself as a candidate for such general acceptance by many different classes of application, so this paper is a preliminary investigation of both how Java code can be linked to REDUCE and what might be done by exploiting such a link.

1 Introduction

The motivation for creating a link between REDUCE and Java is twofold. On the one hand it is to provide other applications with a program-based interface to REDUCE's algebraic capability, using Java as an increasingly widely available mechanism for co-ordinating the transfer of information. On the other it is so that code written in and residing within REDUCE itself can gain access to a wider range of graphical viewers, network services, browsers and general third-party tools.

Use of Java in this way does not address the issues of making algebra system interfaces semantically neutral (as considered by the OpenMath[1] working group), although it does provide the possibility of partitioning the work of providing an OpenMath interface for REDUCE between code internal to REDUCE and a C-like Java veneer. Instead Java is used here to address the practical problems of linking and interfacing disparate packages, where at a lower implementation level a variety of invocations of sockets, pipes, OLE and other arcana may be needed. It also aims to hide all the system-dependent aspects of interface code, so keeping everything as clean and portable as possible.

While it is predicted that Java will grow over the next few years to become a universal portable co-ordinating language, at present it is most directly associated with Web browsers, remote execution of code and the generation of cute animated visual effects. This side of it means that in the short term its most valuable use in association with REDUCE will be in support of operations with that sort of flavour: this is of course close to the collection of ideas presented by Hearn elsewhere[2] at this meeting.

When interfacing two complete programming languages the first big issues to be addressed are how symmetric the link-up should be and how intimate the link-up should be. At one extreme would be the case where one language is totally in change and it invoked the other by passing some neutral data such as strings of characters. At the other would be complete integration of the languages with fully shared data types and representations. The study here attempts to follow the second of these styles, so that REDUCE and Java code can be mixed in a very free manner. As well as the obvious advantages that this brings it causes a number of difficulties both in implementation and use, but before considering these the implementation strategy that was adopted will be described.

Java is defined both by a language definition in the usual style, and by the description of a byte-oriented virtual machine. The official Java Development Kit[4] includes a compiler that maps from the source language into this byte-coded form, and it is expected that this representation (wrapped around with additional symbol table information) will be the way in which Java programs are distributed. Simple implementations of the language can consist of an emulation of the virtual machine together with support for various pre-defined class libraries. Higher performance versions may choose to view the virtual machine as the definition of an architecturally neutral intermediate language, and then provide the back-end of a compiler to map from the byte-coded form of a program into real machine code for the target architecture. The overall structure of the REDUCE implementation used here[3] is in many respects similar. REDUCE packages start off written in their own source language (RLISP), which is then compiled by way of Standard Lisp into a machine independent byte-code representation (which has been optimised for the support of Lisp). An interpreter can then process the byte-codes to run REDUCE. To improve performance it is also possible to compile key parts of REDUCE into C code which is linked in to provide extensions to the normal Lisp run-time environment. The compiled C is typically faster but also typically bulkier than the byte-code version of any given function. But since the implementation allows C coded and byte-coded functions to be mixed without constraint it is possible to gain almost all of the speed of a fully compiled system and almost all the space-saving of a fully byte-compiled one by selecting only the most heavily used functions for compilation into C.

Thus the scheme adopted here is to build an implementation of the Java Virtual Machine as part of the CSL Lisp system, to arrange that it can share at least some data types with REDUCE's Lisp world, and to make it possible for functions that use the Lisp and the Java sub-systems to call each other and otherwise interact.

2 The Bytecode Interpreters

A byte-code interpreter for the Java Virtual Machine has to provide a stack, a garbage-collectable heap and a method area (within which byte-codes are literal tables are stored). Apart from the fact that the exact representation and layout of some data types are rigidly specified (this would, for instance, make it slightly painful to implement the floating point parts of Java on an non-IEEE computer). Otherwise the most complicated part to support involves the resolution of items in literal pools, since the language provide for late binding. Documentation of the

Java release 1.0 suggests that naive implementation of the searching implicit in the basic description of the byte-codes will be inefficient, and described a scheme where once a look-up has been performed the virtual machine instruction that provoked it is overwritten with a "fast" variant that can now access the relevant data directly. This trick is of course thoroughly familiar to Lisp implementers where function calls have been implemented this way in many historical systems — and the consequent problems for dynamic function redefinition within a debugging cycle are also well known.

Java expects to support multi-threaded applications and so its virtual machine has support for mutual exclusion. When implemented as a pure byte-code interpreter the thread implementation does not require much more than making provision for a separate stack for each task and some counting so that after any process has executed a number of bytes it will be pause to allow others to take their share of the real machine. Because the machine accesses items in arrays and members of classes in a fairly abstract way (including the resolution process mentioned above which can sort out the exact layout of class members in memory) there is a lot of flexibility in how these items can be represented.

The virtual machine provided by CSL for Lisp and REDUCE purposes has one big advantage over the Java one. Lisp only has a single generic data type, and all simple data movement and comparison (eg the parts of the machine that support passing arguments to functions, returning values and the SETQ form) can work in terms of this. Furthermore there are only a handful of Lisp special forms that strictly need direct support in terms of byte-codes — all other operations are just calls to library functions. This means that the basic instruction set for Lisp can be very small and simple, while for Java it was necessary to provide (for instance) multiple type-specific versions of the `load` virtual instruction. Thus given the expectation that a byte is 8 bits wide, after having implemented enough instructions to support all Lisp there are still plenty of byte-codes left over. CSL allocates these in three ways:

- To shorter versions of what would otherwise by byte-coded operations that would need to use one or more operand bytes. The Java virtual machine can only provide a handful of load and store instructions of this sort (eg `iload_n`, where locations 0, 1, 2 and 3 of the current stack frame can be accessed especially efficiently), while CSL has a chance to be more generous;
- To combine pairs of otherwise quite separate instructions into one. The pairs treated in this way were selected after measurement of the dynamic use of the original simple instruction set, and include single-byte codes for what would otherwise be combinations such as `LOADLOC 2; LOADLOC 3` and also `LOADLOC 1;EXIT`, where `LOADLOC` loads a value from the current frame;
- To provide in-line support for important Lisp functions. It is very nice that CSL does not need to provide individual byte codes for every arithmetic and logical function, and it attempts to use the 256 distinct operations at its disposal to help with chains of `car` and `cdr` operations, property list access, important simple predicates and the most heavily used arithmetic functions.

The effect is that the CSL byte-code model has a better chance to approach the speed of a "properly" compiled implementation than does the Java one.

CSL stores floating point values boxed in memory with an associated header word. It can handle integers of up to 28-bits as immediate data, and beyond that is changed gear into a bignum representation. Ordinary Lisp vectors that hold numbers will store them in these tagged representations and so will not be at all comfortable for Java purposes. Thus when scalar data is to be exchanged between the Java and Lisp machines various data representation conversions will be needed. However the storage manager in CSL also supports vectors of bytes, half-words, 32-bit words, and of single and double-precision floating point values. These are not used by the main body of REDUCE code (except that of course vectors of bytes are really just a slightly disguised version of strings), but the access functions for them can be called from Reduce. They mean that dynamically allocated Java arrays can be supported within the CSL environment with little trouble and hardly any overhead.

3 Linking with fully compiled code

In CSL calls to some especially common functions are detected by the compiler and converted into short-cut byte-codes (or when Lisp is being compiled into C they may be expanded in-line). Apart from these optimisations all calls are late-bound. Each Lisp symbol has three function cells in its symbol-head, and these hold the entry-points of code to call if that function is invoked with 1, 2 or some other number of arguments (measurements using REDUCE showed that these cases where the most important). A further "environment" call hold Lisp data that is passed to the function that is so called. For a function that expects a fixed number of arguments two of the function calls will point at code that just raises an exception. C coded routines use the environment to gain access to a per-function table of literal Lisp values they may need to refer to. Byte-coded functions are implemented by making the function cell point to the function that interprets byte-codes, and then the environment cell can hold a vector containing the bytes themselves. By indirecting via symbol-heads in this way CSL can apply run-time checks that functions are called with the correct number of arguments, and can mix C and byte-coded functions freely. Calls between Lisp and Java can try to use the same scheme — a Java-coded function can refer to the Java emulator in its function cell and then anybody can call that function and the Java code will be executed. The slight delicacy is the passing of arguments. When Java code is to be called in this way it will be necessary to take the Lisp arguments as passed and both convert them into Java-compatible form, (eg integers must be converted to regular 32-bit un-tagged machine representation), and to place them at the start of a Java stack-frame.

Java can only call functions if they are methods in some class that it knows about, and that means that Lisp code that is to be Java-callable will need to be provided with a veneer that makes it appear to be in a class. The entry in the method block involved can easily indicate that this is a Lisp rather than a Java function, and then calling it can box up floating point values, tag up integers and do whatever conversion seems fitting for other data types. Two convenient loopholes can provide for full generality — firstly because of Java's late binding function calls are essentially by name, so it is possible to provide a general "lisp" class where all possible Lisp (and hence REDUCE) functions are available. Note of course that smaller and better structured class-like interfaces will lead to clearer and more reliable code! Secondly by allowing Java to pass names of Lisp identifiers back to Lisp and giving it access to **eval** and **apply** again its gets universal access to the Lisp world, albeit at the cost of some dynamic data representation conversion at the interface. Note that the universal "lisp" class uses the generality of Java's late binding, while **eval** uses Lisp's.

4 Interfacing with Algebra

The easy parts of a Java interface for REDUCE involve passing integers and strings backwards and forwards. For algebraic formulae things become harder! A first response to this is to ensure that there are fall-back solutions so that calculations become possible, even if clumsy. The most natural hook in this case seems to be to create a Java class that models Lisp prefix representations of formulae, and use this as the lingua franca for the two parties. This in turn means that Java will have to have classes that stand for Lisp identifiers and (arbitrary precision) numbers, and will want access to the main operations that can be performed on same.

Since the Java code will often want to be involved in user interface control it also needs to be able to be involved in input and output operations. By virtual of being a multi-threaded language it is possible to achieve this by providing Lisp with an input stream where a request for characters just invoked code in a Java process, and an output stream that similarly sends individual characters across to the Java world. By selecting these as REDUCE's main working streams it becomes possible for Java code to feed in arbitrary text for REDUCE, and to collect the output (either in Reduce's own 2-D displayed format or using one of the REDUCE TEX-like output packages) and either handle it or pass it on as required.

5 Foreign Class Libraries

One of the big limitations that comes from implementing a fresh Java virtual machine within CSL and hence alongside REDUCE is the problem of other Java libraries and applications. Libraries that use **only** facilities of the Java Virtual Machine can be compiled and loaded into the new implementation easily enough, but many facilities are not like that! To ease the task of building the REDUCE-Java interface only a minimal subset of the official Java library API is supported, and that naturally limits what can be done with it.

The expectation here is that separate Java implementations will be used to host visualisation tools etc, and that it will be sufficient to provide the REDUCE one with just basic IO and communication capabilities, so that complete systems get built in a way that distributes the work between modules that are not obliged even to run on the same machine. Thus for instance the REDUCE-hosted system can avoid supporting the **applet** classes, expecting that it can talk as necessary to some net browser that understands them.

6 Conclusion

At the time of writing release 1.0 of the Java Development Kit represents very fresh news, and the associated documentation of the Java Virtual Machine, class-file format and standard libraries has now started to become firm enough for experiments of this sort to become worthwhile. The eventual usefulness of a Java interface of this sort depends critically on how well the current predictions about general Java acceptance are met. If they are then REDUCE will find itself with an increasing collection of other worthwhile packages for which Java can help provide a platform-independent communication model. So even though Java of itself does not provide a way of ensuring that data passed between different applications is meaningful, it can provide a convenient programming language in which conversions between data representations can be coded and where the behaviour of collections of co-operating packages may be co-ordinated.

References

1. Abbott, J. A., van Leeuwen, A., and Strotmann, A. Tr 12: Objectives of openmath. Tech. rep., RIACS, TU Eindhoven, 1996.
2. Hearn, A. C. Computer algebra and the world wide web. In *Proceedings of DISCO 96* (1996).
3. Norman, A. C. Compact delivery support for REDUCE. *J. Symbolic Computation 19* (1995), 133–143.
4. Sun. Java web site. `ftp://ftp.javasoft.com`, 1996.
5. van Hoff, A., Shaio, S., and Starbuck, O. *Hooked on Java*. Addison Wesley, 1996.

Software Architectures for Computer Algebra: A Case Study

Greg Butler

Centre Interuniversitaire en Calcul Mathématique Algébrique
Department of Computer Science
Concordia University
Montreal, Quebec, H3G 1M8 Canada
Email: gregb@cs.concordia.ca

Abstract. The architectures of the existing computer algebra systems have not been discussed sufficiently in the literature. Instead, the focus has been on the design of the related programming language, or the design of a few key data structures.
We address this deficiency with a case study of the architecture of CAYLEY. Our aim is twofold: to capture this knowledge before the total passing of a system now made obsolete by MAGMA; and to encourage others to describe the architecture of the computer algebra systems with which they are familiar.
The long-term goal is a better understanding of how to construct computer algebra systems in the future.

1 Introduction

The architectures of the existing computer algebra systems have not been discussed sufficiently in the literature. Instead, the focus has been on the user programming language or the mathematical algorithms. The articles on user programming languages discuss its design [7, 11, 20] or key algorithms for its implementation such as pattern matching [19] or type coercion [27]. The articles on mathematical algorithms are often very theoretical, but one can also find discussion of implementations, especially the choice of data structures. A key debate was on vector-based versus list-based data structures [23, 22].

We address this deficiency with a case study of the architecture of CAYLEY [8]. Our aim is twofold:

1. to capture this knowledge before the total passing of a system now made obsolete by MAGMA [4]; and
2. to encourage others to describe the architecture of the computer algebra systems with which they are familiar.

By studying the architecture of existing computer algebra systems we may learn the advantages or limitations of the different approaches as regards a number of issues:

extensibility of its functionality by users;

flexibility in how the system can be used with different interfaces or environments, or in cooperation with other systems;

mathematical rigour of its language, its mathematical toolkit, and of any user extensions to the system;

power of the mathematical toolkit both in terms of breadth of functionality, but also in terms of efficiency of their implementations and the availability of cutting-edge algorithms;

efficiency in terms of both memory and time for the built-in functionality and for the code generated from user language programs; and

usability by novice to expert mathematical users.

In the future, this understanding of architecture may allow us to construct more flexible, extensible, and powerful computer algebra systems and to do so with less effort and higher quality.

The paper first reviews the field of software architecture and then looks at related work from other computer algebra systems before the case study is presented. The case study is of the software architecture of CAYLEY. The case study includes the styles of architecture evident in the design of CAYLEY, and the limitations due to the architecture. The conclusion encourages others to describe the architecture of existing computer algebra systems, because of the impact of architecture.

2 Software Architecture

Software architectures provide an abstract description of the organisational and structural decisions that are evident in a software system.

> "Structural issues include gross organization and global control structure; protocols for communication, synchronization, and data access; assignment of functionality to design elements; physical distribution; composition of design elements; scaling and performance; and selection among design alternatives.
> This is the *software architecture* level of design." [18]

The development [5] of an architecture requires the decomposition of system into subsystems, the distribution of control and responsibility, and the development of the components and their connections or means of communication. General principles of information hiding, such as the use of modules, layers, and abstract machines (API's) help manage the complexity of the task.

Modern methodologies, such as OMT [26], have explicit steps during system design that consider the choice of architecture.

2.1 Architectural Styles

An architecture may be described [1, 17] in terms of its components, connectors, configurations, ports, and roles. Hence a description of an architectural style provides

- a *vocabulary* of the basic design elements (components and connector types),
- a set of *configuration rules* which constrain how components and connectors may be configured,
- a *semantic interpretation* which defines when suitably configured designs have a well-defined meaning as an architecture, and
- *analyses* that may be performed on well-defined designs.

Design patterns [16] can be used to describe an architecture and document the rationale for its development [3].

Mary Shaw and her colleagues [18] are compiling a *catalogue* of existing architectural styles:

- *Pipe-and-Filters* architectures, like that supported by the Unix shell, connect filters in a linear fashion. Each filter has one stream of inputs and one stream of outputs.
- *Data Abstraction* and *Object-Oriented Organization* architectures promote the decomposition of the system into entities (data type variables or objects) that encapsulate their implementation details and present an interface that completely describes their behaviour or functionality.
- *Event-based, Implicit Invocation* architectures are based on components that register an interest in (a class of) events. The components are invoked in response to the occurrence of an event implicitly rather than being called directly by another component.
- *Layered System* architectures are organized as a hierarchy of layers, each layer providing services to the layers above it and each layer being a client for the services provided by layers below it.
- *Repository* architectures have distinct components for a central data store (the repository) and those components which operate upon the repository.
- *Table Driven Interpreter* architectures implement a software virtual machine (the interpreter) by separating the interpretation engine from a table that describes the machine behaviour.
- Heterogeneous architectures combine architectural styles.

The book on OMT [26] also contains very useful sections describing architectures and their use in system design. There is good general agreement of their list with the above list of architectures.

The ultimate aim of such catalogues is to create a *taxonomy* of architectures to organize our knowledge by describing common and distinguishing features of the architectures. This may help designers to appreciate the breadth of choices and trade-offs, and may guide the discovery or invention of new artifacts.

There is a strong relationship between the study of software architectures and the study of programming at the architectural level that is pursued in megaprogramming [28] and the design of module interconnection languages [24, 25].

2.2 Interpreter Architecture

The classic architecture of compilers or interpreters[2] consists of the following *components*: lexical analyser (or scanner), parser, symbol table, and perhaps an optimiser and code generator. The *connectors* are lexical tokens, and various intermediate representations of the program being interpreted. These include abstract syntax trees, three-address code, directed acyclic graphs (DAG's), or polish notation. The *constraints* and *analyses* for interpreters are well-developed due to formal language theory.

Interestingly, as an illustration of their theory on architecture refinement, Moriconi and Qian [21] prove the equivalence between a pipe-and-filter architecture for a compiler and one which uses an abstract data type for the parse tree.

3 Related Work

All computer algebra systems have extensive documentation of their user language and their wealth of mathematical functions in manuals. Most have a summary of these features in the literature. Here we wish to highlight the information related to their architecture.

3.1 MAPLE

There is a reasonable description of the main organisational decisions for MAPLE available in the literature [10, 9]. The system is an interpreter, with a kernel plus a library that provides most of the functionality. The language [11] treats everything as an expression, and hashing provides a unique representation/storage of each entity. The data structures are based on the vector-based model of Norman [23, 22].

Later development separated the user interface from the "algebra-engine". MAPLE is an embedded "algebra-engine" in several mathematical and scientific environments.

3.2 AXIOM

The emphasis of AXIOM is on mathematical rigour, though the development of a strongly typed system language, and through an emphasis on the architecture of mathematical categories [13, 14]. The basic architecture is that of an interpreter (and, later, also a compiler).

3.3 Parallel and Distributed Computer Algebra Systems

The development of parallel or distributed systems for computer algebra, such as DSC [15] and STAR/MPI [12], adds another dimension of architecture to computer algebra systems, namely, physical distribution of tasks/processes, with the cocommittal considerations of communication and synchronization. From algebra's viewpoint, the interest is in the protocols and representations for the communication of mathematical entities.

4 CAYLEY

The CAYLEY project started in 1975 to provide a better user interface to the growing number of group theoretical algorithms available in the GROUP library. Interactive use was the natural mode for exploring properties of groups and the CAYLEY user language was viewed as a means to drive this exploration statement by statement. It was not originally viewed as a programming language per se. In 1980 user-defined procedures were added to the language. In 1987 the implementation language was changed from FORTRAN to C. CAYLEY has now been superseded by MAGMA.

We use the 1980 FORTRAN version of CAYLEY as the basis of our description of the architecture.

4.1 Overview

CAYLEY is primarily a system for computing with groups whose elements are represented as permutations, matrices over finite fields, or words in a set of generators. It computes structural information about groups, as well as doing arithmetic on group elements. The user language adopts Pascal-style commands for iteration and selection, with the **for**-loop extended to iterate over sets and sequences. The basic data types are integer, boolean, groups, group elements, sets and sequences.

The system has a large number of built-in mathematical algorithms, which are accessed as procedures or functions at the user level. Initially, CAYLEY did not allow user-defined procedures or functions. In 1980 user-defined procedures (but not functions) were added.

Variables are not declared before use. There is no explicit typing, except in the two-part definition of algebraic structures, such as groups. The first statement defines the type of group, such as a permutation group of degree 5, and the second statement defines a set of generators for the group.

For educational use, problem libraries can be created as text files and can be accessed from the user language. A library is essential a list of CAYLEY language statements.

4.2 Architecture

The CAYLEY system is composed of an *interpreter*, an *abstract machine* and its *executor*, and implementations of the *mathematical functions*. The mathematical functions are implemented in FORTRAN and called directly from the executor — that is, CAYLEY is a monolithic system which does not use a kernel plus library approach.

The system is built on a platform, called STACKHANDLER, which provides dynamic memory management. The unit of storage it provides is called a STACK (not to be confused with the LIFO data structure called stack). A STACK can be viewed as a one- or two-dimensional dynamic array. Each mathematical object is stored as a STACK. For algebraic structures the STACK accumulates information about the structure as it is computed, either in direct response to user commands or because the information is needed by some algorithm the user calls.

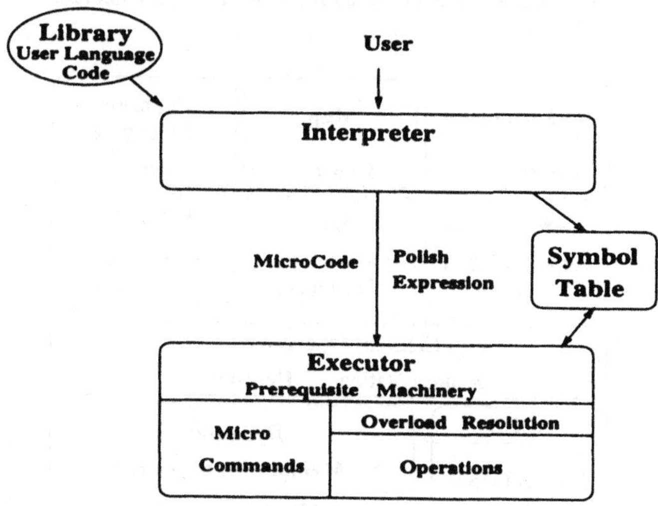

Fig. 1. CAYLEY System Overview

Figure 1 shows the role of the abstract machine. The abstract machine is a stack machine. The interpreter parses a CAYLEY language statement from the user, or from a library. It generates code for the abstract machine that consists of both microcode for the commands and polish notation for the expressions. The microcode is stored in a STACK which contains references to the STACKS holding the polish notation. The basic *microcode* commands are `assign`, `print`, `jump`, and `evaluate` an expression or a list of expressions. The *polish notation* includes the basic operations such as addition and multiplication, calls to each of the built-in mathematical functions, and a `load` operation.

Two major tasks of the executor are ensuring that prerequisite information has been previously computed, and resolving the overloaded operations and functions. A table indicates the required information for each built-in function. This allows the executor to insert the necessary commands into the microcode to test whether the operand's STACK contained the information and to compute it if it is absent. The overload resolution is done at runtime when the type of the operands is known.

The inclusion of user-defined procedures to the CAYLEY language required an additional microcode command `call`, and the implementation of a *copy-on-write* policy [16, p.210] for passing input parameters.

Figure 2 illustrates the layers that implement the mathematical functions. They too build on STACKHANDLER. Some operations can be implemented in *generic* fashion independently of the representation of the mathematical object. The broadly-applicable algorithm that such an operation implements may have steps whose implementation is representation dependent. That is, the generic operations are examples of template methods [16, p.325]

There is a low-level module for input and output. It provides portability across the different character sets by using ASCII within CAYLEY.

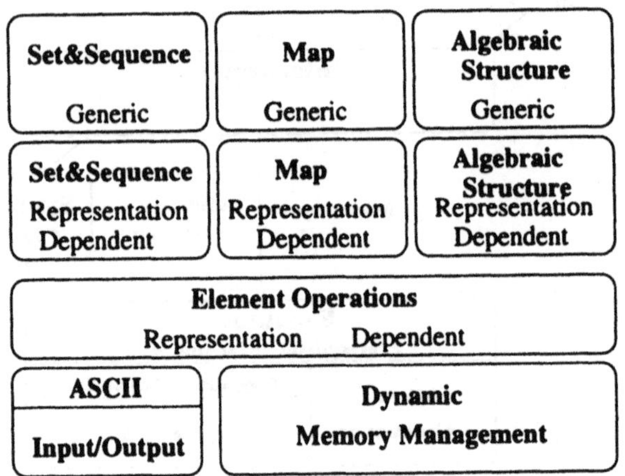

Fig. 2. Some Layers of CAYLEY Mathematical Structures

4.3 Relation to Architectural Styles

There are several architectural styles used in the design of CAYLEY.

Table-Driven Interpreter In many ways, CAYLEY is a classic interpreter: the *components* of lexical analyser, parser, symbol table, code generator, and abstract machine are evident. The *connectors* follow FORTRAN conventions of using `COMMON` blocks to share information between routines: the variable `MICRO` which holds the STACK for the microcode is in the common block `INSTR`; the stack that is used by the abstract machine when evaluating expressions consists of several arrays in the common block `EXPPRM`; and the symbol table is the common block `SYMS`.

Layered Architecture as illustrated in Figure 2.

Data Abstraction The STACK representation of mathematical elements and structures promotes a policy of information hiding and data abstraction. Even when programming in FORTRAN, the developers use pairs of routines `Get`... and `Set`... to manipulate data structures.

4.4 Limitations Due to the Architecture

The architecture seriously limits the development of CAYLEY in several ways.

User-defined functions require the dynamic creation of opcodes for the polish notation for expressions, and the evaluation of these opcodes requires the interpretation of user language code. Within the abstract machine, the command executor for the microcode knows about the CAYLEY user language and how to interpret it. However, the expression evaluator does not. Hence, the addition of user-defined procedures fitted the architecture, while the addition of user-defined functions would necessitate the reorganisation of the abstract machine.

Library calls are restricted to the topmost scope of the CAYLEY statements. This is very serious, as libraries evolved from simple definitions of examples or problems, to be collections of user-language procedures: essentially "modules". There are two ways the architecture limits this: The lexical analyser maintains a line buffer that is used for both user input and for input from a library. Hence, a library call must be on a separate line, since it flushes the line buffer of all other user input on that line. Secondly, the microcode is a single variable in a common block. Neither the microcode, nor its variable are regarded as (LIFO) stacks or as nested. So it is not possible to read a library, create microcode for its statements, and retain the previous microcode that contains the context of the library call.

Optimisation of polish notation is very difficult. Other representations, such as DAG's, are more suited to optimisation.

The CAYLEY user language also poses some limitations, primarily that the two-part definition of mathematical structures leaves a semantic "no man's land" in between the two parts, and that the coercion of values (especially literal values) to their parent mathematical structure is handled in an ad hoc way (without a formal theory).

5 Conclusion

We have presented a case study of the software architecture of CAYLEY. This included positioning it relative to known styles of software architecture, and discussing the limitations due to the choice of architecture.

The limitations of CAYLEY can be laid at the feet of its success. It was conceived as a better, yet still simple, means to interactively control group theoretical computations. No one could have foreseen the uses made: group theorists eagerly wrote large suites of routines, and explored many mathematical structures at the same time. The multi-fold relationships between the structures were of critical importance, as were structures other than groups.

Clearly the choice of architecture has had a very serious impact on the development of CAYLEY. This suggests that more study of the software architectures of computer algebra systems is of vital importance.

Acknowledgements: Tim Nicholson implemented the first version of CAYLEY and is responsible for the design of the interpreter and the abstract machine.

References

1. G. Abowd, R. Allen, D. Garlan, *Using style to give meaning to software architecture*, Proceedings of SIGSOFT'93: Foundations of Software Engineering, ACM Software Engineering Notes **18**, 3 (December 1993) 9–20.
2. A.V. Aho, R. Sethi, J.D. Ullman, **Compilers: Principles, techniques, and Tools**, Addison-Wesley, 1986.
3. Kent Beck and Ralph Johnson, *Patterns generate architectures*, ECOOP'94, M. Tokoro and R. Pareschi (eds), LNCS **821**, Springer-Verlag, pp. 139–149, 1994.
4. W. Bosma, J.J. Cannon, G. Matthews, *Programming with algebraic structures: Design of the* MAGMA *language*, **ISSAC 94**, ACM Press, New York, 1994, pp.52–57.
5. David Budgen, **Software Design**, Addison-Wesley, 1993.
6. W.H. Burge and S.M. Watt, *Infinite structures in Scratchpad II*, **EUROCAL '87**, (Proceedings of the European Conference on Computer Algebra, June 2-5, 1987, Leipzig, DDR), Davenport, J.H. (ed.), Lecture Notes in Computer Science **378**, Springer-Verlag, Berlin, 1989, pp.138-148.
7. G. Butler & John Cannon, *Cayley, version 4: the user language*, **ISSAC'88**, (Proceedings of 1988 International Symposium on Symbolic and Algebraic Computation, Rome, July 4-8), Lecture Notes in Comput. Sci. **358**, 456–466, Springer-Verlag, Berlin, 1989.
8. J.J. Cannon, *An introduction to the group theory language* CAYLEY, **Computational Group Theory**, M.D. Atkinson (ed.), Academic Press, London, 1984, 145–183.
9. B.W. Char, G.J. Fee, K.O. Geddes, G.H. Gonnet, M.B. Monagan, S.M. Watt, *On the design and performance of the Maple system*, Proceedings of the 1984 Macsyma Users' Conference.
10. B.W. Char, K.O. Geddes, W.M. Gentleman, G.H. Gonnet, *The design of Maple: a compact, portable and powerful computer algebra system*, **Computer Algebra**, (Proceedings of EUROCAL '83, European Computer Algebra Conference, March 28-30, 1983, London), van Hulzen, J.A. (ed.), Lecture Notes in Computer Science **162**, Springer-Verlag, Berlin, pp.101–115.
11. B.W. Char, K.O. Geddes, G.H. Gonnet, B.L. Leong, M.B. Monagan, S.M. Watt, **Maple V Languaage Reference Manual**, Springer-Verlag, New York, 1991.
12. G. Cooperman, *Star/MPI: Binding a parallel library to interactive symbolic algebra systems*, **ISSAC 95**, A.H.M. levelt (ed.), ACM Press, New York, 1995, 126–132.
13. J.H. Davenport, B.M. Trager, *Scratchpad's view of algebra I: Basic commutative algebra*, **Design and Implementation of Symbolic Computation Systems**, A. Miola (ed.), Springer LNCS **429**, 1990, 40–54.
14. J.H. Davenport, P. Gianni, B.M. Trager, *Scratchpad's view of algebra II: A categorical view of factorization*, **ISSAC 91**, S.M. Watt (ed.), ACM Press, New York, 1991, 32–38.
15. A. Diaz, E. Kaltofen, K. Schmitz, T. Valente, *DSC: A system for distributed symbolic computation*, **ISSAC 91**, S.M. Watt (ed.), ACM Press, New York, 1991, 323–332.
16. Erich Gamma, Richard Helm, Ralph Johnson, John Vlissides, *Design Patterns: Elements of Reusable Object-Oriented Software*, Addison-Wesley, 1994.
17. D. Garlan, R. Allen, J. Ockerbloom, *Exploiting style in architectural design environments*, Proceedings of SIGSOFT'94: Foundations of Software Engineering, ACM Software Engineering Notes **19**, 5 (December 1994) 175–188.

18. David Garlan and Mary Shaw. *An Introduction to Software Architecture.* Advances in Software Engineering and Knowledge Engineering, Volume 1, World Scientific Publishing Company, 1993.

19. J.M. Greif, *The SMP pattern matcher*, **EUROCAL '85**, (Proceedings of European Conference on Computer Algebra, April 1-3, 1985, Linz, Austria), Caviness, B.F. (ed.), Lecture Notes in Computer Science **204**, pp. 303-314.

20. R..D. Jenks and R.S. Sutor, **Axiom: The Scientific Computation System**, Springer-Verlag, New York, 1992.

21. Mark Moriconi and Xiaolei Qian, *Correctness and composition of software architectures*, Proceedings of SIGSOFT'94: Foundations of Software Engineering, ACM Software Engineering Notes **19**, 5 (December 1994) 164–174.

22. A.C. Norman, *The development of a vector-based algebra system*, **Computer Algebra**, (Proceedings of EUROCAM '82, European Computer Algebra Conference, April 5-7, 1982, Marseille, France), Calmet, J. (ed.), Lecture Notes in Computer Science **144**, Springer-Verlag, Berlin, 1982, pp.237-248.

23. A.C. Norman and P.M.A. Moore, *The initial design of a vector based algebra system*, **Symbolic and Algebraic Computation**, (Proceedings of EUROSAM '79, an International Symposium on Algebraic Manipulation, June 1979, Marseille, France), Ng, E.W. (ed.), Lecture Notes in Computer Science **72**, Springer-Verlag, Berlin, 1979, pp.258-265.

24. R. Prieto-Díaz and J.M. Neighbors, *Module interconnection languages*, J. Systems and Software **6**, 4 (1987) 307–334.

25. J.M. Purtilo, *Applications of a software interconnection system in mathematical problem solving environments*, **SYMSAC '86**, B.W. Char (ed.), ACM, New York, 1986, 16–23.

26. J. Rumbaugh, M. Blaha, W. Premerlani, F. Eddy, W. Lorenson, *Object-Oriented Modelling and Design*, Prentice Hall, 1991.

27. R.S. Sutor and R.D. Jenks, *The type inference and coercion facilities in the Scratchpad II interpreter*, (Proceedings of SIGPLAN '87 Symposium on Interpreters and Interpretive Techniques), SIGPLAN Notices **22**, 7 (1987) pp. 56-63.

28. G. Wiederhold, P.Wegner, S. Ceri, *Towards megaprogramming*, CACM **35** (1992) 89–99.

A Deductive Database for Mathematical Formulas

Stéphane Dalmas[1], Marc Gaëtano[2] and Claude Hurchet[2]

[1] INRIA Sophia Antipolis, projet SAFIR, 2004 route des lucioles, BP 93, 06902 Sophia-Antipolis CEDEX, France
[2] Université de Nice–Sophia-Antipolis, I3S, projet SAFIR

Abstract. In this paper, we present MFD2, a deductive database for mathematical formulas. The database itself is a stand-alone program which can run as a server in a client/server environment and it has been designed to be a powerful assistant for computer algebra systems as well as for other applications. For example, MFD2 could be used in an electronic handbook of mathematical relations or as a lemma database for a theorem prover. The information stored in the database is accessed through a specialized query language. At the heart of MFD2 is a deduction engine based on an algorithm for associate-commutative unification that takes care of the conditions associated with the formulas.

1 Introduction

Today computer algebra systems such as *Axiom*, *Maple*, *Mathematica* or *Macsyma* offer a wide range of mathematical abilities, can operate both symbolically and numerically as well as draw sophisticated graphics. Contrary to what was foreseen in the 60's (in the early days of computer algebra) modern systems do not make use of "artificial intelligence techniques". All of their mathematical knowledge is directly represented in source code. There is however what we can call non-algorithmic knowledge built into these systems as well, in the form of:

- code that expresses properties of elementary functions: the derivative of the sine function is the cosine function, e^x is positive for all real x,
- code to apply various transformations involving special functions, such as $J_{-n}(z) = (-1)^n J_n(z)$ (where J_n is a Bessel function of the first kind).
- code to recognize particular infinite sums or definite integrals.

Having this knowledge directly embedded in the source code of various functions causes consistency, maintainability and extensibility problems. Refining or adding a new property for a known function requires careful examination of the existing code and sometimes several source files must be changed. An alternative approach is to store this kind of information in a separate database and to provide a mechanism to retrieve it. In this paper, we describe such a tool, a deductive database for mathematical formula called MFD2.

MFD2 is a self-contained separate database that can be used in a client/server architecture. The information stored in MFD2 is accessed through a specialized query language designed to express mathematical relations. MFD2 has been designed to be a powerful assistant for computer algebra systems as well as for other applications: to be the heart of an electronic handbook of mathematical relations (such as [2]) or to be used as a lemma database for a theorem prover.

One of the particular application we are aiming at is the use of MFD2 for managing assumptions in computer algebra systems. A growing concern in the computer algebra community is the fact that today systems provide only "generic" results. For example, if in the course of the computation a division by x is performed, x is always assumed to take non-zero values but the system does not keep track of this fact for subsequent computations. More generally, all mathematical computations are performed under a set of conditions that must remain consistent. Only few efforts have been made in this direction. The `assume` functions of *Maple* ([16]) and *Macsyma* (and its proviso feature) are examples of rudimentary attempts to deal with conditions and assumptions, but these systems can still fail to infer very simple facts. For example, declaring `assume(n,integer)` in *Maple* (V release 3) does not cause `sin(n*Pi)` to simplify to zero.

Although checking the consistency of a set of general conditions is undecidable, there are many practical cases where MFD2 can be used to store some assumptions and confirm the validity of the steps that have been taken to produce a result.

We know of no other project with the same goals. [7] presents a program to store integral tables and retrieve user-requested integrals on demand. The techniques used (based on an associative-commutative pattern-matcher) are ad-hoc for the case of integrals. It does not seem to incorporate mechanisms to test or simplify the conditions that can be generated nor that this can be extended to more general formulas. [6] describes a theorem prover that uses the power of a general computer algebra system (*Mathematica*) but there is no provision for storing facts efficiently and it is limited by the generic simplifications of the underlying system. There are quite a few papers dealing with the management of rules and pattern-matching in symbolic computation systems ([9]). These works are not concerned with the management of a large number of rules and the way the rules are applied are generally more limited than in MFD2.

2 Presentation of MFD2

MFD2 is a deductive database for mathematical relations. The retrieval of the information stored in MFD2 is performed through a specialized query language. Its syntax is rather simple and close to the one used in computer algebra systems so we will use regular mathematical notations in the following sections for more readability.

Basically MFD2 can be asked if a mathematical relation is true or if it knows any equivalent form of a given expression. MFD2 can answer queries conditionally i.e. it gives partial solutions together with the conditions under which they are known to be true. Queries can include variables (represented as uppercase letters in our examples), MFD2 then attempts to find a description of the possible values for these variables.

The mathematical facts stored in MFD2 are mainly equalities or inequalities involving all the notions, operators and functions common in most computer algebra systems. MFD2 can typically store facts like:

$$e^x \geq 0 \quad \text{if} \quad x \in \mathbb{R}$$

from which it should be able to infer that $e^x + e^y \geq 0$ if x and y are reals, or formulas such as

$$\sin z = \prod_{k=1}^{\infty}(1 - \frac{z^2}{k^2\pi^2})$$

MFD2 stores a set of mathematical statements together with their conditions of validity. When a query is made, a deduction algorithm tries to find a statement in MFD2 that can be used to solve this query using unification. Assuming the relation "$\sin(2K\pi) = 0$ *for* $K \in \mathbf{Z}$" is stored in MFD2, the answer to the query "*knowing* $n \in \mathbf{Z}$, *is* $\sin((2n+4)\pi)$ *equal to* 0 *?*" is `true`, while solving the query "*is* $\sin(n(n+1)\pi)$ *equal to* 0 *?*" leads to the condition $\frac{n(n+1)}{2} \in \mathbf{Z}$. A condition inferred by MFD2 can be deleted if it appears to be true. For example, assuming MFD2 knows "$x \in \mathbb{R} \Rightarrow \lfloor x \rfloor \in \mathbf{Z}$", the answer to the query "*is* $\sin(2\lfloor x \rfloor \pi)$ *equal to* 0 *?*" is `true`. Notice that in order to solve queries like the ones involving $\sin((2n+4)\pi)$ the deduction algorithm may need to perform the same kind of manipulations a computer algebra system does (factorization and simplification).

Requests for MFD2 can contain variables. For example, we can ask for the values of X and Y such that $\sin(XY) = 0$. Assuming the only relation on the sine function stored in MFD2 is "$\sin(2K\pi) = 0$ for $K \in \mathbf{Z}$", the request "$\sin(XY) =$ *?*" returns multiple values:

$\sin(XY) = 0$ *for* $K \in \mathbf{Z}, X = 1, Y = 2K\pi$
$\sin(XY) = 0$ *for* $K \in \mathbf{Z}, X = 2, Y = K\pi$
$\sin(XY) = 0$ *for* $K \in \mathbf{Z}, X = K, Y = 2\pi$
$\sin(XY) = 0$ *for* $K \in \mathbf{Z}, X = 2K, Y = \pi$
$\sin(XY) = 0$ *for* $K \in \mathbf{Z}, X = 2K\pi, Y = 1$
...

Constraints can be set on variables to restrict their values with some conditions. The request "$\sin(XY) = 0$ *for* $X \in \mathbf{Z}$ *?*" returns only four answers out of the eight produced by the previous request.

Multiple values can be viewed as conditional expressions (if-then-else expressions). If we ask for the value of $\int x^n dx$, MFD2 answers:

$\frac{x^{n+1}}{n+1}$ *for* $n \neq -1$
$\log(|x|)$ *for* $n = -1$

Requests can also specify "syntactic" constraints in the requests. For example, we can ask for particular operators to appear in the answers. The request "*give an equivalent form of* $\log(xy)$, *containing the logarithm operator*", returns (in the domain of the real numbers):

$\log(xy) = \log(x) + \log(y)$ *for* $x > 0,\ y > 0$
$\log(xy) = \log(-x) + \log(-y)$ *for* $x < 0,\ y < 0$

MFD2 can also deal with requests like "*give an equivalent form of* $\cos(x)$ *without the tangent operator*" or "*give an equivalent form of* $\tan(2x^2)$ *with the operators* + *and* tan".

3 The deduction engine

In this section we focus on the original deduction process of MFD2. A simplified version of the Deduction Algorithm is described in figure 1 (as *DA*). It is based on binary resolution, together with a breadth-first search strategy, the formulas and the queries being represented as Horn clauses (a notation $\Sigma \rightarrow q$ is used for clauses). Some parameters, such as the depth of deduction, serve to control the whole procedure. Their default values can be set and modified by the user. At the heart of this algorithm is the extension of an associative-commutative unifier to take into account the relations stored in the database and their conditions of validity. As the unifier manages and introduces conditions, we call it a *conditional unifier*. The unifier is the *CUnifier* function in the algorithm of figure 1.

During the deduction steps, some simplifications are applied to the conditions. They are represented as the *spf* function. For example, if the two conditions $x \geq 0$ and $x > 0$ appear in the same set of conditions c then *spf*(c) is a set of conditions where $x \geq 0$ has been deleted.

The *pivot* function selects a relation from the premisses of a Horn clause, using the following rules: (i) if the head of the clause contains some variables then return it, (ii) if the relation of the head has not been already used then return it, (iii) else return one relation of the premises. Some of the heuristics used to select the relations which are likely to unify with the query (the *fetch* function in the algorithm), are briefly discussed in section 4. As the knowledge of MFD2 can be divided into several parts, only a subset of the relations stored in the database can be involved in a deduction. This subset appears in the algorithm as the parameter named B.

3.1 An example

Here is a simple example involving several facts. It illustrates some of the features of our deduction process and why our conditional unifier is more suitable than an ac-unifier. We want to use MFD2 to check if $x + \sqrt{x^2+1} > 0$ when $x \in \mathbb{R}$ (this is the query), supposing that the following statements about the square root are available (this is the content of the data base):

1. if $X > 0$ and $Y \geq 0$ then $X + Y >$
2. if $X \geq 0$ and $Y \geq 0$ then $X + Y \geq 0$
3. if $X \geq 0$ then $\sqrt{X} \geq 0$
4. $X^2 \geq 0$

$Answer(\Sigma \rightarrow q, B, 0) \longrightarrow$ FAIL
$Answer(\Sigma \rightarrow q, B, depth) \longrightarrow$
let uni the set of all pairs (r, cu) such that
$r \in fetch(q, B)$ and
$cu \in CUnifier(q, head(r))$
and let S the set of clauses $spf(\Sigma\sigma, \Sigma'\sigma, c \rightarrow q\sigma)$ with $(\Sigma' \rightarrow q', (c, \sigma)) \in uni$
then $DA(spf(S), B, depth - 1, \{\})$

$DA(S, B, 0, \{\}) \longrightarrow S$
$DA(\{\}, B, depth, newS) \longrightarrow$
if $depth > 0$ then $DA(spf(newS), B, depth - 1, \{\})$
else $spf(newS)$
$DA(\{\Sigma \rightarrow q\} \cup S, B, depth, newS) \longrightarrow$
let $p = pivot(\Sigma \rightarrow \Delta)$
and let uni the set of all pairs (r, cu) such that
$r \in fetch(q, B)$ and
$cu \in CUnifier(p, head(r))$
if $uni = \{\}$ then $DA(S, B, depth, \{\Sigma \rightarrow q\} \cup newS)$
else let $newS'$ the set of clauses made with $spf((\Sigma \setminus \{p\})\sigma, c, \Sigma'\sigma \rightarrow q)$
for $(\Sigma' \rightarrow q', (c, \sigma)) \in uni$
and return $DA(S, B, depth, newS' \cup newS)$.

Figure 1. Deduction engine

5. if $X \neq 0$ then $X^2 > 0$
6. if $X \geq 0$ then $X = \sqrt{X^2}$
7. if $X < 0$ then $X = -\sqrt{X^2}$
8. if $X \geq 0$ and $Y \geq 0$ and $X > Y$ then $\sqrt{X} > \sqrt{Y}$

Here variables are all supposed to be real, so all the conditions like $X \in \mathbb{R}$ are omitted. Note that we use uppercase letters for the variables in rules. They should not be confused with variables in queries.

If we used only an ac-unifier and with one step of deduction, two answers $\{x>0, \sqrt{x^2+1} \geq 0\} \Rightarrow x + \sqrt{x^2+1} > 0$ and $\{x \geq 0, \sqrt{x^2+1} > 0\} \Rightarrow x + \sqrt{x^2+1} > 0$ are deduced. If more steps are allowed, then the engine is used to try to prove the constraints, and we eventually obtain a single answer with the condition $x \geq 0$ in the premises. Using only associative-commutative unification (together with elementary simplification of mathematical expressions and sets) and these statements, enable to prove $x \in \mathbb{R} \wedge x \geq 0 \Rightarrow x + \sqrt{x^2+1} > 0$ and nothing better. The statement 9 is not used in this deduction because x does not act like a variable whereas X does. x behaves like an "unknown" constant or a parameter, and so no ac-unifier exits between $x + \sqrt{x^2+1} > 0$ and $\sqrt{X} - \sqrt{Y} > 0$.

If we want to go further and verify that $x+\sqrt{x^2+1}>0$ is true for any x real, we could add the *ad-hoc* rule: if $X\leq 0 \wedge Y\geq 0 \wedge Y-X^2>0$ then $X+\sqrt{Y}>0$.

Instead, we have chosen to use the knowledge already stored in the database as much as possible. In this example, the relation 8 implies that the first occurrence of x may be viewed as $-\sqrt{x^2}$, assuming that $x<0$. In this case, the query and the statement 9 have an ac-unifier and the first step of deduction gives three answers with premises: $\{x>0, \sqrt{x^2+1}\geq 0\}$, $\{x\geq 0, \sqrt{x^2+1}>0\}$ and $\{x<0, x^2+1>x^2\}$. If more steps are allowed and with the help of the simplifications, MFD2 is enabled to deduce that $x\in\mathbb{R}\Rightarrow x+\sqrt{x^2+1}>0$.

Translating an occurrence of x into $-\sqrt{x^2}$ is like a paramodulation step [10]. The next subsection shows that this step is tried only when ac-unification fails.

3.2 Conditional unification

As the previous examples show, we need a powerful unifier that may produce new constraints. Unfortunately, it is a very difficult task to develop a complete unifier mixing simple mathematical properties ([3]). Unification modulo associativity and distributivity is undecidable ([10]).

The theory underlying an MFD2 database is made up of the relations stored in it and some mathematical properties that are not explicitely stored relations. For example the commutativity of the addition over the real numbers is a built-in property and the relation $X\in\mathbb{R}, Y\in\mathbb{R}\Rightarrow X+Y=Y+X$ is not stored in MFD2. We thus chose a pragmatic approach to build a general "two-step" unifier. The first step, called **AC-Unify**, is an ordinary ac-unification step (following the algorithm described in [5]) except that it returns all pairs of sub-terms that cause the failure of the unification problem. The second one treats these pairs as equations and tries to solve them, while keeping track of validity conditions for the solutions. That leads to the introduction of conditional substitutions.

One of the main restrictions of MFD2, is that a variable in a query cannot occur twice (variable have been noted as uppercase letters in our examples). This is necessary to have a reasonably efficient algorithm (we use the algorithm described in [5]). In practice, this is not a restriction for the kind of queries we naturally handle.

Here is a definition of conditional unification, using the standard notations of rewriting:

Definition 3.1. A conditional E-unifier for the terms t and t' is a pair (c,σ) consiting of a condition and a substitution such that if c is true in E then $t\sigma =_E t'\sigma$. This is denoted $E\models c\Rightarrow t\sigma = t'\sigma$.

AC-Unify keeps track of a failure as a pair denoted by $x \not\leftarrow y$ (a substitution element is denoted by $x\leftarrow y$). When we try to unify the set of pairs $\{(x_1,y_1),\ldots,(x_n,y_n)\}$, if x_1 and y_1 have no ac-unifier then $x_1\not\leftarrow y_1$ is kept and **AC-Unify** carries on with $\{(x_2,y_2),\ldots,(x_n,y_n)\}$. We call *semi-substitution* $\sigma=\sigma_{ps}\cup\sigma_{ss}$ a set made up of a set of substitution elements (σ_{ps}) and a set (σ_{ss}) of pairs like $x\not\leftarrow y$. Our unifier, **AC-Unify** thus takes two terms x and y

and returns a complete set of unifiers for the equation $x =_{AC} y$ if it exists, or a set of semi-substitutions.

For example **AC-Unify**($x+\sqrt{x^2+1}, \sqrt{X}-\sqrt{Y}$) returns two semi-substitutions: $\{X \leftarrow x^2+1, -\sqrt{Y} \not\leftarrow x\}$ and $\{\sqrt{X} \not\leftarrow x, -\sqrt{Y} \not\leftarrow \sqrt{x^2+1}\}$. And **AC-Unify**($F(X+Y, X-Y), F(a,b)$) returns $\{X+Y \not\leftarrow a, X-Y \not\leftarrow b\}$.

For now, three methods are tried successively to transform semi-substitutions into conditional unifiers. First, simple algebraic transformations, such as applying the distribution law and solving linear parametric equations for a single variable, are tried. Some examples of semi-substitutions that can be reduced at this stage are: $2K \not\leftarrow 2n+4$ or $2(K+\pi) \not\leftarrow 2n+4\pi$. The second transformation consists of the generalization to parametric linear sub-systems of equations, where the semi-substitution element $x \not\leftarrow y$ is considered as the equality $x = y$. This can be done using the algorithm described in [14]. In the third stage, we try to solve a semi-substitution element like $x \not\leftarrow y$ from σ, by ac-unifying $x = y$ with the data stored in MFD2. If a set of complete ac-unifiers exists for $x = y$ and we have an entry $c \Rightarrow u = v$, then for each ac-unifier θ a new conditional semi-substitution is created by removing $x \not\leftarrow y$ from σ, by composing $\theta|_{\mathrm{var}(x=y)}$ with this new semi-substitution and by adding the condition $c\theta$ to the current condition. In the previous example, the semi-substitution $\{X \leftarrow x^2+1, -\sqrt{Y} \not\leftarrow x\}$ is reduced by the third step with the 8th rule and is transformed into the conditional unifier $(x<0, \{X \leftarrow x^2+1, Y \leftarrow x^2\})$. For complexity reasons, the ac-unification is applied only in this third step.

As all these transformations are correct with respect to the underlying mathematical context of the database, the deduction engine always gives correct answers. Of course we cannot hope to have a logically complete engine.

3.3 Normal Form and type

As in computer algebra systems, terms in MFD2 have a normal form (close to the one of *Maple*). As terms are only involved in clauses, we can use conditional simplifications without losing any significant conditions, contrary to what happens in all computer algebra systems. For example, when the term x/x is simplified to 1 in a clause, the condition $x \neq 0$ is added to the premises of this clause.

MFD2 deals with typing through the explicit addition of membership relations in the premises of a clause. This is captured by the $\in$-correct relation :

Definition 3.2. A clause S is *correct for the membership relation* ($\in$-correct), if for each variable or parameter X occurring in S, a single relation $X \in d$ occurs in the premises of S, with d a domain constant, such that the type of each occurrence of X agrees with this condition.

For example, $x \in \mathbb{R}, \mathrm{even}(n) \Rightarrow x^n \geq 0$ is not $\in$-correct (no membership relation for n in the premises), but $x \in \mathbb{R} \wedge x > 0 \Rightarrow \ln(x) \in \mathbb{R}$ is $\in$-correct.

A simple algorithm translates a clause into a $\in$-correct clause. Of course, we use inclusion relations between domains to determine $\in$-correctness and to simplify conditions containing membership expressions. The clauses that do not have a $\in$-correct form, cannot be stored in MFD2.

4 Implementation

We have implemented a first prototype of MFD2, written in SML (using the SML/NJ compiler). We extensively use the facilities provided by the module system of SML to be able to experiment with various design and implementation decisions.

4.1 Entering data

To enter and organize the large sets of formulas that make up the knowledge of MFD2 we have chosen a simple book metaphor. The MFD2 engine can read *books* represented as files. These books consist of chapters that are divided in sections and subsections. Books basically contain symbol declarations and formulas. Formulas are written in a usual syntax (similar, for example, to the syntax of *Maple*). A symbol declaration for a constant (such as π) gives the set it belongs to. For a function, the domain and co-domain are specified.

Keywords, comments and bibliographical references (with support for using different languages) can be attached to symbols, formulas and sectioning units (chapters, sections and subsections).

The MFD2 engine does not read books directly. An intermediate compiler reads the files, does some checking on their contents and generates an intermediate structure that is the true input to the engine. We choose this organization because we plan to provide a "dump and restore" facility for books (writing and reading an intermediate compact format) to achieve a better startup time.

The verifications performed by the compiler are quite simple at this time and mostly consist of a type-checking phase. The symbol declarations are used to check the correctness of each expression. We think that these simple verifications can detect most if not all typing errors. An interesting project would be to use a computer algebra system to try to verify the formulas (the simplest idea would be to perform evaluation at some random points).

The compiler also adds implicit assumptions to the validity conditions of formulas (assumptions related to domains of functions that have not been given explicitly such as a denominator not being 0) and then transforms expressions and conditions into a normal form.

4.2 Storage and retrieval of expressions

The way the formulas are organized in the database is of course of a tremendous importance for the efficiency of our deduction and unification processes. To get a rough idea of how many formulas such a tool would know, we can say that a book like [2] contains about 4000 formulas, involving a little less than 200 functions. The CRC integral table contains approximately 900 entries.

We tried to organized our code carefully to be able to easily experiment with different storage and organization methods for the content of the database (unfortunately this is quite a complicated task).

Following [7], in our first implementation the formulas are organized in a single *semantic discrimination tree* (*cd-tree*). This data structure is an adaptation of the discrimination trees (d-trees) described in [13, 12]. D-trees can store and retrieve data associated with keys that are first order terms. Given a term t, a d-tree is a suitable data structure to fetch a set efficiently that contains all terms t' in the d-tree such that t and t' have a unifier in the empty theory. The data structure we use, the *cd-trees*, can retrieve a set of terms that contain all terms that have an ac-unifier with the input term (experiments showed that this set is not too large in practice). Whereas this extension is suitable when only ac-unification is performed, it is not powerful enough to be the only structure used by our deduction engine as cd-trees are essentially "syntactic" (basically they discriminate on the head symbol of a term). The way semi-substitutions are solved implies that a conditional unifier may be found for two terms with distinct head symbols. The cd-tree structure cannot ensure that it returns all possible entries which can have a conditional unifier with a given query. We have thus implemented another structure on the whole database that is used to find all clauses involving a given set of operators, efficiently. This structure is basically a table that associates each operator with the set of all relations involving this operator. These sets are sorted according to a "cost of extracting". For example, a formula such as $f(x+y) = g(x)g(y)^2$ has a lower cost of extracting with respect to f than to g. The cd-tree structure is used to select a first set of clauses. If they all fail to produce an answer we look for other rules using our additional structure.

5 Conclusion

The current implementation of MFD2 already gives promising results but we still have a lot of improvements to achieve to make it practical for general use.

Experiments that we plan to carry out on the data structures used by MFD2 include using ideas from [4] and making use of hash-coding techniques. We also plan to refine our indexing scheme (linking all rules that contain a given operator) by using sets of operators. For example, we will link together all rules containing the sine function and integral or $\leq$ and the exponential function. Of course, this will be done only for the less discriminating operators (operators that occur in many rules). A little statistical study on the formulas of [2] seems to show that this can be quite effective.

From a very practical point of view, we will make MFD2 use the *OpenMath* protocol ([1]). This should make MFD2 directly available to several computer algebra systems (at least *Maple*, *Axiom* and *Reduce* should be able to access other *OpenMath* programs).

On a more theoretical point of view, we plan to try to evaluate and incorporate more results from rewriting including constrained rewriting ([11]) or dynamically-typed computations as in [8] (which is close to our notion of $\in$-correctness). A way to improve our unifier could be to treat identity at the same level as associative and commutative properties. Some experiments are required to determine the advantages of this choice. The conditional unifier may also be improved by using paramodulation rules for ac-unification (extended and contextual paramodulation rules as in [15]).

One of the original aspects of MFD2 is that our deduction process was designed to handle requests that are "close enough" to some stored facts. The general policy is to trade space for time: if MFD2 cannot infer something simple in reasonable time then this fact should be added. MFD2 is thus more a database of theorems than a real theorem prover.

References

1. The draft OpenMath home page on the World Wide Web. `http://www.can.nl/~abbott/OpenMath/index.html`.
2. Milton Abramowitz and Irene A. Stegun. *Handbook of Mathematical Functions*. Dover, 1970.
3. Alexandre Boudet, Jean-Pierre Jouannaud, and Manfred Schmidt-Schauss. Unification in boolean rings and abelian groups. *Journal of Symbolic Computation*, 8:449–477, 1989.
4. Jim Christian. Flatterms, discrimination nets and fast term rewriting. *Journal of Automated Reasoning*, 10:95–113, 1993.
5. Jim Christian and Patrick Lincoln. Adventures in associative-commutative unification. *Journal of Symbolic Computation*, 8:217–240, 1989.
6. E.M. Clarke and X. Zhao. Analytica - a theorem prover for mathematica. *The Mathematica Journal*, 3(1):56–71, 1993.
7. Richard J. Fateman and T. H. Einwohner. Searching techniques for integral tables. submitted to ISSAC'95, 1995.
8. Claus Hintermeier, Claude Kirchner, and Hélène Kirchner. Dynamically-typed computations for order-sorted equational presentations. Research Report 2208, INRIA, 1994.
9. Richard D. Jenks. A pattern compiler. In *Proceeding of the 1976 Symposium on Symbolic and Algebraic Manipulation*, pages 60–65. ACM Press, August 1976.
10. Jean-Pierre Jouannaud and Claude Kirchner. *Computational logic : essays in honnor of Alan Robinson*, chapter Solving Equations in Abstract Algebras: A Rule-Based Survey of Unification. The MIT press, 1991.
11. Hélène Kirchner. Some extensions of rewriting. In *Term Rewriting, French Spring School of Theoretical Computer Science*, volume 909, pages 54–73. Springer-Verlag, LNCS, 1993.
12. William McCune. Experiments with discrimination-tree indexing and path indexing for term retrieval. *Journal of Automated Reasoning*, 9:147–167, 1992.
13. Peter Norvig. *Paradigms of Artificial Intelligence Programming : Case Studies in Common Lisp*, pages 472–483. Morgan Kaufmann, 1992.
14. William Y. Sit. An algorithm for solving parametric linear systems. *Journal of Symbolic Computation*, 13:353–394, 1992.
15. Laurent Vigneron. *Déduction Automatique avec Contraintes Symboliques dans les Théories Équationnelles*. PhD thesis, Université de Nancy 1, 1994.
16. Trudy Weibel and Gaston H. Gonnet. An assume facility for cas, with a sample implementation for maple. In *Design and Implementation of Symbolic Computation Systems*, 1992.

CASA – A System for Computer Aided Constructive Algebraic Geometry

Michal Mňuk* and Franz Winkler**

Research Institute for Symbolic Computation, Linz, Austria

Abstract. Increasing power of computing devices shed a new light on the role of mathematical experiments. Constructive methods yield new insights into the nature of complicated problems. This paper describes the system CASA which implements basic principles of the classical algebraic geometry. It was created around the notion of the algebraic set and is currently supporting many operations on them while adhering to simplicity and efficiency.

1 Introduction

The old masters of algebraic geometry developed this field in an inherently constructive spirit. This fact together with the increasing power of computers laid a firm basis for implementation of a large portion of this field in software. Thus it becomes possible to automatize many computations hence opening new prospects and possibilities not only to mathematical research but also to industrial applications.

The package CASA – the name standing for *Computer Algebra System for Algebraic geometry* – was designed to perform computations and utilize reasoning about geometrical objects in the classical affine and projective algebraic geometry over algebraically closed fields of characteristic 0. Usually, the underlying ground field is the field of rational numbers or finite algebraic extensions thereof. CASA has been developed at the Research Institute for Symbolic Computation, Linz, Austria, in a research group headed by F. Winkler. The project started in 1990. A first report on development of CASA appeared in [5]. Since that time many people contributed to CASA in one or in the other way. The largest part of work has been done by R. Gebauer, M. Kalkbrener, M. Mňuk, R.J. Sendra, P. Stadelmeyer, B. Wall, and F. Winkler.

CASA has been built around the notion of the algebraic set. Originally, fundamental algorithms were written for algebraic sets in implicit representation, i.e. for sets given by zeros of a set of polynomials. However, due to theoretical and practical reasons the variety of possible representations was extended to

* E-mail: `mmnuk@risc.uni-linz.ac.at`

** E-mail: `winkler@risc.uni-linz.ac.at`

comprise also the parametric, projected and the representation by power series (used for curves).

- *Implicit representation:* An algebraic set is the set of common zeros of a system of polynomial equations. To give an algebraic set in implicit form means to give finitely many polynomials.
- *Projected representation:* As a consequence of the primitive element theorem every irreducible d–dimensional algebraic set in n–dimensional space is, after a suitable linear transformation of coordinates, birationally projectable onto an irreducible d–dimensional algebraic set in $(d+1)$–dimensional space. The image of this projection can be specified by a single polynomial in $d+1$ variables. This can be generalized to unmixed-dimensional algebraic sets. An algebraic set in projected form is given by a polynomial and a tuple of rational functions (specifying the birational mapping).
- *Parametric representation:* Some irreducible algebraic sets can be parametrized by rational functions. An algebraic set in parametric form is given by a tuple of rational functions that parametrizes the algebraic set.
- *Representation by places:* All algebraic curves can be parametrized by a set of power series that are convergent around a point of the curve. An algebraic set is given by places if for each branch passing through a certain point on the algebraic set a tuple of power series that parametrizes the algebraic set around the point is specified.

CASA also works with the polynomial ideals corresponding to these geometric objects.

The operations available in CASA include

- ideal theoretic operations $+, *, \cap, /$,
- creating algebraic sets in different representations,
- generating curves of fixed multiplicities at given points,
- intersection, union, and difference of algebraic sets,
- computing tangent cones and tangent spaces,
- computation of the dimension of an algebraic set,
- decomposition into irreducible components,
- transformations of algebraic sets to hypersurfaces,
- computation of the genus of a curve,
- rational parametrization of curves,
- implicitization of parametrically given algebraic sets,
- Puiseux series expansions,
- plotting both explicitly and implicitly given curves and surfaces.

For more information on the underlying mathematics and software issues we refer to [13, 14]. Some typical sessions of CASA may be found in [8, 17, 18].

The basic philosophy of CASA is to provide a variety of operations on algebraic sets. As the efficiency of these operations is tightly bound to the way algebraic sets are represented, conversion routines have been provided to support various views on one object, to deepen the understanding of its principles, and to speed up algorithms working on algebraic sets. The major goal was to design a system which provides a comfortable, easy to use, efficient, flexible, and mathematically exact working environment for constructive algebraic geometry where all basic theoretical concepts map easily to available data structures. The fact that CASA is based solely on Maple and is fully independent of the operating system, allows it to be used on every hardware where Maple is running.

There are software systems which partially cover some of the fields of CASA, however, to authors' knowledge, there is currently no system with comparable capabilities available.

2 Structure of CASA

CASA evolved over several years from an independent set of Maple programs. From version 2.1 to 2.2 there was a significant change in the internal structure of the system. The code was reorganized into a number of separate modules. Each module contains a set of operations performing related tasks. After this, the code became cleaner and several modules may be now used as stand-alone Maple packages having distinguished name space which avoids conflicts with other software already loaded in the Maple session. The whole system is distributed as a single Maple archive of compiled routines. Unlike previous versions of CASA, the modules and their major parts are loaded into memory using Maple's `readlib`-facility when they get used for the first time. This significantly cuts the amount of memory occupied by the code keeping it free for data.

The documentation of CASA is written using a restricted set of LaTeX commands. The description of every procedure may then be converted to both online Maple help file and to high quality printouts including mathematical formulae. This approach was chosen in order to keep the on-line help consistent with printed manual.

3 Modules

3.1 Basic Operations

The central notion of CASA – the algebraic set – is implemented as a structure (Maple's unevaluated function call) holding the defining entities, a description of the underlying space in which the set is embedded, and some attributes and properties. During a computation, when a new knowledge about an algebraic set is obtained, it is added to its property list to avoid laborious recomputations when this knowledge is needed later. The information already contained in the property list is retrieved whenever some function requires it.

The module for basic operations provides functions to manipulate the structure of the algebraic set, to add new knowledge, and to query about it.

To create an algebraic set given by a polynomial ideal the function `mkImplAlgSet` is called with the generating ideal, the space in form of a list of variables, and some additional information which may be known in advance. The set is implicitly assumed to be affine if it is not explicitly declared projective (by adding the attribute `basespace=projective` to the property list.

```
> A := mkImplAlgSet([x*(x^2+y^2-1),x*z,x^2+y^2+z^2-1],
>                   [x,y,z]);
```

$$A := \text{algebraic\_set}([xz, x(x^2+y^2-1), x^2+y^2+z^2-1], [x,y,z])$$

Parametric sets are given by a list of rational functions in some parameters.

```
> P:=mkParaAlgSet([(t^4-12*t^2-3*t+13)/(t^3-2*t+1),
>                  (5*t^3-6*t^2-4)/(t^3+2*t+1)],[t]);
```

$$P := \text{algebraic\_set}\left(\left[\frac{t^4-12t^2-3t+13}{t^3-2t+1}, \frac{5t^3-6t^2-4}{t^3+2t+1}\right], [t]\right)$$

The representation may be obtained by calling the function `represent`.

```
> represent(P);
```

$$\left[\frac{t^4-12t^2-3t+13}{t^3-2t+1}, \frac{5t^3-6t^2-4}{t^3+2t+1}\right]$$

3.2 Gröbner Bases and Syzygies

The concept of Gröbner bases was originally introduced for polynomial ideals ([1]). But as shown in [9] it can be naturally generalized to submodules of a free module over polynomial rings.

Besides the improved classical Gröbner bases algorithm CASA contains an implementation of Gröbner bases for modules over polynomial rings. It allows to compute Gröbner bases for multivariate polynomial modules and solve systems of linear equations with polynomial coefficients.

Consider the following linear system of equations in $z_1, z_2, z_3 \in K[x,y]$:

$$\begin{pmatrix} x^3-y & xy-1 & -x+y^2 \\ -x+y & x^3-1 & y-1 \end{pmatrix} \cdot \begin{pmatrix} z_1 \\ z_2 \\ z_3 \end{pmatrix} = \begin{pmatrix} xy^2-y^2 \\ -x^5+x^2-x+y^2 \end{pmatrix}$$

The function msolveGB yields the basis of the solution space.

```
> msolveGB([[x^3-y,-x+y],[x*y-1,x^3-1],[-x+y^2,y-1]],
>           [x*y^2-y^2,-x^5+x^2-x+y^2],[x,y],term,tdeg);
```

$$[[y,-x^2,x],[[-1+x+y-y^2+xy-xy^2+x^3y^2-x^4,\\ y-y^2-x^3-x^2+xy+xy^2-y^3+x^3y,\\ x-2y+x^3+xy^2+x^3y-x^2y-x^6]]]$$

3.3 Set Operations

The canonical correspondence of algebraic sets and radical polynomial ideals allows to perform set theoretic operations on algebraic sets (union, difference, intersection, Zariski closure, etc). The basic algorithm used here is the computation of Gröbner bases in polynomial rings.

Let us consider the algebraic set A from Section 3.1 and represent it as an intersection of radical ideals by calling the function computeRadical.

```
> radA := computeRadical(A);
```

$$radA := \text{algebraic\_set}([z,x^2+y^2-1],[x,y,z]),\\ \text{algebraic\_set}([x,y^2+z^2-1],[x,y,z])$$

The union of the two algebraic sets from above is obtained using implUnion.

```
> G:=implUnion(radA);
```

$$G := \text{algebraic\_set}([xz,zy^2+z^3-z,x^3+y^2x-x,\\ x^2y^2+x^2z^2-x^2+y^4+y^2z^2-2y^2-z^2+1],[z,x,y])$$

Using the corresponding operations on ideals, the closure of the set theoretic difference of A and the second component $radA_2$ of $radA$ is computed.

$$A = \text{algebraic\_set}([x(x^2+y^2-1),xz,x^2+y^2+z^2-1],[x,y,z])$$

$$radA_2 = \text{algebraic\_set}([x,y^2+z^2-1],[x,y,z])$$

```
> implDifference(A,radA[2]);
```

$$\text{algebraic\_set}([z,x^2+y^2-1],[x,y,z])$$

3.4 Conversions

To utilize efficient computations over algebraic sets and to support different views, many conversion operations between representations mentioned above are implemented in CASA. The conversion of a one-dimensional set in implicit form to a parametric representation is done by the algorithm described in [12]. To implicitize parametric sets, Gröbner basis computation is used. Algorithms for converting algebraic sets to projected representation are described in [15]. Puiseux series are used to obtain curves represented by places (series).

To convert the set A to the projected representation, `convertRep(A,proj)` is called.

```
> B:=convertRep(A,proj);
```

$$B := \text{algebraic\_set}([[x^2+y^2-1],[x,y,0]],[y,x]),$$
$$\text{algebraic\_set}([[z^2-1+x^2],[0,-x,z]],[x,z])$$

The representation of the first set

```
> represent(B[1]);
```

$$[[x^2+y^2-1],[x,y,0]]$$

consists of a list defining a hypersurface x^2+y^2-1 in the 2-dimensional affine space and a mapping

$$\mathbf{A}^2 \longrightarrow \mathbf{A}^3$$
$$(x,y) \longrightarrow (x,y,0)$$

In this computation, the Gröbner basis of A has been determined and automatically added to the property list by some function called by `convertRep`.

```
> attributes(A);
```

$$[groebnerbasis = [plex,[x^2+y^2+z^2-1,xz,zy^2+z^3-z]]]$$

The basis will be retrieved whenever it is needed in subsequent tasks.

Among others, routines to convert parametric representation to the implicit one are implemented in CASA. The following parametric space curve C may be represented as a zero set of an ideal in $K[x,y,z]$ with 4 generators given below.

```
> C:=mkParaAlgSet([(t^4+3*t^2-7*t+4)/(t^2-3*t+1),
>                  (4*t^2-1)/(2*t^2-3),t],[t]);
```

$$C := \text{algebraic\_set}\left(\left[\frac{t^4+3t^2-7t+4}{t^2-3t+1},\frac{4t^2-1}{2t^2-3},t\right],[t]\right)$$

```
> convertRep(C,impl);
```

$$\begin{aligned}&\text{algebraic\_set}([5010x+9772y-3500z+1500zx-17843+884x^2\\&+21960xy-13421y^2+8076y^3-176x^3+1304x^3y+9548y^2x^2\\&-1072x^3y^2-1688y^3x^2+430y^3x-14280y^2x-12536x^2y\\&+232x^3y^3, 392x+3136y+176x^2+2996xy-2793-2702y^2\\&+2184y^3+1724y^2x^2-768y^3x^2+1564y^3x-4132y^2x-673y^4\\&-1392x^2y+116y^4x^2-148y^4x, 590zy^2-1260zy+910z\\&+116y^3x^2-86x+189y-88x^2-2356xy+653y^2-673y^3\\&-536y^2x^2-148y^3x+1519+1558y^2x+652x^2y, 250z^2\\&-116y^3x^2+196x+1026y+88x^2+1596xy-838y^2+673y^3\\&+536y^2x^2+148y^3x-1268y^2x-1459-652x^2y], [x,y,z])\end{aligned}$$

3.5 Dimension

For the dimension computation the algorithm in [6] was implemented in CASA. Only one Gröbner basis w.r.t. a lexicographic term ordering needs to be computed for determining the dimension of an ideal.

```
> U := mkImplAlgSet([36*z*y^2*x+8*x^3*z^5+16*x^3*z^2*y^3
>           -8*x^5*z -4*x^3*z^2*y^2-2*z*y^2*x^2-z^2*y^4,
>          12*x^6*z*y^2-48*x^4*z-39*z*y^6-12*z*y^2*x,
>          -12*x^3*z-3*z*y^2],[x, y, z]);
```

$$\begin{aligned}U := &\text{algebraic\_set}([12x^6zy^2-48x^4z-39zy^6-12zy^2x,\\&36zy^2x+8x^3z^5+16x^3z^2y^3-8x^5z-4x^3z^2y^2-2zy^2x^2-z^2y^4,\\&-12x^3z-3zy^2],[x,y,z])\end{aligned}$$

```
> dimension(U);
```

$$2$$

3.6 Decomposition

Decomposing algebraic sets into irreducible parts is achieved by constructing the characteristic set for the defining polynomial ideal (cf. [16]). For sets in 3-dimensional space, the projected representation onto unmixed dimensional hypersurfaces may be used to achieve partial splitting. The factors of generating polynomials of hypersurfaces yield then the irreducible components.

```
> N:=mkImplAlgSet([-y^2*z^2 + 10*y^2*z - y*z^2 - 25*y^2
>         + 6*y*z + x - 5*y +1,y^3*z^2 - 10*y^3*z + 25*y^3
>         - y^2*z + 5*y^2 - y*z + y]);
```

```
> U:=decompose(N);
```

U :=algebraic_set($[y, 1+x], [x, y, z]$),
algebraic_set([
$20yz+zx+5z-x^2y-54y-152y^2-22y^2x$
$-19yx+2y^2x^2-4,$
$y^3x^2-76y^3-15y^2-3y-1-11y^3x-15y^2x-3yx-x,$
$4z^2+zx-23z-2y^2x^2+152y^2+106y+36+22y^2x+41yx$
$+16x-x^2y, -16x^2y+65z-44x-438y-1064y^2+2zx$
$-306y^2x-197yx-8y^2x^2+2y^2x^3-x^3y+zx^2-108], [x, y, z]$)

3.7 Curves

Many algorithms for dealing with algebraic curves are implemented – parametrization and implicitization, tangent cones, adjoint curves, singularities, implicit plots, etc. All algorithms use symbolic methods to achieve exact results. In the future, mixed symbolic-numerical methods will be implemented to gain more speed-up while keeping sufficient accuracy.

Let us consider a curve C_1 in the 2-dimensional projective space.

```
> C1:=mkImplAlgSet([-15*y^2*z^3-76*y^3*z^2-z^5-3*y*z^4
>       -15*x*y^2*z^2-11*x*y^3*z-x*z^4-3*x*y*z^3+y^3*x^2],
>       [x,y,z],[basespace=projective]);
```

C_1 :=algebraic_set($[-15y^2z^3-76y^3z^2-z^5-3yz^4-15xy^2z^2-11xy^3z$
$-xz^4-3xyz^3+y^3x^2], [x, y, z]$)

The function `singularities` yields all singular points of C_1 decomposed into classes points having the same multiplicity.

```
> singularities(C1);
```

table([
3 = [[1,0,0]]

$$2=\left[\left[\text{RootOf}(\_Z^2+2\_Z+65),-\frac{1}{4}-\frac{1}{20}\text{RootOf}(\_Z^2+2\_Z+65),1\right],\ [0,1,0]\right]$$

])

The above curve has one triple point and three[1] double points, it has genus 0, and hence may be converted into parametric representation.

```
> convertRep(C1,para);
```

$$\text{algebraic\_set}\left(\left[-t^3-15t^2-75t-126,-\frac{t+5}{26+t^2+11t},1\right],[t]\right)$$

3.8 Plotting

Visualization of algebraic sets is a critical task. Their in many cases complicated nature must be faithfully translated into a picture. For the majority of tasks, only singular or other distinguished points of algebraic sets yield some interesting information. These objects have to be treated with care, requiring special analysis. CASA makes a thorough effort to obtain correct information about the local topology in neighborhoods of singularities and other critical points.

The curve C_1 from Section 3.7 is a projective curve. We may consider the affine pieces of C_1, i.e. $C_{1,x} = C_1 \cap V(x-1)$, $C_{1,y} = C_1 \cap V(y-1)$, and $C_{1,z} = C_1 \cap V(z-1)$:

```
> C1x:=convertSpace(C1,affine,x);
> C1y:=convertSpace(C1,affine,y);
> C1z:=convertSpace(C1,affine,z);
```

$$C_{1,x} := \text{algebraic\_set}([-15y^2z^3-76y^3z^2-z^5-3yz^4-15y^2z^2 -11y^3z-z^4-3yz^3+y^3],[y,z])$$

$$C_{1,y} := \text{algebraic\_set}([-15z^3-76z^2-z^5-3z^4-15xz^2-11xz -xz^4-3xz^3+x^2],[x,z])$$

$$C_{1,z} := \text{algebraic\_set}([-15y^2-76y^3-1-3y-15xy^2-11xy^3 -x-3xy+y^3x^2],[x,y])$$

[1] Note that the first "point" in the list of double points is a class of two points each corresponding to one root of $z^2+2z+65$.

Based on the analysis of certain distinguished points, CASA's function `plotAlgSet` makes exact and topologically correct drawings. See figure 1.

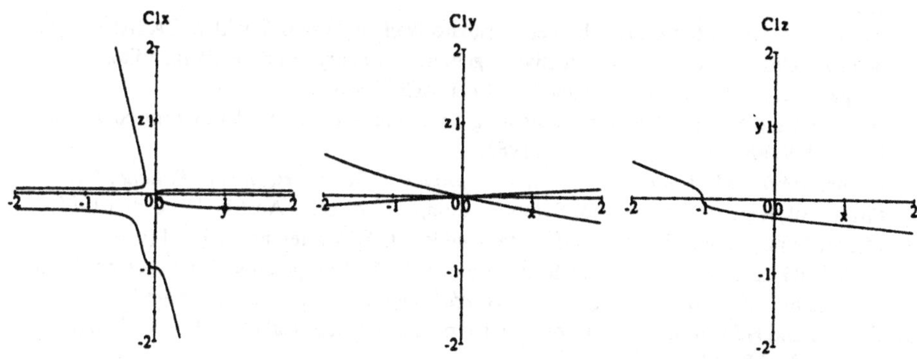

Figure 1. Affine pieces $C_{1,x}$, $C_{1,y}$, and $C_{1,z}$ of C_1

4 Future Developments

The CASA system is undergoing steady development. In next releases, parametrization of surfaces, recent results in parametrization of curves, search for rational points on curves, an improved hybrid symbolic-numerical algorithm for plotting algebraic sets, offset curves and surfaces, and Hilbert polynomial series will be implemented.

Availability

CASA is available for anonymous ftp at *ftp.risc.uni-linz.ac.at* in the directory */pub/CASA*. As of this writing the version 2.2 (patchlevel 1) is available. A WWW page may be accessed at *http://info.risc.uni-linz.ac.at:/labs-info/compal/software/casa/casa.html*.

References

1. Bruno Buchberger. An algorithmic method in polynomial ideal theory. In N.K. Bose, editor, *Multidimensional System Theory*, pages 184–232. Reidel, Dordrecht Boston Lancaster, 1985.

2. Bruce W. Char, Keith O. Geddes, Gaston H. Gonnet, Benton L. Leong, Michael B. Monagan, and Stephen M. Watt. *Maple V Library Reference Manual.* Springer-Verlag, 1991.
3. Bruce W. Char, Keith O. Geddes, Gaston H. Gonnet, Benton L. Leong, Michael B. Monagan, and Stephen M. Watt. *Maple V: Language Reference Manual.* Springer-Verlag, 1991.
4. Bruce W. Char, Keith O. Geddes, Gaston H. Gonnet, Benton L. Leong, Michael B. Monagan, and Stephen M. Watt. *First Leaves. A Tutorial Introduction to Maple V.* Springer-Verlag, 1992.
5. Richard Gebauer, Michael Kalkbrener, Bernhard Wall, and Franz Winkler. CASA: A Computer Algebra Package for Constructive Algebraic Geometry. In S. M. Watt, editor, *ISSAC 91*, pages 403–410, Bonn, Germany, July 1991. ACM Press.
6. H. Kredel and V. Weispfenning. Computing dimension and independent set for polynomial ideals. *J. Symb. Comput.*, 6:231–248, 1983.
7. Michal Mňuk. *Algebraic and Geometric Approach to Parametrization of Rational Curves.* PhD thesis, Research Institute for Symbolic Computation, Linz, Austria, December 1995.
8. Michal Mňuk, Bernhard Wall, and Franz Winkler. CASA reference manual (version 2.2). Technical Report 95-05, Research Institute for Symbolic Computation, Linz, Austria, 1995. See also *http://info.risc.uni-linz.ac.at:/labs-info/compal/software/casa/casa.html.*
9. F. Mora and H.M. Möller. New constructive methods in classical ideal theory. *J. Algebra*, pages 138–178, 1986.
10. Tran Quoc-Nam. A hybrid symbolic-numerical method for tracing surface-to-surface intersections. In A.H.M. Levelt, editor, *Proc. ISSAC'95*, pages 51–58. ACM Press, 1995.
11. Tran Quoc-Nam. On the symbolic-numerical methods for finding the roots of an arbitrary system of non-linear algebraic equations. In *Proc. ATCM'95*, Singapore, 1995. The Assoc. of Math. Eds.
12. J. Rafael Sendra and Franz Winkler. Symbolic parametrization of curves. *J. Symb. Comput.*, 12(6):607–631, 1991.
13. Igor A. Shafarevich. *Basic Algebraic Geometry*, volume 1. Springer Verlag, second edition, 1994.
14. Robert J. Walker. *Algebraic Curves.* Princeton University Press, 1950.
15. Bernhard Wall. *Symbolic Computation with Algebraic Sets.* PhD thesis, Research Institute for Symbolic Computation, Linz, Austria, 1993.
16. Dongming Wang. Irreducible decomposition fo algebraic varieties via characteristic sets and grbner bases. Technical Report 92–55, Research Institute for Symbolic Computation, August 1992.
17. Franz Winkler. Constructive algebraic geometry with CASA. Talk at workshop CoCoA, Cortona, Italy, 1993.
18. Franz Winkler. Algebraic computation in geometry. *Math. and Computers in Simulation*, (1319), 1996.

Making Systems Communicate and Cooperate: The Central Control Approach

Stéphane Dalmas and Marc Gaëtano*

INRIA Sophia Antipolis, projet SAFIR, 2004 route des lucioles, BP 93, 06902 Sophia-Antipolis CEDEX, France

Abstract. In this paper we present the Central Control, a software component designed to be the kernel of environments for scientific computation and which can offer a common and concurrent access to many tools needed by the scientist and the engineer. The Central Control (often abbreviated as CC in the sequel) communicates with servers that can be general purpose or specialized computer algebra systems, numerical systems, visualization programs, graphical interfaces *etc.* The Central Control can abstract the syntaxic and semantic differences of the systems so that, for example, an expression computed by *Mathematica* can be used as input to *Maple.* The CC is in fact an extended *Scheme* interpreter. This enables the dynamic configuration of a network of servers to distribute computations using the full power of the Scheme language. Architectures which support the solution of mathematical problems by linking specialized components are important to the future growth of Computer Algebra. The CC is a tool designed to make experiments in this direction easily. It has been used to build an efficient library for computational algebraic geometry. This library has been used to implement a new algorithm for primary decomposition of ideals.
In this article, we present the main features of the Central Control, illustrated with some examples of actual *Scheme* code. We also discuss the possible uses of the CC, its implementation, its relation with previous work and the problems and benefits of our approach.

1 Introduction

Architectures which support the solution of mathematical problems by linking specialized components are central to the future growth of Computer Algebra. The Central Control is a software component that has been designed to build environments for scientific computations, making various programs (such as computer algebra systems, numerical systems, visualization programs, graphical interfaces *etc*) to communicate and exchange data. An early version of the CC based on Standard ML (see [5]) provided the same facilities through a small command language. This first prototype enabled us to experiment with the basic features but could not be easily extended. The new version of the CC presented in this paper is a *Scheme* interpreter extended with new primitives to deal with

* also I3S, CNRS URA 1376,Université de Nice Sophia–Antipolis

servers (starting a server, sending and receiving data to and from a server, *etc*). This enables the dynamic configuration of a network of servers to distribute computations using the full power of the Scheme language. The CC also permits the transparent use of different servers through a term translation mechanism to provide a convenient way for an application program to access computer algebra facilities independently of a particular computer algebra system.

Since there is a wide variety of mathematical objects and associated representations, our approach is to avoid any predefined meaning on the objects and requests that are exchanged between the CC and the servers. Implementing a new server is fairly easy and does not require particular knowledge about the CC itself, except the protocol to be used as the transport layer. The designer of a new server does not need to cope with any syntactic conventions and he is free to decide what requests and data the server will handle and how they will look like.

In the next section, we review some typical situations where there is a clear need to make several tools cooperate. Then, we describe the Central Control and its most salient features and how it can be used to solve some of the previous problems.

2 Some needs

The need for specialized systems: It is often the case that a general purpose symbolic computation system cannot compete with specialized packages. Often the difference in performance can be tremendous. One typical case is the computation of Gröbner basis for which very specialized and efficient systems exist, such as GB, *Singular* or the PoSSo Gröbner Solver. *Ad hoc* solutions have been implemented for using these programs from general purpose systems, using intermediate files as a simple support of the communication. From a performance point of view, this is not as bad as it might first be expected. In computer algebra the time spent in file system operations is usually negligible compared to the time spent in the computation itself (especially with Gröbner bases...).

Mixing symbolic and numerical computing: Most general purpose symbolic computation systems include some purely numerical algorithms (for linear algebra or ODE solving for example) sometimes both for arbitrary precision "big floats" and machine precision floating-point data, for efficiency reasons. The communication with a server providing the support for native floating-point computations can be a very interesting alternative. Several libraries are available and can be used for building such a server. Some of them are high-quality freeware and most non commercial numerical *Matlab*-like systems use such libraries. Alternatively, we can also imagine other servers providing arbitrary precision floating-point computation. As in the previous example, the efficiency of the communication is not a central problem. The computation time dominates the communication overhead. Splitting a program in several components makes it easier to design. From a software engineering point of view, this certainly enhances the maintainability and ease of development, by enforcing a logical separation and thus avoiding unnecessary or accidental dependencies. The different parts of the program can be developed, tested and debugged concurrently provided that the interfaces are well defined. Generating C or FORTRAN numerical code is another very important issue when optimal performance is needed. Several packages have been written for this purpose, for example MACROFORT or MACROC ([3]) for *Maple* or GENTRAN for *Reduce*. These can also be implemented as special servers.

Distributing computations: Many experiments have been conducted in the past few years in parallel or distributed computing in the field of computer algebra. Of practical interest for many users are facilities for distributing computations that can be done in parallel over a local area network as this is a much more common situation than the availability of parallel computers. Many algorithms can profitably operate at this level of parallelism (see [16]). For this kind of application the efficiency of communications can be very important. Yet there are many algorithms for which this is not crucial. Distributing computation on several machines can also be interesting even without true parallelism. Symbolic computation systems need a lot of resources (time as well as memory) and it can be very desirable to have load sharing facilities. For example, in the case of a system where the numerical operations have been physically separated, the process that implement them can run on another computer. Even if it is not a clear win in term of computation time, an interesting amount of memory could be saved and thus paging activities and the associated performance penalty can be reduced.

Graphical interfaces: The interest to separate the interface from a computing kernel is recognized for quite a long time ([2]) and realized in the major commercial systems like *Maple* ([10]) and *Mathematica*. Unfortunately, we are not at the point where a common interface can be used with several systems or a new interface can be easily build for a commercial system.

Symbolic computation as a tool: Symbolic computation is a natural tool in several areas such as constraints solving ([11]) or CAD. A growing number of industrial products include symbolic computation facilities that are written from scratch, for example in packages for simulations of multi-body systems or control theory. There is an example of a product for the control and simulation of systems, written in Smalltalk, that implements some non-trivial algorithms such as polynomial factorization. There are some rare examples of commercial products that transparently use a commercial symbolic computation system, for example [15] that uses *Maple*. But it seems to be a concern for major vendors of computer algebra: *Maple* distributes a special product (and takes part in the *OpenMath* initiative [1]) and *Mathematica* can be accessed through the *MathLink* protocol.

3 The Central Control

Our first prototype of the Central Control was written in Standard ML (SML). It used its own particular programming language. Our approach for the design of this language was to concentrate on what was really fundamental such as managing servers, promises and translations, and avoid features that were not strictly necessary. The only computation mechanism was through conditional rewrite rules (see [5] for more details). After testing our design with the first prototype it became clear that this approach was not the best in our context. Potential users are not willing to learn a completely new and quite exotic language just for managing their communications. So we took the decision to implement the features of the CC in an existing interpreter for a well-known programming language. Another advantage of this solution is that we can rely on good existing documentation for learning the language.

The Central Control is now a Scheme interpreter extended with a set of new primitive operations. These operations include launching or connecting to a server, sending a computation request and receiving the answer, dealing with exceptional conditions: interrupting a server or asynchronously requesting information such as memory size or CPU time already spent in the current computation, and translating requests and answers to insure the faithful transmission (exchange) of data between servers. This is almost entirely implemented in *Scheme*.

We can illustrate some of the previous items with a short CC session. Here is how to create a MAPLE server, running on the machine `kama`. `maple1` is bound to a new kind of Scheme object, of type `server`.

```
> (define maple1 (server-create "Maple" "kama"))
#<unspecified>
> maple1
#<server>
```

This server belongs to the `Maple` service. A service is an abstraction common to several servers. Services are used for translating requests and results of computations (this mechanism is described below). We can now send a computation request to the server `maple1` and get the result back:

```
> (define prod (server-compute maple1 '("ifactor" 12341234121)))
#<unspecified>
> prod
("*" 3 103 39939269)
```

The `maple1` server interprets the term `("ifactor" 12341234121)` as a request to compute the decomposition of the integer 12341234121. The result is a sequence of prime numbers. The term bound to prod can now be computed on another server:

```
(define math1 (server-create "Mathematica" "ganesa"))
#<unspecified>
> (server-compute math1 prod)
12341234121
```

3.1 Promises

When a computation is expected to take a long time, *promises* can be used to free the CC to wait for the result. A promise is a mutable Scheme object associated with a computation in progress in a given server. There are primitive functions acting on promises, to retrieve the associated server, to check if the computation is done or to get its result when available. When the value of a promise is available we say that the promise is *realized*:

```
> (define nasty (server-compute maple1 '("+" 1 ("factorial" 29))))
#<unspecified>
> nasty
8841761993739701954543616000001
> (define factorized
          (server-compute maple1 (list "ifactor" nasty) 'promise))
#<unspecified>
> factorized
#(promise:tag #f #<server> #f)
```

The keyword **promise** used as an extra argument of the **server-compute** function indicates that the function should return immediately a promise. The function **promise-ready?** checks if a promise has been realized. When the promise is ready, the function **promise-value** can be used to extract its value:

```
> (promise-ready? factorized)
#t
> (promise-value factorized)
("*" 2778942057555023489 218568437 14557)
```

Using **server-compute** with the **promise** flag is similar to **server-compute** except that is returns immediately a *promise* after transmitting the request without waiting for the computation to end in the server. This form allows the user to run a long computation while doing some other activities in the Central Control and actually implements parallel processing. For example, the Scheme function **first-one** sends the same request **req** to two different servers **s1** and **s2** concurrently. As soon as one server terminates the other is interrupted and the result is returned.

```
(define (first-one s1 s2 req)
  (let* ((p1 (server-batch s1 req)) (p2 (server-batch s2 req))
         (pr (wait-for p1 p2)))
    (server-interrupt s2) (server-interrupt s1)
    (promise-value pr)))
```

The primitive **server-interrupt** interrupts a server and **promise-value** extracts the value of a promise. **wait-for** takes promises as arguments and returns the first one that is realized.

If we include a promise in a computation request, in the argument of a regular **server-compute** call, the CC blocks until the promise is realized. If we include a promise in a **server-compute** call with the **promise** flag, the function immediately returns another promise and the request is put on a queue to be sent when the promise is available.

Our promise mechanisms are very similar to the *future* construct of Multilisp ([8]) and to the promises of [7].

3.2 Lazy communication

As the results of the computations can be huge, it is desirable to avoid systematically transmitting them to the CC. The CC can therefore associate a *handle* to a result that is stored in a server:

```
> (define handle
        (server-compute maple1
                        '("expand" ("^" ("+" "x" 1) 10))
                        'handle))
#<unspecified>
> handle
#(handle:tag #<server> 1)
```

Only the request of specific operations on this handle should cause the effective transmission of the associated result. There is *no* transmission of the data associated with the handle when a subsequent computation is addressed to the server that owns the value of the handle:

```
> (server-compute maple1 (list "degree" handle "x"))
10
```

Now the value of **handle** is transmitted to the CC and then to the server **math1** as part of another term:

```
> (server-compute math1 `("Factor" ,handle)))
("^" ("+" 1 "x") 10)
```

This notion of lazy communication can reduce the time of communication, computation (for transforming mathematical objects to the representation used by the communication protocol) and the memory used in the CC. This implies only a few constraints on the server (being able to name and store its results) that are, of course, not mandatory (the CC works with servers that do not support handles).

A difficulty for implementing this lazy communication is how to deal with the termination of servers. When the termination is due to a fault in the server, we cannot do much: every attempt to access the data will result in an error. When the termination is normal, we choose the simplest solution which is to effectively terminate the server only when all the needed handles have been retrieved. In the current implementation we rely on some mechanism in the garbage collector. Another troublesome point is that we cannot access any handles in a single threaded server that is currently computing. The current implementation returns an error in this situation, but an alternative could be to block. In our practice, this has never been a problem, but it should be interesting to find a better solution.

3.3 Translating requests and results

As we don't enforce a standard for encoding all mathematical objects, we need a mechanism to translate from the representation used by one server to another. This does not mean that given n services we need to specify $n(n-1)$ translations. The Central Control can choose its own common representation and thus only n translations are really necessary.

For example, the Central Control understands several representations of arbitrary precision integers (different bases, and byte and digit orderings). We have implemented functions to translate from all representations to a string in base 10 (at least to be able to print them) and to directly translate to and from the most common representations.

The translations are performed for each `server-compute` or `server-batch` call, based on the service of the server that the request is submitted to. It is possible to set translations for requests as well as for results. Nevertheless, we normally use a lazy translation and results are "tagged" with the name of the service they come from. Sometimes this could avoid unnecessary translations between two servers of the same service exchanging data.

Translations are specified with a set of functions manipulating the *translation tables* of a server. A little pattern-matching facility (as well as using general Scheme functions) is used.

3.4 Control in the Central Control

The Central Control is not multi-threaded (contrary to *Glish*, [12]). We can think of situations where a more complicated control could be interesting. For example, we can imagine that a server could ask the CC for some computations (of course to be forwarded to another server). Nothing prevents this in the current implementation (through the mechanism for translating results). The problem is that if the server was called through `server-batch` the request will basically only be honored when a `promise-ready?` call will be issued.

The implementation of a multi-threaded, preemptive version of the Central Control would certainly be a rather tricky task (although the *continuation* mechanism of Scheme can be of some help). We have investigated this possibility and found that it should not be our priority and that it is wiser to first try the simpler approach. Moreover, more flexibility can bring more subtle problems (dealing with exclusion, deadlocks and data protection). Nevertheless, we now plan to make some experiments with a Scheme interpreter that supports native threads.

3.5 Communication protocol

We use the ASAP protocol ([4]) to communicate between a server and the Central Control. The basic objects exchanged are terms or more precisely attributed terms. ASAP also supports transmission of binary data that can be used to provide more efficient specialized encodings. ASAP uses (stream) sockets as the basic communication mechanism. Each ASAP connection uses in fact two sockets, the second one for communicating exceptional conditions (like requests for interruption).

For the moment, the CC only understands the ASAP protocol. Nevertheless, there is no restriction in the design of the CC that will prevent the use of several protocols (a particular protocol can be attached to each service). It would be fairly easy, for instance, to use the OpenMath standard ([1]) as soon as this new protocol is available.

4 Using the Central Control

We provide a convenient interface to the Central Control, using classical input and two dimensional output of mathematical expressions. Through this interface the user feels like he is interacting with a single true symbolic computation system although its computations are transparently done by various servers, running on different machines. This simple interface constitutes an interesting base for someone who wants to develop a specialized package. He can concentrate his work on the essential features of his program to make it a server. The CC can provide an interface and communication with other programs for free.

Furthermore, we plan to enlarge the library of utility functions. For example, we have implemented a few set of utilities to obtain interesting information like average load and memory usage of the machines on the network. They have been used to write some simple functions to provide transparent load sharing on our local area network.

The design of the CC is well suited to be the kernel of an architecture for scientific computation with servers arranged in a "star-like" fashion. The Central Control can be used to combine the power of a general purpose system with the efficiency of a specialized package. It is in fact possible to use the Central Control directly from a general purpose system. This has been done for *Maple*. This is a necessary feature for the acceptance of the CC as many users want to stay with a user interface they know and are used to. This provides an easy access to specialized packages, replacing ad-hoc solutions (some experiments have been done with *Maple* and GB).

The CC can be a very useful tool for applications that need some computer algebra. Such applications can use it as a generic computation server (it can be accessed either through the ASAP protocol or through a textual interface with a special output) and thus removing the dependency on a particular computer algebra system. The application can be distributed with a configuration file that enables the use of different real servers to provide the needed services, depending on which systems are available to the end user.

The previous situation is basically the one of a generic graphical user interface. The CC can be used as an intermediate agent to enable a single graphical interface to communicate with several systems. We are building such an interface (based on a powerful customizable equation editor).

5 *Algeom*, a library for computational algebraic geometry

For his PhD thesis ([14]), Alain Sausse developed *Algeom*, a fairly complete library for computational algebraic geometry. This library uses *Maple*, *Macaulay* and GB to perform its computations.

Algeom was used to implement a new algorithm for primary decomposition of ideals. This application uses non-trivial programming at the Scheme level to mix different methods with a complicated "heuristic" control, for example, stopping a process when too much time has elapsed and trying something else. The Central Control allows to exploit some very useful concurrency in these algorithms as well as to use the most efficient server for each sub-task.

6 Some problems with the Central Control approach

The Central Control is not as useful in cases where the communication is really intensive or when efficiency is the main concern. Its centralized communication model is restrictive but it is also very simple to manage and understand.

Using the CC to access general purpose systems can be frustrating for two reasons: there is no natural way to use the language of the system from the CC and it can be difficult to know how much of the semantics of the system is supported.

Of course, it is possible to load files written in the language of the system in such servers (by means of a generic `load` operator that is translated to the correct function for each service) but we cannot write programs from the Central Control. Although it could be possible to map every construct of the language of the system in the Central Control, we think that the languages of computer algebra systems have too many peculiarities for this to be manageable.

One of the goals of the Central Control is to enable the exchange of data between various servers. For that task, it needs to perform translations. A general purpose computer algebra system has several hundreds of symbols with special meanings but the Central Control is likely to know only part of these. So we face a possible semantic gap: it is possible that a symbol with a special meaning to one server cannot be translated to another one. This is, of course, unavoidable. But it is, nevertheless, a safe situation. The translation mechanism can ensure that a symbol with no special meaning for the Central Control will not be interpreted by the server it is sent to. This prevents unintended computations, for example when sending `W(x)` as a formal expression to *Maple*, without knowing that `W` as a special meaning for this system.

7 Implementation

The Central Control is implemented on top of the SCM Scheme interpreter. SCM has been ported to many computers (and incidentally seems to have been chosen to be the base of the GNU extension language). We plan to port our primitives and Scheme code to at least another Scheme interpreter to demonstrate that our design is not dependent on a particular implementation. Our current implementation is small: approximately 1200 lines of C code and 600 lines of Scheme code. The process occupies approximately 700K (including 400K of sharable code) on a SparcStation at startup (the size of a *Maple* 5.3 kernel).

We have used some parts of the JACAL program (a symbolic mathematics system which runs under Scheme language implementations, written by Aubrey Jaffer, the author of SCM) to create an interface with classical input and two dimensional output of expressions.

8 Joining the Central Control

We have developed several tools to ease the building of servers. These include C program templates, a library and a compiler of ASAP specifications that generates a C function from patterns of terms and associated actions.

The amount of work required to obtain a sufficiently robust server (starting from a program that was not designed as a server) can be estimated from the experience we have now gained after more than a year of experiments. If the parts of the program that manage input and output are sufficiently structured, the job seems fairly easy: a week for PARI, done by one of the authors of this paper with no prior knowledge of this program, including the understanding of the internals of PARI (the most time consuming part). The amount of extra code is rather small: something like 160 lines for sending results and 200 lines for handling communication in the case of the PARI server.

The task can be more difficult in the case of a commercial system that is not accessible at the source level. Two options are possible that involve using an intermediate process to make a translation. Both *Maple* and *Mathematica* can offer a non-interactive access through a dedicated communication protocol: *Mathematica* through the *MathLink* protocol and *Maple* with the *MathEdge* product. Unfortunately *MathEdge* is not delivered with every distribution, it is a separate product. Hence, we had to resort to a more adventurous method: encapsulating *Maple* with a process that emulates a real terminal and parses its output. It can be difficult to obtain a robust server with this approach as our experience shows. The output format is not specified at the limits (especially how long tokens are dealt with) and error detection and recovery is problematic.

9 Existing works and approaches

Distributed computation and distributed software architecture have developed to become topics of great interest. Many tools have been proposed and there are some emerging standards (*Tooltalk*, *OpenDoc*, CORBA, HP *softbench* to name but a few). CAS/PI ([9]) and SUI ([6]) are two examples of software architectures designed for symbolic computation. None of these are really programmable in the way the Central Control is. SUI seems to focus on the interface and is not programmable at the user level. Purtillo explicitly addressed the issue of communication between numerical and symbolic programs with *Polylith* in [13]. But his software bus approach (similar to the one of CAS/PI) renders the addition of new facilities and the access to new operations inside servers more difficult and not as dynamic as in the Central Control, where you can add a new type of server on the fly and change any interface dynamically.

Glish is an array-oriented interpreted language for building distributed systems from event-oriented programs ([12]). It is similar to the Central Control as it uses a centralized communication model where all exchanged data pass through the *Glish* interpreter (but it also support point-to-point links when more efficiency is needed).

Even if most concepts such as central data repository, lazy communications or use of rules to translate expressions were already implemented in previous prototypes, the major advance of our work is probably to make all these concepts available within a compact and very flexible software component.

10 Conclusion

The main advantage of the Central Control is certainly its flexibility. We thus believe that the Central Control can be a useful tool for easily making various experiments as well as designing complete architectures. Our approach seems promising for building a powerful scientific computation environment from specialized components, both from a software engineering and a pragmatic point of view. That is one of our aims.

For now, we have successfully built servers for PARI, *Maple*, *Mathematica*, *Macaulay*, GB. We decided to try to make as many servers as possible for validating our work. Of course, not all of these servers are at the same level of robustness but most of them are perfectly usable. A sophisticated graphical user interface that directly communicates with the Central Control is currently under development.

Our experiments with *Algeom* showed that the Central Control can be a very effective tool to prototype complex algorithms.

Acknowledgments

This work was partially supported by the ESPRIT POSSO project (6846). The Central Control is used as the heart of the so-called *comprehensive solver* which provides a common access to the various tools developed inside the project. We thank Alain Sausse for his work on the first version of the Central Control and the *Algeom* library.

References

1. The draft OpenMath home page on the World Wide Web. `http://www.can.nl/~abbott/OpenMath/index.html`.
2. Dennis Arnon, Richard Beach, Kevin McIsaac, and Carl Waldspurger. Caminoreal: an interactive mathematical notebook. In *EP'88 International Conference on Electronic Publishing, Document Manipulation and Typography*, Nice, France, April 1988. Cambridge University Press.
3. Patrick Capolsini. MacroC : C code generation within Maple. Rapport technique 151, INRIA-I3S, February 1993.
4. Stéphane Dalmas, Marc Gaëtano, and Alain Sausse. ASAP : a protocol for symbolic computation systems. Rapport Technique 162, INRIA, March 1994.
5. Stéphane Dalmas, Marc Gaëtano, and Alain Sausse. Distributed computer algebra: the Central Control approach. In *PASCO'94*, pages 104–113, Linz, Austria, September 1994. World Scientific.
6. Y. Doleh and P.S. Wang. SUI : A system independant user interface for an integrated scientific computing environment. In *ISSAC'90*, pages 88–94, August 1990.
7. Daniel P. Friedman and David S. Wise. The impact of applicative programming on multiprocessing. In *Proc. 1976 International Conference on Parallel Processing*, pages 269–272, August 1976.
8. Robert H. Halstead. Multilisp: A language for concurrent symbolic computation. *ACM Transactions on Programming Languages and Systems*, pages 501–538, October 1985.
9. Norbert Kajler. CAS/PI: a Portable and Extensible Interface for Computer Algebra Systems. In *Proc. of ISSAC'92*, pages 376–386, Berkeley, USA, July 1992. ACM Press.
10. B.L. Leong. Iris: Design of a user interface program for symbolic algebra. In *Proceedings of the 1986 Symposium on Symbolic and Algebraic Computation*, pages 1–6, July 1986.
11. Philippe Marti and Michel Rueher. A cooperative scheme for solving constraints over the reals. In *PASCO'94*, pages 284–293, Linz, Austria, September 1994. World Scientific.
12. Vern Paxon. Glish : A software bus for high-level control. Paper submitted to ICALEPCS '93, 1993.
13. James M. Purtillo. Applications of a software interconnection system in mathematical problem solving environments. In *Proceedings of the 1986 ACM-SIGSAM Symposium on Symbolic and Algebraic Computation*, pages 16–23, July 1986.
14. Alain Sausse. *Architecture logicielle distribuée pour le calcul formel. Application à la Décomposition Primaire d'Idéaux.* PhD thesis, Université de Nice-Sophia Antipolis, December 1995.
15. Philip H. Todd, Robin J. McLeod, and Marcia Harris. A system for the symbolic analysis of problems in engineering mechanics. In *Proc. ISSAC'94*, pages 84–89. ACM Press, July 1994.
16. Stephen M. Watt. A system for parallel computer algebra programs. In *Proc. EUROCAL'85*, Linz, Austria, April 1985. Springer Verlag, LNCS 204.

A Database for Number Fields

Mario Daberkow[1] and Andreas Weber*[2]

[1] Technische Universität Berlin
Fachbereich Mathematik MA 8-1
10623 Berlin, Germany
E-mail: daberkow@math.tu-berlin.de
[2] Department of Computer Science
Cornell University
Ithaca, NY 14853, U. S. A.
E-mail: aweber@cs.cornell.edu

Abstract. We describe a database for number fields that has been integrated into the algebraic number theory system Kant. The database gives efficient access to the tables of number fields that have been computed during the last years and is easily extended.
A set of functions that are specific for a number field database has been integrated into the user interface Kash of Kant. The user has thus the possibility to create queries which involve special functions on number fields provided by Kant.

1 Introduction

In the last years several extensive lists of number fields were calculated, containing data for hundreds of thousands of number fields (see e.g. [22, 2, 11, 3], or [5, App. B] for other references). These lists have been generated with the help of programs for algebraic number theory such as Kant [7, 15, 6] and Pari [1] and most of this data is available in formats that can be read by those programs. However, no mechanisms had been implemented that allow an efficient access to the data.

In this paper we describe a database for number fields that has been integrated into Kant and that gives efficient access to the computed number fields. The database has been designed to use one of the available relational database management systems that support SQL [8, 13] as a query language. It can thus use the technologies that have been developed for relational databases to handle large amounts of data. A set of functions that are specific to a number field database have been integrated into the user interface Kash [15] of Kant. Thus the user can create queries that involve special functions on number fields provided by Kant.

For the connection to the database management system that stores the data we use several layers of functionality and we utilize standards as often as possible. It is thus possible to change the underlying database management system without too much effort, should such an undertaking be necessary.

* Supported by the *Deutsche Forschungsgemeinschaft.*

The organization of the paper is as follows. The basic design considerations for the database are described in Sec. 2. The layers of functionality we have used are given in Sec. 3 A discussion of some special topics that arise in connection with a number field database is contained in Sec. 4.

In Sec. 5 we give some examples of the use of the database. Queries that can be handled by the database system alone are described in Sec. 5.1. However, the examples given in Sec. 5.2 involve the combination of the database management system and the number theory system in an essential way. This section exemplifies that the combination of the involved systems is more powerful than the "sum of its parts".

2 Design of the Database

Several partially contradicting goals had to be balanced when designing the database.

- The database has to allow efficient access to hundreds of thousands or even millions of number fields.
- The database should support access for all platforms supported by Kash.
- The format of the data in the database should be independent of the platform used and possible changes to the binary representation of number fields in the Kant kernel.
- Since Kash is available for academic institutions free of charge, the database system that is used to store number field data should be available under the same conditions.
- The database has to be extensible.

For these reasons we have chosen to use a database management system (DBMS) that supports the SQL standard [8, 13]. By utilizing a standard it is possible to support several databases — ones that are free of charge for academic institutions and also highly optimized ones if needed by users.

2.1 Currently Used Tables

Currently the database contains tables in a form roughly corresponding to the file output format for orders in Kash [15, 6].[3]

Within these tables we use the generating polynomial as a key. This is possible because two non-isomorphic number fields cannot have the same generating polynomial. However, two isomorphic number fields may have different generating polynomials, see the discussion given in Sec 4.

[3] In order to speed up common queries we store the full signature of the number fields and not only its real part. We also store the regulator both with the full given precision but also as a floating-point number of the database, because this enables fast queries which use a bound on the regulator as a criterion. These additions cause only a negligible storage overhead.

If we want to ensure that we do not multiply store information of isomorphic number fields, we have to use some functionality that is provided by the surrounding Kash system, cf. Sec. 4.

3 Layering the Database Functions

3.1 System Independent Connection Functions

Although the query language is standardized by SQL, the C-APIs[4] used to access the database are not. Some databases support standards also in this respect like Microsoft's "Open Database Connectivity" (ODBC) [20], but most of the databases available under UNIX do not. However, the functionality that is needed is similar for all databases. It is necessary that there are APIs for

- opening a connection to the DBMS;
- sending SQL queries to the DBMS;
- retrieving the result of a query from the DBMS.

Using several layers of functionality it is thus possible to introduce a layer of connection functions that is independent of the used system and through which all of the access of the higher layers to the database management system can go through.

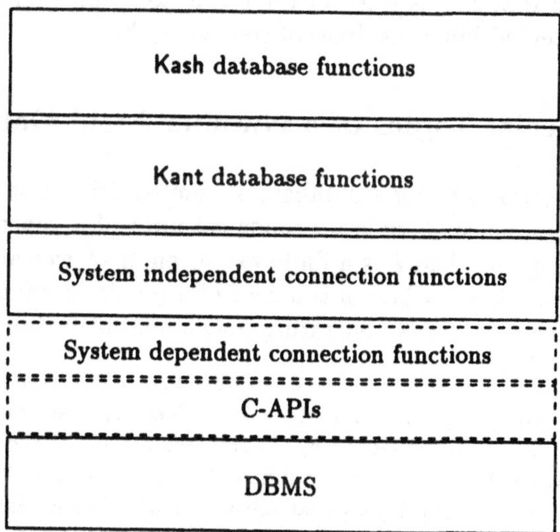

Fig. 1. Layers of the number field database

[4] We use the usual acronym API for *application programming interface*. Almost all database management systems provide APIs for the C programming language [16].

The use of the layer "system dependent connection functions" are similar in spirit to the "Z classes" of D. Fuqua [12][5] or of DBI [4].

In the current implementation we have provided system dependent connection functions for the shareware database mSQL [14], which is available free of charge for academic institutions. Extensions of this layer to support other database systems are relatively easy.[6] Implementations of the necessary system dependent connection functions for databases supporting ODBC [20] and Postgres95 [24] are currently under development.

3.2 Kant Database Functions and Kash Database Functions

Only a basic set of functions that is necessary to deal with the number field database had to be implemented as a C-library, a library extending the Kant library functions. Basically, following functionality is given by this C-library.

- Opening and closing connections to a number field database.
- Inserting new number fields into the database.
- Deleting number fields from the database.
- Sending a query to the database and successively retrieving the matching number fields. The retrieved number fields are parsed into the internally used binary format.
- Counting the number of matches of a query.

With the help of these functions it is easily possible to write a set of more sophisticated functions in Kash that facilitate the use of the database. For instance a function that checks whether there is a isomorphic number field in the database is written within a few lines of code in Kash.

4 Some Special Topics of a Number Field Database

One of the major problems for a database of number fields is the representation of the stored data, since there is no canonical form of a general number field known. Since a number field k is a finite extension of $\mathbb{Q}$, we can assume that k is given in the form $\mathbb{Q}(\rho)$, where ρ is a root of a monic, irreducible polynomial $f(t) \in \mathbb{Z}[t]$, which is called a generating polynomial of k. Hence, k is a finite

[5] The "Z classes" are implemented in C++ [23] to provide consistent C++ APIs to commonly used database. Our system dependent layer are C functions, whose implementation has been influenced by the implementation of the "Z classes".

[6] In Fig. 1 the layers that vary with the different database managements systems are given as dashed boxes. The layers that are invariant to such changes are given as framed boxes. Since the query language is the standardized SQL for the database management systems, the "user visible part" is independent of the used system and thus the layer "DBMS" is given as a framed box. Only the C-APIs are varying and have to be matched by a corresponding set of "system dependent connection functions". All other layers are not affected by changes of the underlying database management system.

vector space over $\mathbb{Q}$ and every element α in k can be uniquely represented in the form

$$\alpha = a_1 + a_2\rho + \ldots + a_n\rho^{n-1},$$

with $a_i \in \mathbb{Q}$ $(1 \leq i \leq n)$ and $n := \deg(f)$. Moreover, using this representation we can perform all kinds of arithmetic in k and therefore the description of a generating polynomial is an important part of the system we describe. But unfortunately, the polynomial f is not uniquely determined, as the next example shows:

Example. Consider the polynomials $f_1(t) = t^3 - 21t + 35$ and $f_2(t) = t^3 + 3t^2 - 18t + 15$ and the number fields k_i generated by root ρ_i of f_i $(i = 1, 2)$. These two number fields are isomorphic and an embedding of k_1 into k_2 is given by

$$\tau : k_1 \to k_2 \ : \rho_1 \mapsto 10 - 5\rho_2 - \rho_2^2.$$

Note, however, that this embedding is not unique, since

$$\tilde{\tau} : k_1 \to k_2 \ : \rho_1 \mapsto 1 + \rho_2$$

is also a proper embedding. We would like to remark that by ordering the roots of each polynomial, we could define a unique embedding from k_1 into k_2.

4.1 Test for Isomorphisms and Constructing Embeddings

As a consequence of the above representation of number fields using a primitive element, a test whether or not a certain field is already contained in the database is not trivial and in addition it is not a straight–forward procedure to extend the stored information of a number field, if the additional data is given in a different representation. Before we consider the situation that we want to extend the database, we put our focus on the situation that only two number fields k_1 and k_2 are given. Here we have to solve the following two problems:

1. decide, whether or not these two fields are isomorphic,
2. if k_1 and k_2 are isomorphic, find an embedding $\tau : k_1 \to k_2$.

We will see that these two problems are closely connected and that all algorithms we state here will prove that two fields are isomorphic by constructing an embedding from one field into the other.

Before we construct an embedding $\tau : k_1 \to k_2$ we perform some easy test to make sure that k_1 and k_2 are not definitely non isomorphic fields. Checking the invariants of a number field are a good way to do so, since two isomorphic number fields have the same invariants [21, 17]. Note, that the reverse conclusion is not true, and hence the invariants are only a test, but not a proof.

For our purposes the most important invariants of a number field k are the signature and the discriminant k. There are certainly more invariants, like the regulator R_k, the class group Cl_k, the Galois group $\mathrm{Gal}(k)$ or the ζ–function of k, but these are much harder to compute than the two mentioned above and long computations have shown that these two invariants are very good indicators for non isomorphic fields. For example there are only 2253 fields of degree 5 and signature $(3,1)$ with a non unique discriminant (for 2139 of these fields the number of non isomorphic fields is exactly 2) in the set of all fields of this signature and a discriminant bounded by 5000000. The total number of fields with this property is 79394 [22]. We would like to mention, that some additional fast tests, like the verification of the decomposition behavior of small rational primes in a number field, are possible if the pseudo tests need to be improved. We will now sketch two algorithms to construct an embedding of k_1 into k_2 if the two fields passed the pseudo tests. In the following let k_i be generated by a root ρ_i of a monic, irreducible polynomial $f_i(t) \in \mathbb{Z}[t]$ $(i = 1,2)$ of degree $n = \deg(f_1) = \deg(f_2)$.

Embedding by factorization. The first algorithm uses the factorization of polynomials over number fields [18, 9]. The number field k_1 is isomorphic to k_2 if and only if the polynomial f_1 has a linear factor in $k_2[t]$, e.g. there is an $\alpha \in k_2$ satisfying $(t-\alpha) \mid f_1(t)$. Using α we define an embedding of k_1 into k_2 by

$$\tau : k_1 \to k_2 \; : \rho_1 \mapsto \alpha.$$

Embedding by enumeration. This second method uses the LLL–algorithm [19] combined with the Fincke–Pohst algorithm for determining all lattice points in an ellipsoid. The basic idea is the same as in the first algorithm. We try to find a root α of the polynomial f_1 in the number field k_2. The roots of f_1 are algebraic integers, and since we already know the discriminant of the number field k_2, we can assume that we know an integral basis $\omega_1, \ldots, \omega_n$ of k_2 [21, 5], i.e. a $\mathbb{Q}$-basis of k_2 such that an element $\alpha = \sum_{i=1}^n a_i\omega_i \in k_2$ is an algebraic integer if and only if $a_i \in \mathbb{Z}$ $(1 \le i \le n)$ and we denote the set of the algebraic integers of k_2 by $\mathcal{O}_{k_2}$.

To sketch the method, let $\sigma_1, \ldots, \sigma_n$ be the set of all $\mathbb{Q}$-isomorphisms from k_2 into subfields of $\mathbb{C}$. By

$$\langle \cdot, \cdot \rangle \; : k_2 \times k_2 \to \mathbb{R} \; : (x,y) \mapsto \sum_{i=1}^{n} \sigma_i(x)\overline{\sigma_i(y)}$$

we define a scalar product on k_2, such that $(\mathcal{O}_{k_2}, \langle \cdot, \cdot \rangle)$ is a lattice. The Fincke–Pohst algorithm [10] computes all points in this lattice with a bounded or given value of the norm function T_2 associated to the scalar product. The enumeration procedure can be improved dramatically if an LLL reduced basis of L is used. Since we can compute the value of T_2 for all roots of f_1 by evaluating $\sum_{i=1}^n \rho_1^{(i)}\overline{\rho_1^{(i)}}$, where $\rho_1^{(i)}$ $(1 \le i \le n)$ are the roots of f_1, we are able to prove whether or not the polynomial f_1 has a root in k_2 by enumerating the finite number of points in $\mathcal{O}_{k_2}$ with the correct T_2 value. For each of the evaluated points we check if it is a root of f_1. Once such an element is found, we can define an embedding of k_1 into k_2 analog to the last algorithm. If no such element is found, k_1 is not isomorphic to k_2.

There has been a tremendous improvement on algorithms for the factorization of polynomials over number fields in the recent past, so that this method seems to be the best choice in general at the moment. Nonetheless the enumeration method has proven to be better for fields of degrees up to 5 with moderate discriminants. Moreover the second method has advantages over the factorization method, if we have to do several isomorphism tests for a single number field, since we can precompute and store a lot of the needed data.

4.2 Extending the Stored Information

We are now coming back to the problem of extending an existing database. Here we have to check whether for a given number field k an isomorphic one is already stored and if we can extend the data of the stored field. We proceed quite similar to the above case. First, we extract all fields from the database having the same signature and discriminant (if more informations are without any additional computations available, we extend the search criterion) as k and then we test if one of the extracted fields is isomorphic to k. In case this is true, the isomorphism test has computed an embedding from k into the field stored in the database as well, so if informations are known for k that are not already stored, we can transfer these data easily.

5 Use of the Database

5.1 Queries Involving the DBMS Only

Several queries can be handled by the underlying database management system alone. The Kash functions are only needed to parse the number fields from their representation in the relational database to their representation in Kash.

The queries that can be handled by the database management system alone are the ones that consist of logical connections of atomic queries with the

- generating polynomial,
- degree,
- discriminant,
- signature,
- regulator of the unit system,
- Galois group,
- class number,
- number of cyclic factors of the class group,
- the transformation matrix for an integral basis,

to give the most important ones.

In Table 1 we summarize the results of some queries. The timings that are given heavily depend on the used database system. The given timings show that for a subset of the currently computed number field data the shareware database mSQL is sufficiently efficient. Notice that the time for *counting* all matching number fields is generally smaller than the one for returning even one matching number field, because in the latter case the DBMS has to collect data that has to be returned. So the timings depend on the size of the database and the number of matches of the query.

Table 1. Summary of some queries

The timings where on a HP 735 using mSQL. The sample database contained 83806 number fields.
Query 1: `"degree=5 and [signature real]=3 and [class number]>1"`
Query 2: `"degree=5 and [signature real]=5 and regulator>99.5"`
Query 3: `"[signature real]=5 and [KANT name Galois group] LIKE '%D5%'"`

	Query 1	Query 2	Query 3
Time for returning first match	0.51 sec	4.95 sec	0.00 sec
Avg. time for returning a consecutive match	0.00 sec	0.00 sec	0.00 sec
Time for counting all matches	0.05 sec	0.76 sec	0.00 sec
Number of matches	1284	17939	26

5.2 Queries Involving Kash Functions

We will give some examples of queries that involve the combination of a database query with functions of Kash.

Example 1. We want to extract all fields k of degree 5 and class number 2 from the database, such that there are exactly two prime ideals above 3 and 5 in k. There have been 343 fields with this property stored in the sample database. The total number of fields of degree 5 and class number 2 in the database were 1244. Here is the Kash sample session:

```
kash> DbOpen ("donald:kantnf");;
kash> DbQueryFLDTable ("degree = 5 and [class number] = 2");
true
kash> L := [];
[ ]
kash>
kash> repeat
>    o := DbNextOrderFromQuery ();
```

```
>    if o <> false then
>      if (Length (Factor (3*o)) = 2) and (Length (Factor (5*o)) = 2)
>         then Add (L, o);
>      fi;
>    fi;
> until o = false;
kash> Length (L);
343
kash> Factor (3*L[100]);
[ [
        [3 1 2 2 0]
        [0 1 0 0 0]
        [0 0 1 0 0]
        [0 0 0 1 0]
        [0 0 0 0 1]
        , 1 ], [
        [3 2 2 0 1]
        [0 1 0 0 0]
        [0 0 1 0 0]
        [0 0 0 3 0]
        [0 0 0 0 1]
        , 2 ] ]
kash> Factor (5*L[100]);
[ [ <5, [2, 1, 0, 0, 0]>, 1 ], [ <5, [15, 10, 9, 15, 3]>, 1 ] ]
kash> DbCountMatchesQueryFLDTable
> ("degree = 5 and [class number] = 2");
1244
```

Example 2. In our second example we are interested in all totally real fields k of degree 3, such that there exists a totally real quadratic extension generated by a unit of k, which is unramified at all places. We know that for such an extension the class number of k has to be divisible by 2. The procedure given below found 146 fields.

The database request alone extracted 612 fields and has thus limited the search space for the Kash procedure tremendously.

```
kash> DbOpen ("donald:kantnf");;
kash> DbQueryFLDTable
> ("degree=3 and [signature real]=3 and [class number]>1");
true
kash> L := [];
[  ]
kash> repeat
>    o := DbNextOrderFromQuery ();
>    if o <> false then
>      if (OrderClassGroup (o) [1] mod 2 = 0) then
>        U := OrderUnitsFund (o);
>        for u in [U[1],U[2],U[1]*U[2],-U[1],-U[2],-U[1]*U[2]] do
>           tmp := Flat (MatToColList (EltCon (u)));
```

```
>           if ForAll (tmp, n -> Re (n) > 0) then
>             O := Order (o,2,u);
>             if (Norm (OrderKextDisc (O)) = 1) then
>               Add (L, [o,u]);
>             fi;
>           fi;
>        od;
>      fi;
>    fi;
> until o = false;
kash> Length (L);
146
kash> DbCountMatchesQueryFLDTable
> ("degree=3 and [signature real]=3 and [class number]>1");
612
```

Acknowledgments. We are grateful to A. Myka for suggesting to consider mSQL. The second author is grateful to R. Zippel for awaking his interest in databases for algebraic objects.

References

1. BATUT, C., BERNARDI, D., COHEN, H., AND OLIVIER, M. *User's Guide to PARI-GP — Version 1.39.03.* Université Bordeaux I, 33405 Talence Cedex, France, 1995. Available at ftp://megrez.math.u-bordeaux.fr/.
2. BUCHMANN, J., FORD, D., AND POHST, M. Enumeration of quartic fields of small discriminant. *Mathematics of Computation 61* (1993), 873–879.
3. BUCHMANN, J., AND FORD, D. J. On the computation of totally real quartic fields of small discriminant. *Mathematics of Computation 52* (1989), 161–174.
4. BUNCE, T. Perl 5 database interface (DBI) API specification — version 0.6, 1994. Available at ftp://ftp.mcqueen.com/pub/dbperl/DBI/.
5. COHEN, H. *A Course in Computational Algebraic Number Theory*, vol. 138 of *Graduate Texts in Mathematics.* Springer-Verlag, Berlin, 1993.
6. DABERKOW, M., JÜNTGEN, M., JURK, A., POHST, M. E., AND VON SCHMETTOW, J. KANT. In *Computeralgebra in Deutschland — Bestandsaufnahme, Möglichkeiten, Perspektiven*, Fachgruppe Computeralgebra der GI, DMV, GAMM, Ed. Passau, 1993, pp. 212–218.
7. DABERKOW, M., POHST, M., ET AL. KANT V4. Submitted to *Journal of Symbolic Computation.*
8. DATE, C. J., AND DARWEN, H. *A Guide to The SQL Standard*, 3rd ed. Addison-Wesley, 1993.
9. ENCARNACIÓN, M. J. *Fast algorithms for reconstructing rationals, computing polynomial GCDs and factoring polynomials.* PhD thesis, RISC Linz, 1995.
10. FINCKE, U., AND POHST, M. A procedure for determining algebraic integers of given norm. In *Proc. Eurosam 83* (1983), vol. 162 of *Lecture Notes in Computer Science*, Springer-Verlag, pp. 194–202.

11. FORD, D. Enumeration of totally complex quartic fields of small discriminant. In *Proceedings of the Colloquium on Computational Number Theory* (Debrecen, Hungary, 1989), A. Pethö, M. E. Pohst, H. C. Williams, and H. G. Zimmer, Eds., de Gruyter, pp. 129–138.
12. FUQUA, D. Z classes for database access, 1995. Available at `http://alfred.nih.gov/Dean/ZDB.html`.
13. GRUBER, M. *SQL Instant Reference.* Sybex, 1993.
14. HUGHES, D. J. *Mini SQL — A Lightweigth Database Engine*, 1995. Available at `ftp://Bond.edu.au/pub/Minerva/msql/`.
15. KANT GROUP. *KASH — A User's Guide.* Straße des 17. Juni 136, 10623 Berlin, Germany, 1995. Available at `ftp://ftp.math.tu-berlin.de/pub/algebra/Kant/`.
16. KERNIGHAN, B. W., AND RITCHIE, D. M. *The C Programming Language*, second ed. Prentice Hall, Englewood Cliffs, NJ 07632, 1988.
17. LANG, S. *Algebraic Number Theory*, vol. 110 of *Graduate Texts in Mathematics.* Springer-Verlag, 1986.
18. LENSTRA, A. K. Factoring polynomials over algebraic number fields. In *Proceedings of the 1983 European Computer Algebra Conference* (1982), vol. 144 of *Lecture Notes in Computer Science*, Springer-Verlag, pp. 32–39.
19. LENSTRA, A. K., LENSTRA, H. W., AND LOVÁSZ, L. Factoring polynomials with rational coefficients. *Mathematische Annalen 261* (1982), 515–534.
20. MICROSOFT CORPORATION. Accessing the world of information: Open Database Connectivity (ODBC), 1995. Available at `http://www.microsoft.com/DEVONLY/STATEGY/ODBC/ODBCBG.HTM`.
21. POHST, M. E., AND ZASSENHAUS, H. *Algorithmic Algebraic Number Theory.* Cambridge University Press, Cambridge, 1989.
22. SCHWARZ, A., POHST, M., AND DIAZ Y DIAZ, F. A table of quintic fields. *Mathematics of Computation 63* (1994), 361–376.
23. STROUSTRUP, B. *The C++ Programming Language*, second ed. Addison-Wesley, Reading, MA, 1991.
24. YU, A., AND CHEN, J. *The Postgres95 User Manual.* Computer Science Division, University of California at Berkeley. Available at `http://epoch.cs.berkeley.edu:8000/postgres95/`.

Compiling Residuation for a Multiparadigm Symbolic Programming Language

Georgios Grivas[1] and Alexios Palinginis
Department of Computer Science
ETH Zürich
{grivas@inf|apalingi@iiic}.ethz.ch

Abstract. This paper describes a new compilation method for the residuation mechanism underlying a multiparadigm language. Furthermore, it proposes the constraint functional and functional logic paradigms for the programming language of symbolic computation systems.The integration of functional and logic programming for the programming language of the symbolic computation system AlgBench is demonstrated using the residuation evaluation mechanism. The proposed multiparadigm language LAB could serve as a stand alone programming language. We present a residuation compiler and a collection of new commands and structures for a uniform abstract machine, an extension of the Warren's Abstract Machine (WAM), with a common memory for both functional and logic programms. Our implementation is done in an object-oriented way and it shows that the proposed approach and optimizations leads to a better execution and evaluation performance.

1 Introduction

As computers are becoming indispensable to scientists and engineers, advanced features of programming are needed for symbolic computation systems. Nowadays, the most widely used symbolic computation systems own either a procedural or a functional programming language. However, we are sure that the various kinds of logics will be an important part of the next generation symbolic computation systems. The need for logic and proving capabilities in the programming language of systems doing mathematics (especially in Mathematica) is also highlighted in [4]. A step towards this direction is the integration of the most pragmatic logic at the moment, the Horn-clause logic with SLD-resolution [12]. We have integrated the functional language of AlgBench with a constraint logic language using the residuation theory as evaluation mechanism. In this paper we present a compiler for an integrated language where functions, predicates and constraints can be combined to form a programming language of a symbolic computation system. We have implemented a new compiler for the resulting multiparadigm language LAB (**Logic** + **AlgBench**) and integrated it with the AlgBench interpreter.

In the area of programming languages although no widely spread research or commercial functional logic system exists, the amalgamation of functional and logic paradigms has been a field of great research interest on the last decade. It combines the features and flexibility of both paradigms, introducing new powerful programming and teaching advantages. First, the similarities of both paradigms and second, the expressive power and the efficient operational behavior due to their integration make the use of such programming languages attractive.

The LAB compiler, written in C++, supports standard Prolog functionality and an equation constraints solver, both using a refinement implementation of the WAM, also in C++, which can be easily and efficiently emulated on standard hardware. The functional part of the compiler makes possible the integration with the logic part through an extension of the exist-

[1]Research supported by the Swiss National Science Foundation

ing logic abstract machine keeping the same concepts with those of the interpreter's functions and adding the suspension *residuation mechanism*. To combine both paradigms we introduce special function calls (residuations or delayed constraints), which can be called with insufficiently instantiated arguments (residuals). This gives the freedom and the chance to the programmer to produce terms not depended on the data order, combining to a program logic and functional statements. Such calls are promising parallel execution with terms, possible equation constraints, which may share the same variables. The communication between those processes takes place through this insufficiently instantiated variables under control of unification. The synchronization of those processes is been realized through the suspended functions. In this way a multilevel processes communication can be obtained. Each residuation is being suspended and gives the control to the next disjunction goal, having the possibility to resolve the residuation.

Our main contributions presented are: (1) Enrichment of the programming language of symbolic computation systems with constraint logic and constraint functional programming using the residuation evaluation mechanism, (2) compiler for a declarative symbolic computation system, and (3) compilation of the residuation mechanism in order to provide an efficient full integrated language.

The rest of the paper is organized as follows. Section 2 describes the notion of multiparadigm programming using delay constraints and gives some examples presenting at the same time informally the LAB syntax and semantics. In thesection 3 the basic concepts of compiling delay constraints are introduced and their implementation is discussed in detail. We assume though that the reader is familiar with the WAM concepts and programming primitives. For a complete overview see [1]. Section 4 presents some optimization techniques on the current program structures. In Section 5 we show the results of the comparison between the LAB compiler with other functional or logic languages. Finally, we summarize our work and conclude.

2 Combining functions and predicates by delay constraints

Under multiparadigm programming languages we understand not only that a language supports more than one paradigm, but also the combination of these paradigms enriching its expressive power. The difficulty in the design of such a language is to keep the semantics of mixed programs simple and comprehensible. In our view LAB achieves this goal without sacrificing expressiveness.

For the integration of functional and logic paradigms we have to make design decisions regarding the semantics of the *foreign calls*. These are calls to functions in a logic program or to predicates in a functional program. On the other hand, the *native calls* are function calls in functional programs or predicates calls in logic programs. Furthermore, we distinguish between ground and non-ground function calls in predicates, because non-ground function calls may lead to residuation and consequently to concurrency. We deal with the following combinations of functional, logic and constraint paradigms (A detailed description can be found in [10]): predicate calls in functions, ground function calls in predicates, suspension of non-ground function calls allowing concurrency and constraint functional programming (CFP) and arithmetic constraints in predicates leading to constraint logic programming (CLP).

Term rewriting is not as general as SLD-resolution. *Narrowing* [5] is a generalization of term rewriting. Programs augmented with constraints may also achieve narrowing behavior e.g., the computation of inverse of functions. Furthermore, the residuation mechanism in conjunction with some constraint solvers is just expressive as narrowing and easier to achieve an efficient compilation. For these reasons, we have not tried to compile the general narrowing mechanism.

Now, assume odd is a boolean function which returns `True` if its argument is an odd number. The function call `odd(x_)` at the first goal position of the RHS of the predicate `q` below can be replaced by a `True` or `False`:

```
equal(x_, x_).
odd(x_) := equal(Mod(x, 2), 1)
q(x_) :- odd(x_), r(x_).
```

The query `q(3)?` succeeds, if the fact `r(3)` exists, since `odd(3)` is `True`. But what will happen if `q(z_)?` is to be evaluated, which is completely legitimate in logic queries? The function call `odd(x_)` becomes `odd(z_)`. In such case, it is desired to skip this execution point (i.e., the next argument or in case of logic programs the next goal and try the next one, if one exists. In the predicate `q`, it is the goal `r`. Every function call with argument variables may residuate. For such cases, we introduce a special kind of function calls. We consider such calls as *delay constraints* or *residuations* or *constraint functions* following the special evaluation scheme of residuation (see Figure 1). The residuated argument variables are called *residuals*.

Such delay constraints can be nested in logic rules where the unbound variables may take a value. If the arguments are still uninstantiated at the end of the rule evaluation, then the residuation is unresolved and the rule fails. In the previous example, we skip the `odd(x_)` execution point and try the next existing one. If `x_` becomes instantiated in the goal `r`, then the residuated function currently sleeping and waiting for `x_` must immediately be informed and entered into the current processing.

```
FUNCTION eval(e: term): term
BEGIN
    evaluate function symbol and function arguments;
  L: choose first rule l ⟶ r with the same root symbol as e;        term rewriting
     IF no such rule exists THEN fail
     ELSE IF match(e, l) THEN eval(r)
     pick the next rule which defines e;
     IF there are no more rules THEN fail ELSE GOTO L END
     ELSE
     suspend(e);
     IF any argument of e is refined THEN goto L END
     END
END
```

Figure 1. The residuation mechanism evaluating the functional term *e*

Let us show some residuation examples starting with the case of residuation in a fact, trying to describe a constraint between values: `isDouble(2*x_, x_)`. In Table 2 we show the difference in the behaviour with respect to a logic language giving the results of LAB and PROLOG to the queries of the `isDouble` predicate. The function call is the term `2*x_` and the function symbol is the multiplication operator `*`. In the first query, the unification of `10` with 2*x_, will be suspended and the next execution point will be tried out (here the next argument), which bounds the value 5 to x_. As soon as it takes place the suspended constraint is released, and the evaluation at this residuation point is finished. The program flow returns to the second unification and to the end of the fact finding both arguments successfully unified and returns success. The PROLOG result is not correct because of syntactic unification. The only solution to make the function call 2*x_ in PROLOG is the use of the non-logical is/2 built-in function.

	LAB	*PROLOG*
`isDouble(10, 5)?`	`yes`	`no`
`isDouble(y_, 3)?`	`[y=6]`	`Y = 2*3`
`isDouble(6, z_)?`	`[z=3]`	`no`

Table 1. A simple residuation example in LAB and in PROLOG

The next query shows that residuation is a constraint solving process. The variable y_ in the beginning is suspended and the execution continues to the next point where the variable x_ gets by unification the value 3. Then the residual y_ is awaken and takes the double value of 3, i.e. 6. PROLOG gives a correct but not satisfactory result because of syntactic unification. Finally, in the third query the variable z_ takes half the value of 6 using the incorporated constraint solver and modifying appropriately the predicate `isDouble`.

In the example below non-ground function calls occur in the facts q. The query `p(s_, t_)` causes the execution of the equality constraints `s_+t_ = s_*t_`. However, in predicate p during the execution of the first goal q in the RHS, the unification with the fact q tries to unify the residuals x_ and y_ with v_ and w_. This makes v_ and w_ residuals as well although they are not variable arguments of a function call. We call such residuals *foreign residuals*.

```
r(3, 5).
r(2, 2).
r(4, 6).
q(x_, y_, x_+y_, x_*y_).
q(x_, y_, x_+y_, (x_*y_)-14).
p(v_, w_) :- q(v_, w_, z_, z_), r(v_, w_).
```

Example 1. Another case of residuation proposed in [7]

In this case, the residuation mechanism suspends the execution of this goal and continues the execution to the next goal `r(v_, w_)`. This succeeds and since the residuals have become bound the residuation is resolved. But the constraints fail. Backtracking resets the bindings to unbound and tries another fact for r. The residuation is resolved again, but this time the constraint in fact q is satisfied as shown below. If we force backtracking, we get an alternative solution to the constraint problem:

```
In(1):= p(s_, t_)?
Out(1)= [s=2, t=2]
In(2):= ;
Out(2)= [s=4, t=6]
```

Another case of residuation is when a non-ground function call is nested in a function constraint or has already seen variables as arguments. Consider the following example `isHalf` which is analogous to `isDouble` with the difference that the function call occurs in the second argument.

```
f(x_) := x*2
isHalf(x_, f(x_)).
```

This makes the situation in certain cases easier. In the first three queries below x_ becomes instantiated before the function call due to the unification of the first arguments. So the non-ground function call becomes ground and we get the correct result. However, in the last query the residuation remains unresolved. A solution to this problem is to use the constraint solver described in the next section.

```
In(1):= isHalf(4, 8)?
Out(1)= []
In(2):= isHalf(4,3)?
Out(2)=  ***   failed   ***
In(3):= isHalf(4, x_)?
Out(3)= [x=8]
```

Another form of residuation occurs in constraints with undefined root symbols which we call *dynamic residuation*. For example the variable root symbol `h(x_)` of the next function rule may refer to a function with arity 2 at runtime: `h(x_)(z_, 5) := x * z`

3 Compiling Residuation

The functional and logic paradigms have been widely implemented in an interpreter platform. Compilation of such symbolic and declarative languages is used in order to increase their efficiency. We can split the LAB compiler in three major phases: translation phase, register allocation, and code generation. The LAB compiler makes use of a temporary data structure to get information for the code generation. The structure is built by the compiler preprocessing the expressions produced by the interpreter's scanner and parser. The class hierarchy of this structure is shown in Figure 2 (we use the name of the pointer to a class).

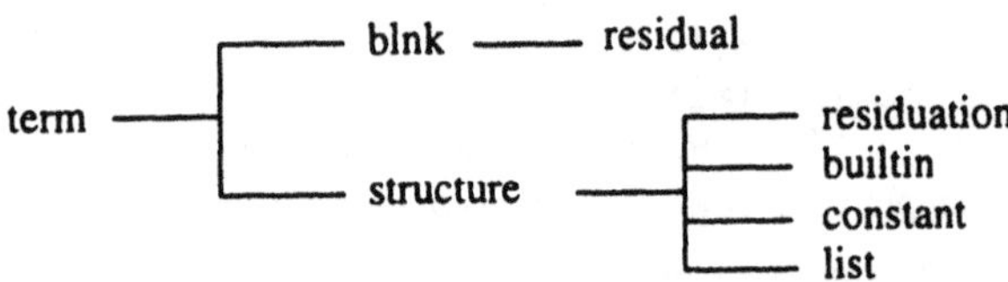

Figure 2. Compiler's Temporary Structure Class Hierarchy

The class `term` is on the top of the hierarchy with subclasses `blnk` and `structure`. Here we distinguish between `constants`, `structures`, `blnks`, `lists` and `builtin`-functions. The class `blnk` includes all variables, bound or unbound with their corresponding `get`, `put`, `set` and `unify` methods. We concentrate on the class `structure` which includes all the terms having a composite structure, such as the constraints. This class contains the implementation of the basic methods for code generation. For a detailed analysis of the LAB compiler see [10].

The mechanism of residuation is introduced by means of two additional classes, called `residual` and `residuation`. As discussed in the previous section, residuating function calls are a special case of constraints with partially instantiated arguments. Consequently, they are clearly a subclass of the class `structure`. The new class `residuation` provides methods for creating the appropriate code in case of residuation. The class `residual` provides operations for creating and maintaining some new extended resources such as a residual list and an update list.

The constructor of the `residuation` class creates an object initializing its fields shown in Figure 3, finding the variables that are possible to residuate and building in this way the residual list (see below) which consists of objects variables of class `residual`. A residual variable, is responsible to send an update message to the residuation as soon as it gets a value. From that point on the residual will generate a `branch` instruction after every `get`, `set` or `unify` instruction, which gives control to the suspended residuation.

At that early point of compilation the variable is marked as residual, keeping a link to the residuation enclosing this residual. Figure 3 shows the compiler's temporary data structure just after calling the residuation initiator. The example illustrates a case of two residuation calls those of `f(x_, z_)` and `g(y_, x_)` taking place in the same rule. At this point of compilation we insert the residuals into a temporary global residual list, to allow the recognition of foreign residuals at the currently nested level residual list. This is needed for a later optimization.

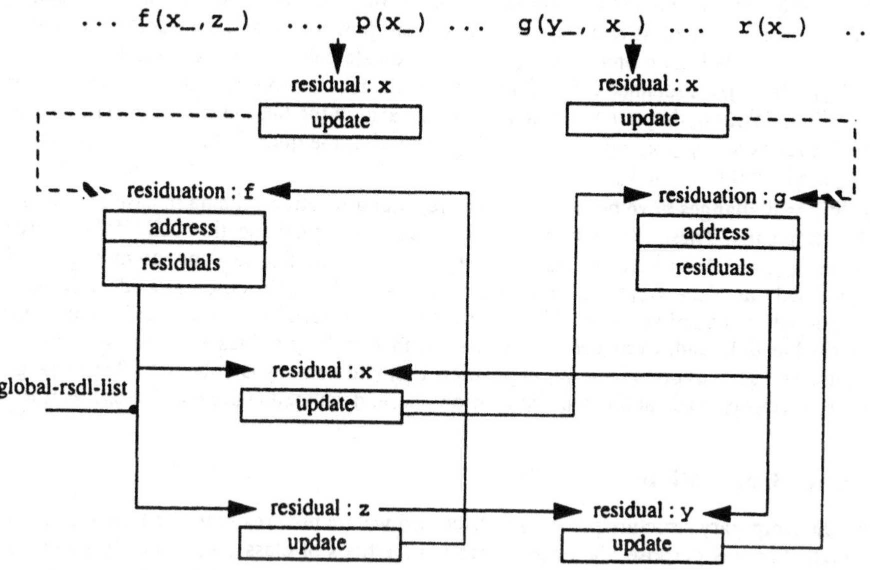

Figure 3. Temporary data structure in the residuation code generation

Every residual that occurs in a predicate will make a copy of the corresponding residual hanging currently on the global residual list. In the example of Figure 3 , `x_` in `p` should update only `f`, while `r` should update `f` and `g`. When the temporary data structure of the entire rule is finished, the rule head is searched to determine whether it has any residuals in argument position. In this case, the residual field `argument_position` is set with the corresponding position and the residual is moved from the temporary global residual list to the rule residual list in the symbol table. We keep such information, to be able to check later in the compilation if variable arguments in predicate calls are nested in an argument position referring to a residual. This foreign residual will also update the residuation in case of a definition even if the rule enclosing the residuation has been compiled earlier. Figure 4 shows the symbol table entry of a fact `q`.

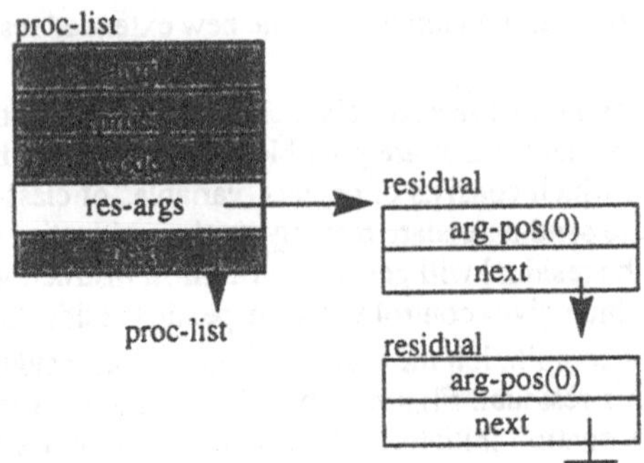

Figure 4. Symbol table entry for foreign residuals recognition

Predicate calls with variable argument use the information stored in the symbol table to examine references to a residual argument position. In Example 1 the arguments `x_` and `y_` of the rule `q` are marked as residuals with argument_position, `0` and `1` respectively, and inserted in the symbol table under fact `q`. Every predicate call (in our example `q(v_, w_, z_, z_)` in rule `p`) is scanned for variable arguments. If a variable such as `v_` and `w_` occurs in a residual position, they are marked as residuals also in the later predicate `p`. Thus we get the residuations `x_+y_`, `x_*y_`, etc., to be updated after the query `r(v_, w_)` because `v_` and `w_` are bound to `x_` and `y_`.

Additional changes have been made to the register allocation schema in case of residuating function calls. These function calls must be seen as normal function calls if the instruction `check_residuation` finds all arguments instantiated. So, the function must generate code normally as if the variables were bound to a value. For that reason a residuation refers to its residuals as bound variables. Moreover the first seen variable, not in a residuation, must be handled as unbound. Consider the `isdouble` fact of the previous example, where `x_` in `2*x_` must be seen as bound and `x_` in position `1` should be seen as unbound. These changes are more precisely made at the code generation phase, discussed at the end of next section.

3.1 Code Generation

Using the temporary structure previously discussed we are now ready to generate the emulator's code. For this purpose, the code generation methods of class `residuation` and `residual`, have been overriden from their superclasses, `structure` and `blnk` respectively.

As discussed previously, residuated functions are suspended waiting for a resume signal. This means, that the local variable stack must be kept for a future use. For that purpose we extended the memory of the LAB abstract machine (LABAM) by adding a memory partition called *residuation environment*. This memory area has the same structure as the local environment stack of the LABAM. Thus, we keep these two stacks compatible in order to achieve a smooth switch between them. The management of this memory region is done through two new LABAM instructions: `alloc_res` and `dealloc_res`. We must point out that no actual deallocation of memory is possible for this partition, since a residuation can be resumed at any time in the future, and more than once in case of backtracking. The need for the `dealloc_res` instruction is only to reorganize the special registers. The generated residuation code is the one following the functional call compilation scheme [10] with the insertion of some new emulators instructions at the beginning and end of the residuation code. The init_residuation and check_residuation instructions are inserted at the beginning to check and modify the program flow when entering the residuation. The end_residuation instruction modifies the flow when leaving the residuation.

All three commands use a residuation global enviroment memory to communicate and keep track of the residuation status. Each residuation holds a list of the register numbers of its variable arguments, which is an index of the corresponding general register of the variable. Typically for arguments of function calls the indexes are references to the stack pointer. Since the check_residuation instruction can be called from another environment, the current stack pointer is also needed.

The residuals keep a list of all addresses to branch to after an instantiation. After the generation of the instruction check_residuation its code address is known. This address (stored in the residuation class field address) will be sent to all of its residuals, creating so their address list. Actually, no list is needed but a piece of code with all the branch commands. This is stored in field update_code of class residual.

A problem arise at a foreign residuation, where check_residuation will only update the residuals of the current rule. Who is going to update the foreign residuals? The branch address of a foreign residual is of course known, since we generate the corresponding check_residuation instruction in an earlier compilation. We need to keep those addresses for every residual occuring in argument position on the LHS of a rule. Each time we find a foreign residual during the compilers translation phase, we can add a branch instruction with the code address, from the symbol table entry to the residual's update_code field. Unfortunately, not all problems are solved in this way.

Consider, for example, the existence of more than one rule with same root symbol and arity. It is possible that in some of them, Residuals may appear in an argument position of a rule, or in another position of a rule or not appear at all. Which information are we going to store in the symbol table? We need namely all of them in case of backtracking. We store a list with the residuals for each rule and we introduce two new LABAM instructions for dealing with branch points in case of backtracking in a predicate rule with residual arguments. The instruction init_rsdl_args initializes the first rule by setting the references to the possible residuals of the first rule. It is inserted after a try_me instruction. The second instruction, nxt_rsdl_args, sets the symbol table references to the possible residuals of the next rule. It is inserted after a trust_me or retry_me_else instruction. Both instruction have as single argument a pointer to the corresponding symbol table entry of the rules.

Furthermore, for the foreign residuals in order to branch always to the correct point in the code, we are creating a slotted interface to access the residuals update code without bothering about backtracking inconsistencies. The slots are a single instruction code making the decision of whether a residual for the current rule exists or not. It is also important for this instruction to keep always the same address, slot property, for objects that reference this argument position. The instruction name is nnbranch and stands for not null branch. It accepts one argument, the branch address. If the address is Null a proceed is executed, otherwise it jumps to the given address.

Residuals generate code according to the code generation schema for variables. After a set, get or unify instruction, where residuals may take a value, a branch instruction is generated to jump to the check_residuation instruction of the residual's code. In the case of a predicate call with residual arguments, a put instruction is normally generated for each argument and a call instruction gives control to the unify phase of the LHS of the corresponding rule or fact. We need the residuation's update calls right after the call instruction of the unifier query. To illustrate this, let us reconsider the fact `isdouble(2*x_, x_)`. The LAB compiler generates for this fact the following code:

```
(1)        alloc_args 2
(2)        put_constant 2 X1
(3)        init_residuation RE L
(4)        R1 check_residuation RE
(5)        put_value X3 X2
(6)        put_structure Times/2 X0
(7)        set_constant 2
(8)        set_local_value_x X3
(9)        eval_fun
(A)        ealloc_args=20
(B)        get_function X1
(C)        end_residuation RE
(D)    L   get_variable X3 X2
(E)        branch B1
(F)        deallocate
(G)        proceed
(H)        B1 branch R1
(I)        deallocate
(J)        proceed
```

After the register allocation, the variable `x_` is linked to register `X3`. The code generator of class `residuation` generate the code as for the class `structure`. The first two instructions reflect this. As soon as the code generator detects a residual, by looking up the residual list, it inserts the `init_residuation` and `check_residuation` instructions referring to the next continuation point (`L`) and to the residuation enviroment (`RE`). The mark `L` is set to the next statement having the possibility to instantiate the residual. In the case of a fact this is the next argument, where a unification takes place. The next instructions (5 to 8) are generated according to the class `structure` code generation method. The residuation code ends with the `end_residuation` instruction.

When a residual is instantiated an effort is made to resume the residuation. At line (`D`), the residual `x_` is used to generate a `get` command. The residual must generate right afterwards a `branch` command to update the residuation `2*x_`.

4 Optimizations

The residuation mechanism causes some overhead for trying to keep the residuations up-to-date. We would like here to propose some optimizations, handling some special cases to avoid unnecessary delays. With a simple static analysis of the expression, we can decide at compilation time the form of the residuation. Consider, for example the case of a residuation call in a function rule or as a query. It is definitely sure that this residuation will stay unresolved if no constraint operates on the residual. The same happens if a function call residuates on a variable that never occurs again in the expression. For example, the residuation `f` in rule `p(x_,y_):-m(f(z_)),l(x_),t(y_)` will stay unresolved waiting for a value for variable `z_`. Such cases can be detected with little effort at compilation time.

Nested residuations could handle unbound variables as residuals only if this variables don't occur in residuation at upper levels. For example in `f(x_,g(x_, z_))` `g` may consider its first argument x_ as a bound variable since a check_residuation instruction for x_ is already associated to f.

The order, in which the residual list and update list is to be constructed affects also the residuation mechanism overhead. The best order in which the residuals are checked is according to their instantiation order. The residuation f(3,x_,y_,z_,w_) has four residuals to be checked at a check_residuation execution. It will fail as long as it finds an uninstantiated variable and it will succeed only if all residuals are resolved. When succeeds on a residual, it is marked 'succeeded' and moved to the end of the list. Thus, the check will succeed when at the top of the list a 'succeeded' residual occurs. In our example the starting order of the residuals list is x, y, z, w. At the first update call, suppose that y is instantiated, the list will be traversed until y, where it will be marked as defined, moved to the end and give control passed further, while x (the top of the list) was skipped since it is undefined. Afterwards, the residuals list has the order x, z, w, y′. The same will happen twice (for example with x and w), and the list will be x, y′, z′, w′. At the next check, x gets a value and all residuals are instantiated, allowing f to be processed. With this strategy we can decide faster whether a residuation is resolved or not by checking only the variables that are not yet instantiated. With this optimization we increase performance, by doing less checking, i.e., less dereferencing. This is especially useful for residuations waiting for many residuals. We encapsulate this optimazation in init_residuation.

As mentioned before, after every get or unify instruction an update to the corresponding residuation occurs. The need of such an update message can be obtained also by optimizing the compiler. In the residuation f(x_, g(x_)) we do not need to update the second residuation g for the variable x_ as long as it is processed after f allows it, since f is waiting also for variable x_. For the implementation of this optimization on nested residuations we use a global residual list for the current nested term. For example, in f(x_,g(x_,z_))...l(z_)...p(x_)... we can have z_'s update message to g, to be inserted after the update message of x_ to f. In this way, we minimize the update calls by sending them only at the time when they are "useful".

The last proposal is based on the update code of the residuals. Normally, a branch instruction is referring to a code part with branches to the check_residuation addresses. In the case of only one residuation update, we can reduce the amount of branch and proceed instructions by using one branch directly to the check_residuation address. Consider the isdouble example discussed in the previous section. At line (E), instead of branching to B1 we can branch directly to R1. Nevertheless, on the symbol table the update code must be inserted, since x_ is at argument position.

5 Benchmarks

We tried to compare our LAB system with LIFE, a functional logic interpreter using also the residuation mechanism. The programs tried were small but typical functional logic programs in the sense that functions are called with non_ground arguments so that the residuation rules must be applied in order to evaluate this functions. All programs were executed on a Sparc 20 with 64MB main memory under SunOS 5.3. In the nat programs, natural numbers are represented by two constructs: the constant symbol 0 and the successor function s of arity 1. The naive reverese programs use explicitly the cons operator.

We were surprised to see that although LIFE is based on the residuation mechanism, the complicated residuation examples, such as the Example 1 proposed in [1], did not return a proper result. A comparison of our residuation compiler is therefore at the moment not possible. We tried to compare the additional time needed for the residuation mechanism with those of AlgBench, but since the compiler adds only the new command and mechanisms for residuation cases, the execution times were essential the same. Moreover the compilation time has been decreased thanks to a better design of the model of the compiler compared to an early version [8].

Furthermore, we compare the results received from uniparadigm LAB programms with those of the interpretersAlgBench, Mathematica and LIFE (Wild_Life version 1.02) and of the WAM-based compiler SICStus Prolog (emulated version 2.1 #7) as well as the execution times of the functional and logic versions to each other. Table 5.1 summarizes the functional benchmark results of LAB, AlgBench, Mathematica, ALF and LIFE. Table 5.2 compares the execution times of their logic counterparts for LAB, SICStus Prolog, and LIFE. We make no detailed attempt to explain all individual results, which is always a difficult task in benchmarking, but rather we intend to give an indication of the comparative performance of the five languages.

Benchmark	LAB	AlgBench	Mathematica	LIFE
nrev_cons120	0.637 (1.0)	15.28 (24)	1.09 (1.7)	3.133 (4.9)
fact_nat7	1.382 (1.0)	1755 (1270)	2.47 (1.8)	no termination
Ack_nat(3,3)	2.169 (1.0)	22.46 (10)	3.91 (1.8)	9.21 (4.25)

These comparative tests show that LAB is significantly faster for programs not involving non-numeric computation than either AlgBench or Mathematica or LIFE; and comparable or sometimes even a little faster, than the emulated version of SICStus Prolog. Both the functional and the logic version of the naive reverse benchmark programs of LAB are faster than any of the other systems compared.

Benchmark	LAB	SICStus Prolog	LIFE
nrev30	0.035 (1.0)	0.113 (3.0)	0.100 (2.7)
fact_nat6	0.091 (1.0)	0.154 (1.7)	0.483 (5.3)
Ack_nat(3,3)	0.196 (1.0)	0.166 (0.9)	0.633 (3.2)

6 Conclusions

We have presented a multiparadigm language LAB for symbolic computation which first, enables teaching all paradigms within the same language, the construction of very high-level expression of constrained data structures and constraint programming at user level for symbolic computation systems. Additionally, it is more expressive than functional imperative programs due to unification and backtracking and more efficient than logic programs due to deterministic functional programming (no need for non-declarative control constructs like is/2, Cut !, freeze etc.). Finally, the concurrent programming via residuation allows a parallel implementation. The communication between processes is done through unification of shared variables and the synchronization through residuation. We have proposed an object-oriented compiler which generates code for an extended WAM. Our implementations show that the WAM seems to be is a suitable vehicle for implementation of integrated languages and that the compilation of residuation is feasible.

Bibliography

1 H. Aït-Kaci, P. Lincoln, and R. Nasr. Le Fun: Logic, Equations, and Functions. In *Proceedings of IEEE Symposioum on Logic Programming*, 1987.

2 H. Aït-Kaci. *Warren's Abstract Machine. A Tutorial Reconstruction.* MIT Press, 1991.

3 H. Aït-Kaci and A. Podelski. *Functions as Passive Constraints in LIFE.* DEC Paris Research Laboratory, Research Report 13, June 1991.

4 H. Aït-Kaci. Towards a Meaning of LIFE. In *Proc of the 3rd International Symposium on Programming Language Implementation and Logic Programming (PLILP'91)*, Springer-Verlag, LNCS 528, Passau, Germany, August 1991.

5 B. Buchberger. *Mathematica*: A System for doing Mathematics by Computer? Invited Talk.. In *Proc. of the 3rd Intern. Symposium on Design and Implementation of Symbolic Computation Systems (DISCO'93)*, Springer Verlag, LNCS 722, Gmunden, Austria, September 1993.

6 M. Fay. First order unification in equational theories. In *Proc. of the 4th Intern. Conference on Automated Deduction (CADE-4)*, Academic Press, Austin, Texas, 1979.

7 G. Grivas. A Unification- and Object-Based Symbolic Computation System. In *Proceedings of the 6th International Workshop on Unification (UNIF'92).* BUCS Tech. Report #93-004, Computer Science Department, Boston University, July 1992.

8 G. Grivas and R. E. Maeder. Matching and Unification for the Object-Oriented Symbolic Computation System *AlgBench.* In *Proceedings of the 3rd International Symposium on Design and Implementation of Symbolic Computation Systems (DISCO'93)*, Springer-Verlag, LNCS 722, Gmunden, Austria, September 1993.

9 G. Grivas. Efficient Integration of Declarative Paradigms into Symbolic Computation Systems. In *Proc. of ICLP'94 Post-Conference Workshop on Integration of Declarative Paradigms*, MPI-I-94-224, Max-Planck-Institut für Informatik, Santa Margherita, Italy, June 1994.

10 G. Grivas. Towards a Constraint Functional Logic Compiler for Symbolic Computation Systems. In *Proceedings of the 6th International Symposium on Programming Language Implementation and Logic Programming (PLILP'94).* Madrid, Spain, September 1994.

11 G. Grivas. *The Design and Implementation of a Multiparadigm Symbolic Programming Language.* PhD Thesis, Department of Computer Science, ETH Zürich, July 1996 (to be published).

12 M. Hanus. The Integration of Functions into Logic Programming: From Theory to Practice. *Journal of Logic Programming*, Vol. 19/20, pages 583-628, May 1994.

13 J. W. Lloyd. *Foundations of Logic Programming.* Springer-Verlag, second edition, 1987.

14 R. E. Maeder. AlgBench: An Object-Oriented Symbolic Core System. In Proc. of the 2nd International Symposium on Design and Implementation of Symbolic Computation Systems (DISCO'92), Springer Verlag, LNCS 721, Bath, England, April 1992.

Pluggability Issues in the Multi Protocol*

Simon Gray[1], Norbert Kajler[2], Paul S. Wang[1]

[1] Kent State University, Kent, OH 44242-4001, USA
{sgray, pwang}@mcs.kent.edu
[2] Ecole des Mines de Paris, 60 Bd. St-Michel, 75006 Paris, France
kajler@paris.ensmp.fr

Abstract. There are several advantages to providing communication links between independent scientific applications. Important problems to solve include data and control integration. Solving these problems separately is an essential aspect of the design and implementation of a protocol for mathematics. The Multi Protocol (MP) specification addresses the exchange of mathematical data by focusing only on the data encoding issues. In this way, MP can be plugged into various existing data transport mechanisms addressing control integration, or augmented by a higher control-related protocol layer. Our implementation of MP is independent of the data transport mechanism and can work with several devices. An application puts/gets data to/from MP buffers which communicate with the transport device through an abstract device interface. This paper describes the general design of the interface between MP and a transport device and the lessons we have learned during its implementation.

1 Introduction and Motivation

There has been increasing interest, over the last few years, in the ability to integrate independent software packages to work cooperatively on the solution to a problem. Indeed, the design of scientific computing packages is moving away from single image monolithic systems and toward interconnected components that can run independently (see for instance [10, 20, 9, 11] and the proceedings of PASCO'94 [18]). Advantages to this approach include:

1. Researchers gain the freedom to choose applications most suited to their needs and, with a plug-and-play style of interoperability, can experiment with different applications to determine suitability.
2. Autonomous components can be developed and maintained separately, providing access to a wealth of software resources.
3. These components can run on different platforms for convenience, better performance, or to meet license restrictions.
4. The components can be *reused* independently of each other.

* Work reported herein has been supported in part by the National Science Foundation under Grant CCR-9503650

Three dimensions to the problem of tool integration are readily identifiable: data, control, and user interface integration [22, 19]. Data integration involves the exchange of data between separate tools, including the definition of a mechanism allowing the tools to share a common format (and possibly a shared understanding of the *meaning* of the data).

Control integration concerns the establishment, management, and coordination of inter-tool communications. Finally, the aim of user interface integration is to provide the user with a logical and consistent style of interaction with each of the components of the integrated system.

Furthermore, it is important within the context of integration to be able to support different computational paradigms. Especially promising areas for research are parallel symbolic computation [7, 10] and distributed problem solving environments (PSEs) [12, 21]. Figure 1 gives a sampling of the possibilities and further illustrates that these paradigms are not mutually exclusive.

The Multi Project at Kent State is part of an ongoing research effort into the integration of software tools for scientific computing. A key philosophy of the project has been to view the dimensions of tool integration as *individual* problems to be solved *separately*, recognizing that each of these problems may have multiple solutions. The focus of our first efforts has been on data integra-

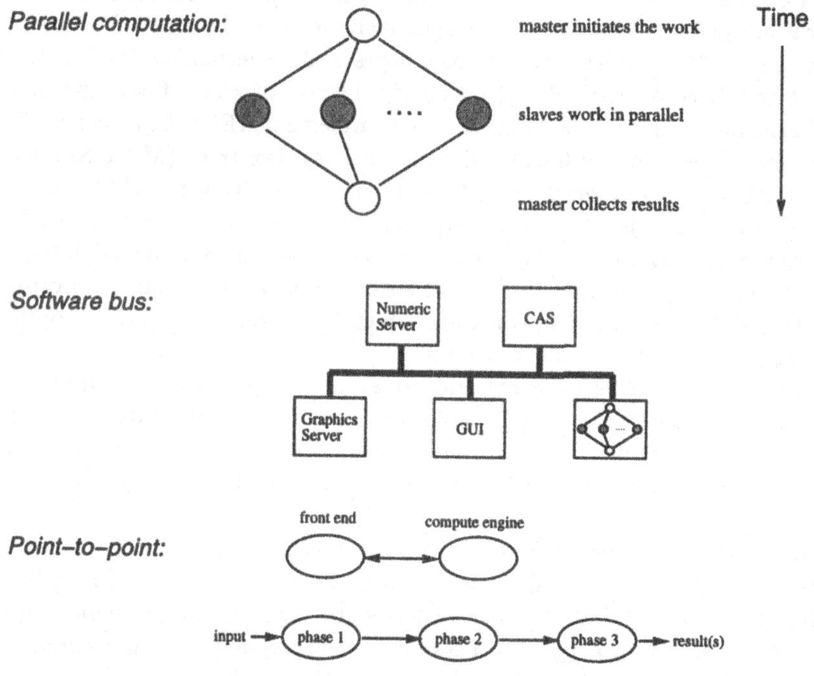

Fig. 1. Computing paradigms to be supported

tion and has produced the Multi Protocol (MP) [15], a specification for encoding mathematical expressions for efficient communication among scientific computing systems. A library of C routines based on the MP specification has been implemented and is publicly available. In implementing MP we had two goals in mind with respect to computing paradigms[3]:

1. That we be able to transmit MP-formatted data using a variety of transport devices, promoting plug-and-play at the communication level.
2. That the integration of MP and transport devices be as seamless as possible. That is, we wanted to hide details about the transport device.

While we have had good success with the first goal, our second goal has been more elusive and our experiences have suggested not just changes in our implementation, but also a rethinking of the goals themselves. This paper describes the lessons learned while trying to do control integration in a generic way through an abstract device interface contained in a separate layer of software.

2 MP Overview

The MP specification focuses on the efficient encoding of mathematical expressions and numerical data. The major design and implementation decisions made in MP are given in [14] and [15] (respectively versions 0.5 and 1.0 of MP).

In short, MP defines a set of basic types and a mechanism for constructing structured data. Numeric data (fixed and arbitrary precision floats and integers) are transmitted in a binary format (2's complement, IEEE float, etc.). Expressions are represented as linearized, annotated syntax trees (MP trees) and are transmitted as a sequence of *node* and *annotation packets*, where each node packet transmits a node from the expression's syntax tree. The node packet has fields giving the *type* of the data carried in the packet, the number of children (for operators) that follow, the number of annotations, some semantic information, and the data. Operators and constants which occur frequently have an optimized encoding and are known as "common".

Annotations efficiently carry additional information which is either supplementary and can be safely ignored by the receiver, or may contain information essential to the proper decoding of the data. In any case, each annotation is tagged in such a way that the receiver always knows whether it can safely ignore the annotation content or not.

In a layer above the data exchange portion of the protocol, MP supports collections of definitions for annotations and mathematical symbols (operators and symbolic constants) in *dictionaries*. Dictionaries address the problem of application heterogeneity by supplying a standardized representation and semantics for mathematical objects. They are identified within packets through a dictionary tag field. Applications that communicate according to definitions provided in dictionaries do not need to have direct knowledge of each other, promoting a

[3] We use "MP" here to refer to both the specification and the implementation.

plug-and-play style of inter-operation at the *application* level. A dictionary for polynomials has been built on top of MP in this manner [3].

Applications send and receive *messages* containing one or more MP trees which are created by calling routines from the MP Application Programming Interface (API). Logically, the format and representation of the data are completely separate from the mechanism used to transmit it. We have maintained this separation in the implementation of MP and thus are able to support a diverse collection of delivery mechanisms.

3 MP Links and Device Independence

Within MP, an application communicates with other applications through an *MP link*, which is simply an abstraction of an underlying data transport mechanism that is bound to the link at the time of its creation. The link sits on top of a transport device. Data is exchanged between the link and the transport through an *abstract device interface*. This section discusses the issues involved in maintaining device independence.

3.1 MP Links

A link is created within an MP *environment*. The environment contains a set of resources to be shared by a collection of links created within an application, including a list of the transport devices available to links and a pool of memory buffers. The programmer may customize the environment by setting options that reset the buffer size and determine how many buffers are initially in the pool.

As illustrated in Fig. 2, a link has two layers: a transport layer and a buffer layer. The buffering layer lies between the MP API and the transport mechanism and is where messages are assembled and disassembled. The programmer's view

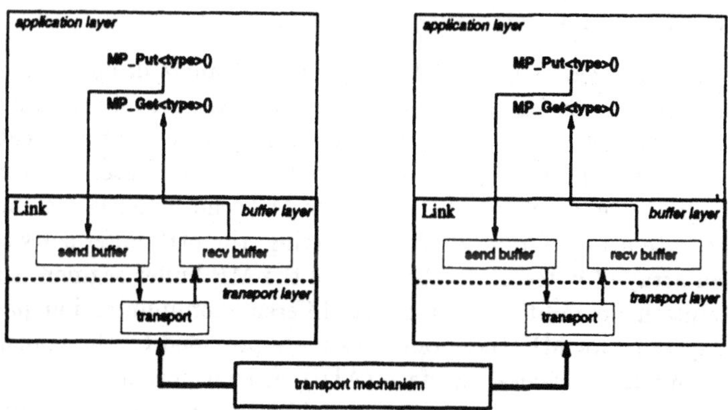

Fig. 2. Link connection organization

is that each **put** and **get** operation accesses a link connecting the application with one or more other applications. In reality, all accesses are to one of two buffers: **puts** access a link's send buffer and **gets** the receive buffer. These buffers are implemented as a linked list of buffer segments.

Information is exchanged between applications in the form of *messages*. A message is composed of one or more *message fragments*, where each fragment corresponds to the data in a single buffer segment. For some transport devices, the sending side has the option of choosing between building an entire message in the link's send buffer before actually sending it, or to send each buffer segment when it becomes full. Similarly on the receiving side. This allows the sender and receiver to overlap I/O with processing. A sending application builds a message through a series of **puts**. The sender marks the end of a message with a call to **MP_EndMsg()** which sets a bit in the message segment header indicating that this is the last fragment in a complete message. It then flushes the link's send buffer to the underlying transport mechanism. One or more MP trees may be contained within a single message. A receiving application parses the incoming message with a series of **gets**. The receiver must perform some message alignment by calling **MP_SkipMsg()** before reading in *each* message. It is the receiver's responsibility to ensure that the entire contents of a message are consumed before proceeding to the next message. This can be done with a call to **MP_TestEofMsg()**, which returns true if the end of the current message has been reached. A useful convention is to transmit a single expression per message. This provides a level of error recovery and allows the receiving side to efficiently skip to the end of an expression it cannot process.

A link is created and initialized with a call to **MP_OpenLink (MP_Env_pt env, int argc, char *argv[])** which returns a pointer to an MP link structure on success and NULL on failure. Arguments passed in the **argv** vector identify the transport to use, what mode to open it in, etc. These arguments may be given on the command line when the application is launched, hard coded within the application, or, for interactive applications, provided by user input.

3.2 The Abstract Device Interface

Transmitting data generally consists of two steps: *marshalling* (or *packing*) the data, producing a linearized encoding of the data suitable for transmission, and sending the data. Receiving follows this process in reverse. Communication packages that can send and receive an array of *bytes* are candidates for transmitting MP trees. On the sending side, data from the MP send buffer is presented to the device as a byte array. Similarly, when receiving, the MP buffer software simply requests an array of bytes up to the limit of the MP buffer segment size.

To maintain flexibility and support different communication packages as transport systems for MP, these operations are provided by the transport layer through an Abstract Device Interface (ADI). A transport device is represented by a *generic* transport device structure which contains fields common to all devices, including a device operations structure, and a pointer to an opaque structure that is specific to the transport device bound to the link.

The operations in the device interface are:

1. **dev_open_connection()** - Allocate memory for the device structure, assign the appropriate device operations structure, and perform device-specific open operations to make a connection.
2. **dev_close_connection()** - Break down the connection and release all memory allocated for the device structure.
3. **dev_read()** - Read a specified number of bytes from the device into the link's receive buffer. This routine is called by the buffering layer when the application has attempted to **get** a data item and the link's receive buffer is empty.
4. **dev_write()** - Write a specified number of bytes from the link's send buffer to the device. This routine is invoked by the buffering layer either to empty the link's send buffer to free it for another message *fragment*, or when the message has been completed and needs to be transmitted.
5. **dev_get_status()** - Determine the status of the device.

The intended behavior of this interface made three assumptions about the capabilities of the actual transport device:

1. The device provides buffering of the messages it delivers, allowing the MP buffering layer to read and write its messages in fragments.
2. The read/write routines in the ADI simply need to move data between the device's buffers and MP's buffers.
3. The source and destination information can be easily kept in the transport-specific device structure.

Under these assumptions, adding a new transport requires providing only the device-specific structure and the functions that implement the operations in the abstract device interface. Figure 3 shows the relationship between the MP API and the operations in the ADI. As we will see, the degree to which this can be done seamlessly varies with the characteristics of the device. Additional interface routines were sometimes necessary.

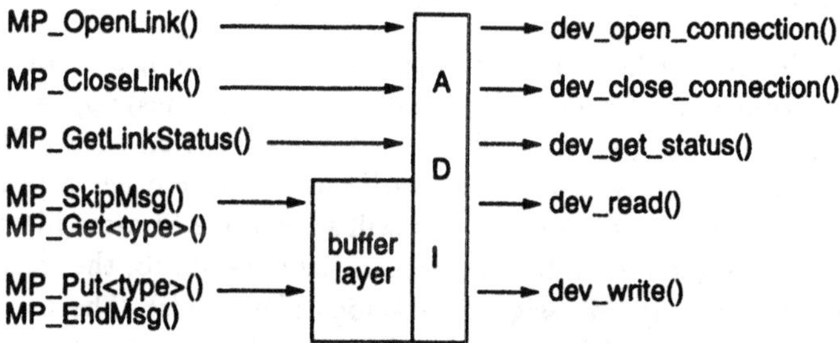

Fig. 3. MP API to device operations

4 Current Transports

This section briefly describes the transport interfaces currently supported by MP. These devices represent the range of computing paradigms illustrated in Fig. 1. The devices discussed here also provide contrasting features and requirements, and demonstrate that the ADI is no panacea – it cannot elegantly solve all the problems set out for it.

4.1 Files and TCP Sockets

The FILE and TCP transport devices are the simplest and adhere most closely to the blueprint for a generic transport device. Both rely on a Unix I/O descriptor for access to the "device", and use standard Unix system calls to open, close, read, and write on the descriptor. The read/write calls block, but it is possible to test for device readiness first through a call to the `MP_GetLinkStatus()` routine.

Discussion. These devices meet all the assumptions we made for the abstract device interface. The reason is not difficult to see. As a system that communicates via messages, the MP buffering layer fits most naturally with a stream-oriented device that provides its own buffering. We expect that a shared memory device would also fall into this category.

The FILE device is useful for archiving and was also used in some early experimentation for systems that originally only communicated using files. The TCP device provides reliable, ordered delivery of data for point-to-point communication. The best throughput is gained when the MP environment's buffer option is used to reset the buffer segment size to that of the TCP packet size. A smaller segment size favors better response times.

We have used these devices most extensively. Both were used in our experiments to connect Maxima with the graphing package IZIC [2] and to provide point-to-point communication between the packages SINGULAR, Mathematica, and FACTORY [3].

4.2 PVM

The Parallel Virtual Machine (PVM) [13] is a message-passing system that makes a collection of computers (workstations and parallel machines) appear to the user as a single multiprocessor system. When used in conjunction with PVM, an MP link is an endpoint for communication with one *or more* other processes. In a typical PVM application, a single master will spawn one or more identical slave *tasks* which are assigned work by, and return their results to, the master. Tasks are identified through PVM-assigned task identifiers (*tids*). The user's data is *pack*ed into the PVM send buffer and, when complete, is sent to one or more recipients who are identified in a list of tids. Also, messages may be tagged so that the receiver may specify exactly which messages it is willing to accept. The receiver performs a receive operation to read data into the PVM receive buffers, from which the application may *unpack* the data as needed into the user's data structures. Arguments in the receive routine allow the receiver to specify the kind of message it will accept and from which source.

Because PVM is quite flexible with regard to the packing/unpacking of data, it fits fairly nicely into our abstract device model. PVM allows you to pack data in pieces; that is, there is no requirement that you have all your data ready and, in a single call, pack it into PVM buffers. Instead, the sender can pack data items individually until the message is complete and *then* issue a send command, which transmits the entire message. Moving the data from a link's send buffer to the PVM buffer is done with the `pvm_pkbyte()` routine, which treats its data simply as an array of bytes. This call amounts to a `memcpy()` from the MP buffer space to the PVM buffer space. On the receiving side, the transport read routine performs both the receive and unpack operations. A receive call reads an entire message into the PVM read buffer. The unpack routine `pvm_upkbyte()` moves the requested number of bytes from the PVM buffer to the MP receive buffer.

Discussion. PVM's ability to do packing (unpacking) independently of sending (receiving) simplifies the interface and supplies the behavior expected by the buffering layer, allowing MP trees to be exchanged in fragments.

However, several factors complicate fitting PVM seamlessly into the abstract device interface. First, typically a master *spawns* a set of slave processes and the connection between them is established by the PVM daemon, so there is little work for the device open and close routines to do. Second, PVM supports send/receive routines for point-to-point, multicast, and broadcast communication.[4] So the device interface write/read routines should be able to determine at execution time which is appropriate. Third, the PVM send routines take an argument specifying the intended receiver. Similarly, the receiver may specify the source from which it expects data. These tids could be placed inside the PVM device structure by an auxiliary routine in the interface and accessed by the device write/read routines. Finally, recall that messages may be tagged. A tag could also be stored in the PVM device structure. But keeping this kind of information in the device structure is really only viable as long as it doesn't have to be changed often. If an application uses different sets of tids or message tags, having to constantly reset them in the device structure is awkward, and, for regular PVM users, unnatural.

4.3 ToolBus

ToolBus [4] is one of several new software bus architectures that can be used to glue independent software components together (either on a single machine or over a network).

Like ToolTalk [24], ToolBus requires that all data travel through a central manager process, but it is unique in that it uses a *script* to completely define and control the set of interactions that can occur between tools connected by the

[4] Point-to-point communication is one-to-one, multicasting is one-to-many, and broadcasting is one-to-all.

bus. Data is exchanged between tools in the form of typed *terms*. Among these is a binary type which treats the data as an array of bytes. Through patterns in a script, a tool specifies what kinds of service requests it can carry out and which types of terms it will accept. Tools open *ports* to the toolbus through which they exchange terms. Each port has a *handler* associated with it that is invoked whenever a term arrives on the corresponding port. The handler's job is to determine if the incoming request matches any of the services offered by the tool, and, if so, to invoke the routine that provides the service. The service routine returns a (possibly NULL) term as its result, which the handler passes back to the toolbus (and from there possibly to other tools).

The user's data is packed into a term by `TBmake()`, which takes a scanf-like argument list containing a pattern describing the data to be packed in the term. The term is sent to the toolbus either by returning it as the value of the handler function or by making an explicit call to `TBsend()`. It is also possible to incrementally build a recursive list of terms stored within a single term. We use this feature to pack a link's send buffer fragments into a single toolbus term. Unpacking terms is done through `TBmatch()`, which takes the incoming term, a pattern against which to match the term, and the address of the program's variables where the data from the term will be unpacked (if the match succeeds).

Discussion. ToolBus required more work to adapt to the ADI paradigm than the other transports we have used, but with the addition of some auxiliary routines, it fit very nicely[5].

The device write routine simply packs the link's send buffer fragments into a special term (`write_term` - a recursive list of binary type terms) stored in the device structure. When the sending tool is ready to send the completed message, an auxiliary routine provides access to `write_term`, which the sender can then return to the toolbus. This was simple to implement, but requires that the sender build a complete message (MP tree) in its memory before transmitting (so sender and receiver cannot overlap I/O and processing), and requires more memory copies.

The receiving side was slightly more complicated. To meet the requirements of the MP buffering layer, a simple buffering layer had to be incorporated into the device read routine to move the MP data packed in the incoming ToolBus term to a buffer fragment in the link's receive buffer. The auxiliary routine `MP_SetTbTerm()` places the incoming term into `read_term` (the MP link's reading-side complement to `write_term`). Recall that this term is a recursive list of binary type terms, each of which stores a single sender buffer fragment. The device read routine unpacks a single binary term (sender buffer fragment) from `read_term` into a buffer kept in the device structure. Requests for data made by the MP buffering layer are satisfied from this device buffer. When the device's buffer is empty, another binary term (sender buffer fragment) is unpacked from `read_term`.

[5] We anticipate that other software bus packages such as ToolTalk [24] will present similar problems.

Our experience is that the auxiliary routines do not complicate the interface and that despite the additional memory copies, performance is quite good. Certainly the cost of the extra copies does not outweigh the advantages gained by the *connectivity* that is provided.

4.4 Mail and MIME

MP trees can also be sent using the *Multipurpose Internet Mail Extensions* (MIME) standard [5], which allows textual and non-textual data to be transported together through the network by SMTP (Simple Mail Transfer Protocol) without loss of information. At first appearance, mail would not seem to fit any computing paradigm. However, the reader program need not simply be a display or conversion package. It could be associated with, for example, a computation package. In this scenario, the reader would queue the incoming requests, taking note of the address of the sender, launch an application (possibly one of a selection of computation tools) to service the request and return an MP formatted result in another MIME message. This model nicely fits the point-to-point paradigm, but it also fits the multicast paradigm if the request is sent to multiple servers.

Discussion. The MIME standard defines several new header fields including **`Content-Type`** to specify the type and subtype of the data that follows, and **`Content-Transfer-Encoding`** to specify the encoding used to make the message contents acceptable to SMTP. In our use, we specify the **`application`** content type with an **`mp`** subtype and **`base64`** as the encoding type. Our **`.mailcap`** customization file specifies a reader program for processing MP trees sent via email. Currently we simply use one of the MP test routines to write a human readable version of the MP tree to a file. It is a simple matter to write a more elaborate display program to format the output on the screen. Also, the MP content received via email is readable directly by any MP-compliant compute engine.

5 Related Work

MathLink [25] is a commercial mathematics protocol packaged with Mathematica. Although it has a set of Mathematica-specific routines, it is a general protocol that can be used independently of Mathematica. MathLink's notion of a link is very similar to MP's. At the time a link is created, a transport device is bound to it. Currently MathLink supports several built-in devices: TCP sockets (Macintosh and Unix), Unix pipes, the Macintosh System 7 program-to-program (PPC) communication mechanism, and, under Macintosh System 6, the Apple Talk Data Stream Protocol (ADSP). At this time MathLink is the only other implementation of a mathematics protocol which has this kind of flexibility.

ASAP (A Simple ASCII Protocol) [8] is a public domain mathematics protocol. It uses a pair of TCP socket connections, called *channels*, to establish a point-to-point connection between two processes. One socket is dedicated to the *data channel* for transmitting expressions. The other is dedicated to the *urgent channel* for handling exceptional conditions and uses the Unix SIGIO signal to interrupt the receiving process.

The POlynomial System SOlver (PoSSo) project includes a protocol, PossoXDR [1], defining the external representation of the data types manipulated inside PoSSo processes. The encoding extends the XDR technology [23] with some new types and constructors, and by tagging each data element with its type. PossoXDR communicates through a TCP socket or a file.

6 Lessons Learned and Future Work

Clearly some devices fit into our device abstraction more neatly than others. But it is very important to remember that the MP put and get routines an application uses to send and receive MP data are *completely indifferent* to which transport device is used to communicate the data. Indeed, the development and testing of these devices was done with the same suite of test routines and the *only* changes made were isolated to the few incompatibilities described for PVM and ToolBus. The clear advantage of this approach is that by focusing on the MP API, one can reuse most of the interface code developed for an application when plugging in a different transport, protecting the initial investment in the interface. This means, for example, that the interface we developed for Singular [16], may be largely reused when plugging it in as a specialized server (in a distributed PSE or a point-to-point connection to augment another system, as we did with Mathematica) or as a subsystem for parallel computation (as we want to do with PVM for parallel Gröbner bases computations).

Our initial goal was to be able to plug different transport devices *into* MP. Our perspective was that of a programmer who is familiar with MP and simply wants to use some specific device as a data transport mechanism. In such a case, we want to assume knowledge of MP and offer transparent use of the selected device. But there is a second perspective to consider, that of a programmer accustomed to the particulars of the transport device who wants to use MP for transmitting mathematical data. In this second case, we want to assume knowledge of the transport device and its interface, and little or no knowledge of MP. This usually requires extending the device's API with a series of functions built on top of the MP API (and trying to make these functions as coherent as possible with the functions in the device's API). For example, we have `pvm_mp_pkList()` and `pvm_mp_pkPoly()` to pack a list and polynomial respectively for PVM-MP. But for systems such as ToolTalk that have no notion of data types and simply provide the delivery of messages marshalled using an independent mechanism, the issue of whether to provide an MP-style or package-style API is largely irrelevant. How best to resolve these issues is not immediately clear and is a subject of ongoing investigation.

The flexibility gained through our approach opens a wide range of possibilities. We are especially interested in pursuing two areas.

1. Parallel distributed symbolic computation. When done within the context of p4 [6], MPI [17], or PVM, this work is readily portable to tightly-coupled, shared memory machines such as the T3D which have optimized implementations of both MPI and PVM. In this scenario, the applications typically know each other quite well, allowing data exchange performance optimizations.
2. Distributed problem solving environments. We want to explore using MP in conjunction with software buses such as ToolBus and ToolTalk to integrate symbolic, numeric, and graphics processing. Clearly here the challenges are greater. Exchanging data is more complicated, as are the MP-application interfaces. Solving the problems inherent in this kind of integration should be useful for designing and implementing larger scale problem solving environments.

We continue to add new transport devices and to work on cross-platform communication. In particular, the point-to-point mechanism should translate well to other platforms (Macintosh PPC, Windows WinSocket). Work is underway on a shared memory device for Unix-based machines.

In our opinion (see [15]), commands providing inter-tool *control* should be kept outside the definition of a mathematics protocol. This includes being able to stop a remote computation, get the status of a remote computation, determine if a link is still alive, and so on. Clearly, such commands are not provided by MP. Instead, they are are expected to be available from the device used in complement of MP – at least when such commands make sense (which is not always the case, see MIME for instance). Still, we may provide such a capability as part of a separate, higher software layer, to be used with the core of MP when no other technology provides its own control mechanism. Along these lines we want to enhance the `MP_GetLinkStatus()` routine to accept requests for those control aspects that relate to the device itself and to have the device-specific `dev_get_status()` routine handle them as provided for by the device.

7 Conclusion

A key enabling technology for distributed and parallel mathematical computation is a standard protocol for exchanging mathematical objects. MP is our attempt to contribute to a standard, non-proprietary mathematics protocol. An essential feature of MP's design and implementation is its independence from the mechanism used to transmit the data. When designing MP, we focused on not interfering with communication-level issues in order to ensure the highest possible degree of "pluggability" for our protocol. So, instead of solving both data and control problems inside a single unique protocol, we decided to address the data encoding aspect *only* and make sure that the resulting technology fit well inside a variety of communication and computing paradigms.

At the implementation level, MP attempts to provide this independence by communicating through an abstract device interface which is mapped to an actual device structure at the time a transport link is created. Some devices map well to this abstract interface, while others require auxiliary routines.

We believe this approach allows a better separation of problems and leads to a highly pluggable protocol which can be used together with well-known communication and computing technologies such as files, TCP sockets, Mail, ToolBus, or PVM, and hopefully others such as CORBA, ToolTalk, or WWW.

8 Availability

The source for the MP library is available from `ftp.mcs.kent.edu` in `/pub/MP`. Also see `http://SymbolicNet.mcs.kent.edu/areas/protocols/mp.html`.

Acknowledgments

The authors would like to thank Olaf Bachmann and Hans Schönemann for their insightful comments on earlier drafts of this paper.

References

1. J. Abbott and C. Traverso. Specification of the POSSO External Data Representation. Technical report, September 1995.
2. R. Avitzur, O. Bachmann, and N. Kajler. From Honest to Intelligent Plotting. In A. H. M. Levelt, editor, *Proc. of the International Symposium on Symbolic and Algebraic Computation (ISSAC'95), Montreal, Canada*, pages 32 – 41. ACM Press, July 1995.
3. O. Bachmann, H. Schönemann, and S. Gray. A Framework for Distributed Polynomial Systems Based on MP. To appear in the Proceedings of ISSAC'96.
4. J.A. Bergstra and P. Klint. The Discrete Time ToolBus. Technical Report P9502, Programming Research Group, University of Amsterdam, 1995.
5. N. Borenstein and N. Freed. MIME (Multipurpose Internet Mail Extensions) Part One: Mechanisms for Specifying and Describing the Format of Internet Message Bodies. RFC 1521, September 1993.
6. R. Butler and E. Lusk. Monitors, messages, and clusters: the p4 parallel programming system. *Parallel Computing*, 1994.
7. G. Cooperman. STAR/MPI: Binding a Parallel Library to Interactive Symbolic Algebra Systems. In A. H. M. Levelt, editor, *Proc. of the International Symposium on Symbolic and Algebraic Computation (ISSAC'95), Montreal, Canada*, pages 126 – 132. ACM Press, July 1995.
8. S. Dalmas, M. Gaëtano, and A. Sausse. ASAP: a protocol for symbolic computation systems. INRIA Technical Report 162, March 1994.
9. M. C. Dewar. Manipulating Fortran Code in AXIOM and the AXIOM-NAG Link. In H. Apiola, M. Laine, and E. Valkeila, editors, *Proceedings of the Workshop on Symbolic and Numeric Computing*, pages 1–12. University of Helsinki, Finland, 1994. Available as Technical Report B10, Rolf Nevanlinna Institute.

10. A. Diaz, E. Kaltofen, K. Schmitz, T. Valente, M. Hitz, A. Lobo, and P. Smyth. DSC: A System for Distributed Symbolic Computation. In S. M. Watt, editor, *Proc. of the International Symposium on Symbolic and Algebraic Computation (ISSAC'91), Bonn, Germany*, pages 323–332. ACM Press, July 1991.
11. Y. Doleh. *SUI: A system Independent User Interface for an Integrated Scientific Computing Environment*. PhD thesis, Kent State University, May 1995.
12. E. Gallopoulos, E. Houstis, and J. Rice. Computer as Thinker/Doer: Problem-Solving Environments for Computational Science. *IEEE Computational Science and Engineering*, pages 11–23, 1994.
13. A. Geist, A. Beguelin, J. Dongarra, W. Jiang, R. Manchek, and V. Sunderam. PVM3 User's Guide and Reference Manual. Technical Report ORNL/TM-12187, Oak Ridge National Laboratory, Septempber 1994.
14. S. Gray, N. Kajler, and P. S. Wang. MP: A Protocol for Efficient Exchange of Mathematical Expressions. In M. Giesbrecht, editor, *Proc. of the International Symposium on Symbolic and Algebraic Computation (ISSAC'94), Oxford, GB*, pages 330–335. ACM Press, July 1994.
15. S. Gray, N. Kajler, and P. S. Wang. Design and Implementation of MP, a Protocol for Efficient Exchange of Mathematical Expressions. 1996. Forthcoming in *Journal of Symbolic Computing*.
16. G.-M. Greuel, G. Pfister, and H. Schönemann. *Singular*: A system for computation in algebraic geometry and singularity theory. University of Kaiserslautern, Dept. of Mathematics, 1995. Available via anonymous ftp from `helios.mathematik.uni-kl.de`.
17. W. Gropp, R. Lusk, and A. Skjellum. *Using MPI*. MIT Press, 1994.
18. H. Hong, editor. *Proc. of the 1st Intl. Symp. on Parallel Symbolic Computation (PASCO'94)*, volume 5. World Scientific, September 1994.
19. N. Kajler. Building a Computer Algebra Environment by Composition of Collaborative Tools. In J. P. Fitch, editor, *Proc. of DISCO'92, Bath, GB*, volume 721 of *LNCS*, pages 85–94. Springer-Verlag, April 1992.
20. N. Kajler. CAS/PI: a Portable and Extensible Interface for Computer Algebra Systems. In P. S. Wang, editor, *Proc. of the International Symposium on Symbolic and Algebraic Computation (ISSAC'92), Berkeley, USA*, pages 376–386. ACM Press, July 1992.
21. John Rice. Scalable Scientific Software Libraries and Problem Solving Environments. Technical Report CSD TR-96-001, Department of Computer Science, Purdue University, January 1996.
22. D. Schefström and G. van den Broek, editors. *Tool Integration*. Wiley Press, 1993.
23. Sun Microsystems, Inc., Mountain View, CA. *Network Programming Guide (revision A)*, 1990. Part number 800-3850-10.
24. SunSoft Press. *The ToolTalk Service: An Interoperability Solution*. 1992.
25. Wolfram Research, Inc. MathLink Reference Guide (version 2.2). Mathematica Technical Report, 1993.

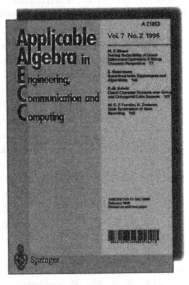

Springer-Verlag and the Environment

We at Springer-Verlag firmly believe that an international science publisher has a special obligation to the environment, and our corporate policies consistently reflect this conviction.

We also expect our business partners – paper mills, printers, packaging manufacturers, etc. – to commit themselves to using environmentally friendly materials and production processes.

The paper in this book is made from low- or no-chlorine pulp and is acid free, in conformance with international standards for paper permanency.

Lecture Notes in Computer Science

For information about Vols. 1–1071

please contact your bookseller or Springer-Verlag

Vol. 1106: M. Jampel, E. Freuder, M. Maher (Eds.), Over-Constrained Systems. X, 309 pages. 1996.

Vol. 1107: J.-P. Briot, J.-M. Geib, A. Yonezawa (Eds.), Object-Based Parallel and Distributed Computation. Proceedings, 1995. X, 349 pages. 1996.

Vol. 1108: A. Díaz de Ilarraza Sánchez, I. Fernández de Castro (Eds.), Computer Aided Learning and Instruction in Science and Engineering. Proceedings, 1996. XIV, 480 pages. 1996.

Vol. 1109: N. Koblitz (Ed.), Advances in Cryptology – Crypto '96. Proceedings, 1996. XII, 417 pages. 1996.

Vol. 1110: O. Danvy, R. Glück, P. Thiemann (Eds.), Partial Evaluation. Proceedings, 1996. XII, 514 pages. 1996.

Vol. 1111: J.J. Alferes, L. Moniz Pereira, Reasoning with Logic Programming. XXI, 326 pages. 1996. (Subseries LNAI).

Vol. 1112: C. von der Malsburg, W. von Seelen, J.C. Vorbrüggen, B. Sendhoff (Eds.), Artificial Neural Networks – ICANN 96. Proceedings, 1996. XXV, 922 pages. 1996.

Vol. 1113: W. Penczek, A. Szałas (Eds.), Mathematical Foundations of Computer Science 1996. Proceedings, 1996. X, 592 pages. 1996.

Vol. 1114: N. Foo, R. Goebel (Eds.), PRICAI'96: Topics in Artificial Intelligence. Proceedings, 1996. XXI, 658 pages. 1996. (Subseries LNAI).

Vol. 1115: P.W. Eklund, G. Ellis, G. Mann (Eds.), Conceptual Structures: Knowledge Representation as Interlingua. Proceedings, 1996. XIII, 321 pages. 1996. (Subseries LNAI).

Vol. 1116: J. Hall (Ed.), Management of Telecommunication Systems and Services. XXI, 229 pages. 1996.

Vol. 1117: A. Ferreira, J. Rolim, Y. Saad, T. Yang (Eds.), Parallel Algorithms for Irregularly Structured Problems. Proceedings, 1996. IX, 358 pages. 1996.

Vol. 1118: E.C. Freuder (Ed.), Principles and Practice of Constraint Programming — CP 96. Proceedings, 1996. XIX, 574 pages. 1996.

Vol. 1119: U. Montanari, V. Sassone (Eds.), CONCUR '96: Concurrency Theory. Proceedings, 1996. XII, 751 pages. 1996.

Vol. 1120: M. Deza. R. Euler, I. Manoussakis (Eds.), Combinatorics and Computer Science. Proceedings, 1995. IX, 415 pages. 1996.

Vol. 1121: P. Perner, P. Wang, A. Rosenfeld (Eds.), Advances in Structural and Syntactical Pattern Recognition. Proceedings, 1996. X, 393 pages. 1996.

Vol. 1122: H. Cohen (Ed.), Algorithmic Number Theory. Proceedings, 1996. IX, 405 pages. 1996.

Vol. 1123: L. Bougé, P. Fraigniaud, A. Mignotte, Y. Robert (Eds.), Euro-Par'96. Parallel Processing. Proceedings, 1996, Vol. I. XXXIII, 842 pages. 1996.

Vol. 1124: L. Bougé, P. Fraigniaud, A. Mignotte, Y. Robert (Eds.), Euro-Par'96. Parallel Processing. Proceedings, 1996, Vol. II. XXXIII, 926 pages. 1996.

Vol. 1125: J. von Wright, J. Grundy, J. Harrison (Eds.), Theorem Proving in Higher Order Logics. Proceedings, 1996. VIII, 447 pages. 1996.

Vol. 1126: J.J. Alferes, L. Moniz Pereira, E. Orlowska (Eds.), Logics in Artificial Intelligence. Proceedings, 1996. IX, 417 pages. 1996. (Subseries LNAI).

Vol. 1127: L. Böszörményi (Ed.), Parallel Computation. Proceedings, 1996. XI, 235 pages. 1996.

Vol. 1128: J. Calmet, C. Limongelli (Eds.), Design and Implementation of Symbolic Computation Systems. Proceedings, 1996. IX, 356 pages. 1996.

Vol. 1129: J. Launchbury, E. Meijer, T. Sheard (Eds.), Advanced Functional Programming. Proceedings, 1996. VII, 238 pages. 1996.

Vol. 1130: M. Haveraaen, O. Owe, O.-J. Dahl (Eds.), Recent Trends in Data Type Specification. Proceedings, 1995. VIII, 551 pages. 1996.

Vol. 1131: K.H. Höhne, R. Kikinis (Eds.), Visualization in Biomedical Computing. Proceedings, 1996. XII, 610 pages. 1996.

Vol. 1132: G.-R. Perrin, A. Darte (Eds.), The Data Parallel Programming Model. XV, 284 pages. 1996.

Vol. 1133: J.-Y. Chouinard, P. Fortier, T.A. Gulliver (Eds.), Information Theory and Applications. Proceedings, 1995. XII, 309 pages. 1996.

Vol. 1134: R. Wagner, H. Thoma (Eds.), Database and Expert Systems Applications. Proceedings, 1996. XV, 921 pages. 1996.

Vol. 1135: B. Jonsson, J. Parrow (Eds.), Formal Techniques in Real-Time and Fault-Tolerant Systems. Proceedings, 1996. X, 479 pages. 1996.

Vol. 1136: J. Diaz, M. Serna (Eds.), Algorithms - ESA '96. Proceedings, 1996. XII, 566 pages. 1996.

Vol. 1137: G. Görz, S. Hölldobler (Eds.), KI-96: Advances in Artificial Intelligence. Proceedings, 1996. XI, 387 pages. 1996. (Subseries LNAI).

Vol. 1138: J. Calmet, J.A. Campbell, J. Pfalzgraf (Eds.), Artificial Intelligence and Symbolic Mathematical Computation. Proceedings, 1996. VIII, 381 pages. 1996.

Vol. 1139: M. Hanus, M. Rogriguez-Artalejo (Eds.), Algebraic and Logic Programming. Proceedings, 1996. VIII, 345 pages. 1996.

Vol. 1140: H. R. Kuchen, S. Doaitse Swierstra (Eds.), Programming Languages: Implementations, Logics, and Programs. Proceedings, 1996. XI, 479 pages. 1996.

Vol. 1141: W. Ebeling, I. Rechenberg, H.-P. Schwefel, H.-M. Voigt (Eds.), Parallel Problem Solving from Nature — PPSN IV. Proceedings, 1996. XVII, 1.050 pages. 1996.

Vol. 1142: R.W. Hartenstein, M. Glesner (Eds.), Field-Programmable Logic. Proceedings, 1996. X, 432 pages. 1996.

Vol. 1143: T.C. Fogarty (Ed.), Evolutionary Computing. Proceedings, 1996. VIII, 305 pages. 1996.

Vol. 1144: J. Ponce, A. Zisserman, M. Hebert (Eds.), Object Representation in Computer Vision. Proceedings, 1996. VIII, 403 pages. 1996.

Vol. 1145: R. Cousot, D.A. Schmidt (Eds.), Static Analysis. Proceedings, 1996. IX, 389 pages. 1996.